T0201382

Design of Experiments for Reliability Achievement

WILEY SERIES IN PROBABILITY AND STATISTICS

Established by Walter A. Shewhart and Samuel S. Wilks

The **Wiley Series in Probability and Statistics** is well established and authoritative. It covers many topics of current research interest in both pure and applied statistics and probability theory. Written by leading statisticians and institutions, the titles span both state-of-the-art developments in the field and classical methods.

Reflecting the wide range of current research in statistics, the series encompasses applied, methodological and theoretical statistics, ranging from applications and new techniques made possible by advances in computerized practice to rigorous treatment of theoretical approaches. This series provides essential and invaluable reading for all statisticians, whether in academia, industry, government, or research.

Design of Experiments for Reliability Achievement

Steven E. Rigdon
Saint Louis University
Epidemiology and Biostatistics
Saint Louis, MO

Rong Pan
Arizona State University
School of Computing and Augmented Intelligence
Tempe, AZ

Douglas C. Montgomery
Arizona State University
School of Computing and Augmented Intelligence
Phoenix, AZ

Laura J. Freeman
Statistics
Virginia Tech
Blacksburg, VA

This edition first published 2022

© 2022 John Wiley & Sons, Inc.

The right of Steven E. Rigdon, Rong Pan, Douglas C. Montgomery, and Laura J. Freeman to be identified as the authors of this work has been asserted in accordance with law.

Registered Office
John Wiley & Sons, Inc., 111 River Street, Hoboken, NJ 07030, USA

Editorial Office
111 River Street, Hoboken, NJ 07030, USA

For details of our global editorial offices, customer services, and more information about Wiley products visit us at www .wiley.com.

Wiley also publishes its books in a variety of electronic formats and by print-on-demand. Some content that appears in standard print versions of this book may not be available in other formats.

Library of Congress Cataloging-in-Publication Data Applied for:

ISBN: 9781119237693

Cover image: Wiley
Cover design by © Eoneren/Getty Images

Set in 9.5/12.5pt STIXTwoText by Straive, Chennai, India

SKY10034162_042222

In memory of our friend and colleague, Connie Margaret Borror (1966–2016)

Contents

Preface

Techniques of design of experiments (DOE) have for decades been used in industry to achieve quality products and processes. These methods often involve analyzing models that assume, at least approximately, that the outcome is normally distributed. When the outcome is a lifetime, similar techniques can be applied, although the methods require more complicated models. Reliability experiments are special in two respects: (i) there is almost always *censoring*, i.e. the termination of the experiment before all units have failed, and (ii) lifetime distributions are usually not well approximated by the normal. This book is about designing experiments and analyzing data when the outcome is a lifetime.

Condra (2001b) suggests three aspects of reliability methods:

1. Methods for measuring and predicting failures
2. Methods for accommodating failures
3. Methods for preventing failures

The first, measuring and predicting failures, usually involves fitting models to lifetime data in order to assess the reliability of a system. The second, accommodating failures, involves concepts like parallel redundancy (where failure of a single component does not cause failure of the system), repairability (the ability to quickly fix a problem and return the system to working condition), maintainability (the ability to keep a system in working condition), and others. The last, methods for preventing failures, is potentially the most useful. DOE methods can be used to find characteristics of the product, or maybe the process used to make the product, that lead to the highest possible reliability. Of course, this involves methods for measuring and predicting failures (the first item earlier) and it could involve the second (accommodating failures), but the idea of designing experiments to improve reliability is a powerful one. DOE has been used successfully in a number of areas where a normally distributed response is reasonable, but applications in reliability are rather sparse.

We have divided the book into four parts:

I **Reliability** Here we cover the basic concepts and definitions of reliability. We present models for lifetimes, including the exponential, Weibull, gamma, and log-normal. In addition, we discuss log-location-scale distributions, such as the smallest extreme value (SEV) distribution, which is a general class of distributions that can be used to model the logarithm of lifetimes. Inference for lifetime distributions, or log lifetime distributions, is the topic of Chapter 3. There, we develop point and interval estimate of model parameters and ways we could test hypotheses regarding those parameters.

II Design of Experiments In the second part we present the basic ideas of experimental design and analysis. We cover the DOE for linear and generalized linear models. Chapter 4 covers factorial designs in general, the 2^k design, and fractional two-level designs. Chapter 5 covers designs for response surfaces.

III Regression Models for Reliability Studies This part consists of two chapters. Chapter 6 covers parametric regression models. This includes models on transformed data, exponential regression, and Weibull regression. Chapter 7 covers semi-parametric regression models including the Cox proportional hazards model.

IV Experimental Design for Reliability Studies The final part addresses experimental designs for reliability studies. Chapter 8 covers tests done under a single test condition. Chapter 9 covers multiple-factor experiments, including accelerated life tests.

The material in this book requires a one- or two-semester course in probability and statistics that uses some calculus. Readers with a background in reliability but not DOE can skip Part I and proceed to Part II on experimental design, and then to Parts III and IV. Readers with a background in experimental design but not reliability can begin with Part I, skip Part II, and proceed to Parts III and IV. Those who are well versed in both reliability and DOE can proceed directly to Parts III and IV.

The book's companion web site contains the data sets used in the book, along with the R and JMP code used to obtain the analyses. The web site also contains lists of known errors in the book.

Steven E. Rigdon
Saint Louis
26 November 2021

Rong Pan, Douglas C. Montgomery
Tempe
26 November 2021

Laura J. Freeman
Arlington
26 November 2021

About the Companion Website

This book is accompanied by a companion website:

www.wiley.com/go/rigdon/designexperiments

The website includes data sets and computer code.

Part I

Reliability

1

Reliability Concepts

1.1 Definitions of Reliability

It is difficult to define reliability precisely because this term evokes many different meanings in different contexts. In the field of reliability engineering, we primarily deal with engineered devices and systems. Single-word descriptions may depict one or two aspects of reliability in an engineering application context, but they are inadequate for a technical definition of *engineering reliability*. So, how do engineers and technical experts define *reliability*?

- Radio Electronics Television Manufacturers Association (1955) – "Reliability is the probability of a device performing its purpose adequately for the period of time intended under the operating conditions encountered."
- ASQ (2020) – "Reliability is defined as the probability that a product, system, or service will perform its intended function adequately for a specified period of time, or will operate in a defined environment without failure."
- Meeker and Escobar (1998a) – "Reliability is often defined as the probability that a system, vehicle, machine, device, and so on will perform its intended function under operating conditions, for a specified period of time."
- Condra (2001a) – "Reliability is quality over time."
- Yang (2007) - "Reliability is defined as the probability that a product performs its intended function without failure under specified conditions for a specified period of time."

There are some variations in the aforementioned definitions, but they all either explicitly or implicitly state the following characteristics of reliability:

- Reliability is a probabilistic measure – the probability of a functioning product, service, or system.
- Reliability is a function of time – the probability function of successfully performing tasks, as designed, over time.
- Reliability is defined under specified or intended operating conditions.

We define a function, $S(t)$, to be the survival, or reliability, function, which is the probability of the product, service, or system being successfully operated under its normal operating condition at time t; in other words, the unit survived past time t.

Design of Experiments for Reliability Achievement, First Edition.
Steven E. Rigdon, Rong Pan, Douglas C. Montgomery, and Laura J. Freeman.
Companion website: www.wiley.com/go/rigdon/designexperiments

1.2 Concepts for Lifetimes

When an item fails, the "fix" sometimes involves making a repair to bring it back to a working condition. Another possibility is to discard the item and replace it with a working item. In general, the more complex a system is, the more likely we are to repair it, and the simpler it is the more likely we are to scrap it and replace it with a new item. For example, if the starter on our automobile fails, we would probably take out the old starter and replace it with a new one. In a case like this, the automobile is a *repairable* system, but the starter is *nonrepairable* since our fix has been to replace it entirely.

Since complex systems, which are usually repairable, are made up of component parts that are nonrepairable, we will focus in this book on nonrepairable items. If these nonrepairable items are designed and built to have high reliability, then the system should be reliable as well. For nonrepairable systems we are interested in studying the distribution of the time to the first (and only) failure, or more generally, the effect of predictor variables on this lifetime. This lifetime need not be measured in calendar time; it could be measured in operating time (for an item that is switched on and off periodically), miles driven (for a motor vehicle like a car or truck), copies made (for a copier or printer), or cycles (for an industrial machine). For nonrepairable systems, we study the occurrence of events in time, such as failures (and subsequent repairs) or recurrence of a disease or its symptoms. See Rigdon and Basu (2000) for a treatment of repairable systems.

The lifetime T of a unit is a random variable that necessarily takes on nonnegative values. Usually, but not always, we think of T as a continuous random variable taking on values in the interval $[0, \infty)$. There are various forms that the distribution may take, many of which, including the exponential, Weibull and gamma, are presented in detail Chapter 2. Here we present the fundamental ideas and terms for continuous random variables.

Definition 1.1 *The probability density function (PDF) of a continuous random variable T is a function $f(t)$ with the property that*

$$P(a < T < b) = \int_a^b f(t) \ dt.$$

Thus, probabilities for a continuous random variable are found as areas under the PDF. (See Figure 1.1a.) Note that a and b can be $-\infty$ or ∞. Since $\int_a^a f(t) \ dt = 0$, the probability that $T = a$, that is, the probability that T *equals* a particular value a, is equal to zero. This also implies that

$$P(a < T < b) = P(a < T \leq b) = P(a \leq T < b) = P(a \leq T \leq b) = \int_a^b f(t) \ dt.$$

See Figure 1.1b.

Since the probability is 1 that T is between $-\infty$ and ∞, we have the property that

$$\int_{-\infty}^{\infty} f(t) \ dt = 1. \tag{1.1}$$

See Figure 1.1c. Also, since all probabilities must be nonnegative, the PDF $f(t)$ must satisfy

$$f(t) \geq 0, \quad \text{all } t. \tag{1.2}$$

The results in (1.1) and (1.2) are the fundamental properties for a PDF.

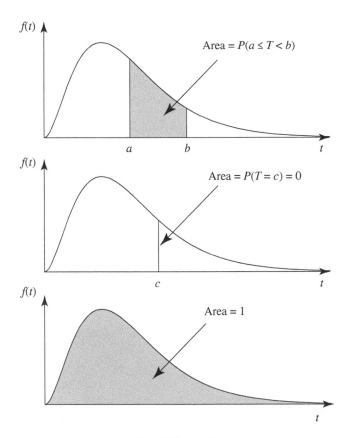

Figure 1.1 Properties of the PDF for a lifetime distribution.

The development earlier makes no assumption about the possible values that the random variable T can take on. For lifetimes, which must be nonnegative, we have $f(t) = 0$ for $t < 0$. Thus, for lifetimes, the PDF must satisfy

$$f(t) \geq 0, \quad \text{all } t,$$
$$f(t) = 0, \quad t < 0,$$
$$\int_0^\infty f(t) \, dt = 1.$$

The set of values for which the PDF of the random variable T is positive is called the *support* of T. The support for a lifetime distribution is $[0, \infty)$, although for some distributions we exclude the possibility of $t = 0$.

Note that the PDF does not give probabilities directly; for example, $f(4)$ does not give the probability that $T = 4$. Rather, as an approximation we can write

$$P(4 < T < 4 + \Delta t) = \int_4^{4+\Delta t} f(t) \, dt \approx \Delta t \, f(4).$$

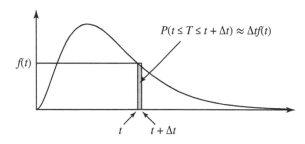

Figure 1.2 Approximation $P(t \le T \le t + \Delta t) \approx \Delta t f(t)$.

(See Figure 1.2.) Thus, the PDF can be interpreted as

$$\Delta t \, f(t) \approx P(t < T < t + \Delta t), \qquad (1.3)$$

or equivalently

$$f(t) \approx \frac{P(t < T < t + \Delta t)}{\Delta t}.$$

To be precise, the PDF is equal to the limit of the right side earlier as $\Delta t \to 0^+$:

$$f(t) = \lim_{\Delta t \to 0^+} \frac{P(t < T < t + \Delta t)}{\Delta t}.$$

Definition 1.2 *The cumulative distribution function (CDF) of the random variable T is defined as*

$$F(t) = P(T \le t) = \int_{-\infty}^{t} f(x) \, dx, \qquad (1.4)$$

where $f(x)$ is the PDF for T.

Note that we have changed the variable of integration from t to x, in order to avoid confusion with the upper limit on the integral. For a lifetime distribution with support $(0, \infty)$, we have the result

$$F(t) = P(T \le t) = P(0 \le T \le t) = \int_{0}^{t} f(x) \, dx. \qquad (1.5)$$

As $t \to \infty$ on the right side earlier, the integral goes to $\int_{0}^{\infty} f(x) \, dx$, which equals 1. Also, since the probability of having a negative lifetime is 0, the CDF must be zero for all $t < 0$. Finally, since the CDF "accumulates" probability up to and including t, increasing t can only increase (or hold constant) the CDF. Thus, for a lifetime distribution, the CDF must satisfy

$$F(t) = 0, \quad \text{for } t \le 0,$$

$$\lim_{t \to \infty} F(t) = 1,$$

$$F(t) \text{ is nondecreasing.}$$

Equation (1.4) shows how to get the CDF given the PDF. A formula for the reverse (getting the PDF from the CDF) can be obtained by differentiating both sides of (1.4) with respect to t and applying the fundamental theorem of calculus:

$$F'(t) = \frac{d}{dt} \int_{-\infty}^{t} f(x) \, dx = f(t). \qquad (1.6)$$

In other words, the PDF is the derivative of the CDF.

Definition 1.3 *The survival function, or reliability function, is defined to be*

$$S(t) = P(T > t) = \int_t^\infty f(x) \, dx.$$

In other words, $S(t)$ is the probability that an item survives past time t, while $F(t)$ is the probability that it fails at or before time t (that is, that it *doesn't* survive past time t). Thus, $S(t)$ and $F(t)$ are related by

$$S(t) = 1 - F(t).$$

One of the most important concepts in lifetime analysis is the hazard function.

Definition 1.4 *The hazard function is*

$$h(t) = \lim_{\Delta t \to 0^+} \frac{P(t \le T \le t + \Delta t | T > t)}{\Delta t} \tag{1.7}$$

As an approximation, we can write

$$\Delta t \, h(t) \approx P(t \le T \le t + \Delta t | T > t) \tag{1.8}$$

analogous to (1.3).

The probability in the definition of the hazard is a *conditional* probability; it is conditioned on survival to the beginning of the interval. This is a natural quantity to consider because it makes intuitive sense to talk about the failure probability of an item that is still working. It is conceptually more difficult to talk about the probability of an item failing if the item might or might not be working. If we replace the conditional probability in the definition of the hazard function with an unconditional probability, we get

$$\lim_{\Delta t \to 0^+} \frac{P(t \le T \le t + \Delta t)}{\Delta t} \tag{1.9}$$

which is equal to the PDF $f(t)$. Thus, the PDF is the (limit of) the probability of failing in a small interval when viewed *before* testing begins. The hazard is the (limit of) the probability of failing in a small interval for a unit that is known to be working.

The hazard function can be written as

$$
\begin{aligned}
h(t) &= \lim_{\Delta t \to 0^+} \frac{P(t < T \le t + \Delta t | T > t)}{\Delta t} \\
&= \lim_{\Delta t \to 0^+} \frac{P(t < T \le t + \Delta t \wedge T > t)}{\Delta t P(T > t)} \\
&= \lim_{\Delta t \to 0^+} \frac{P(t < T \le t + \Delta t)/\Delta t}{P(T > t)} \\
&= \frac{f(t)}{S(t)}.
\end{aligned} \tag{1.10}
$$

Indeed, many books *define* the hazard function in this way. We choose to define the hazard as the limit of a conditional probability because this intuitive concept is helpful for understanding the failure mechanism.

To illustrate the difference between hazard and density, consider a discrete case, say, where items are placed on test and are observed every 1000 hours. Let h_i denote the probability that an item fails in the ith interval $(1000(i-1), 1000i]$. Suppose first that $h_i = \frac{1}{10}$, so that there is a probability of $\frac{1}{10} = 0.1$ that a working unit will fail in any time interval. Thus, at the end of a time interval when we inspect those units still operating, we would expect that about one-tenth of them would fail. We could naturally ask the question "What is the probability that a unit fails in the ith interval?" This is different from the question "What is the probability that a unit that is currently operating fails in the ith interval?" The difference is that the latter is a conditional probability (conditioned on the unit still operating), whereas the former is an unconditional probability. The answer to the latter question is: the hazard h_i. To get at the answer to the latter question, we can observe that

$$p_1 = P(\text{failure in } (0, 1000]) = \frac{1}{10}$$

$$p_2 = P(\text{failure in } (1000, 2000])$$

$$= P(\text{no failure in } (0, 1000]) P(\text{failure in } (1000, 2000] | \text{no failure in } (0, 1000])$$

$$= \frac{9}{10} \frac{1}{10}$$

$$p_3 = P(\text{failure in } (2000, 3000])$$

$$= P(\text{no failure in } (0, 1000])$$

$$\times P(\text{no failure in } (1000, 2000] | \text{no failure in } (0, 1000])$$

$$\times P(\text{failure in } (2000, 3000] | \text{no failure in } (0, 2000])$$

$$= \frac{9}{10} \frac{9}{10} \frac{1}{10},$$

and in general,

$$p_i = P(\text{failure in } (1000(i-1), 1000i])$$

$$= P(\text{no failure in } (0, 1000])$$

$$\times P(\text{no failure in } (1000, 2000] | \text{no failure in } (0, 1000])$$

$$\times \cdots$$

$$\times P(\text{failure in } (1000(i-1), 1000i] | \text{no failure in } (0, 1000(i-1)])$$

$$= \left(\frac{9}{10}\right)^{i-1} \frac{1}{10}, \quad i = 1, 2, \ldots$$

This is, of course, the geometric distribution. Plots of h_i and p_i for the case of a constant (discrete) hazard are shown in Figure 1.3.

Suppose now that the probability mass function (rather than the hazard function) is constant, with $p_i = 0.1$. Since $\sum p_i = 1$, we conclude that $p_i = 0.1$ for $i = 1, 2, \ldots, 10$. The hazard is then

$$h_i = P(\text{failure in interval } i | \text{survival to beginning of interval } i)$$

$$= \frac{P(\text{failure in interval } i)}{P(\text{survival to beginning of interval } i)}$$

$$= \frac{P(\text{failure in interval } i)}{1 - P(\text{death before beginning of interval } i)}$$

$$= \frac{1/10}{1 - (i-1)/10}$$

$$= \frac{1}{10 - i + 1}.$$

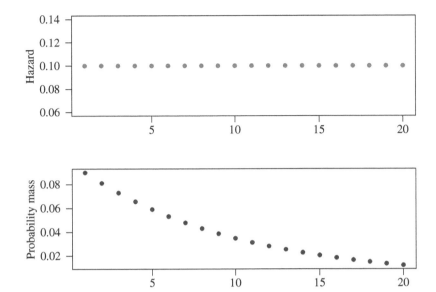

Figure 1.3 Hazard and probability mass function for the case of constant hazard.

Thus, $h_1 = \frac{1}{10}, h_2 = \frac{1}{9}, \ldots, h_{10} = \frac{1}{1} = 1$. The last number, $h_{10} = 1$, may seem a little surprising, but if $p_i = 0$ for $i > 10$, and if an item hasn't failed up through interval $i = 9$, then it *must* fail at time $i = 10$. The hazard and probability mass for the constant probability case are shown in Figure 1.4.

The cumulative hazard is defined to be the accumulated area under the hazard function. To be precise, the cumulative hazard is defined to be

$$H(t) = \int_0^t h(u)\, du. \tag{1.11}$$

Any one of the PDF, CDF, survival function, hazard, or cumulative hazard function is enough to determine the lifetime distribution. In other words, knowing any one of these can get you all of the others. For example, Eq. (1.5) shows how you can get the CDF if you know the PDF. If we know the hazard function $h(t)$, we can use the relationship

$$h(t) = \frac{f(t)}{S(t)} = -\frac{S'(t)}{S(t)}$$

to find $S(t)$. To see this, notice that this is a simple first-order linear differential equation with initial condition $S(0) = 1$, which can be solved by integrating both sides from $u = 0$ to $u = t$. This yields

$$\int_0^t h(u)\, du = \int_0^t -\frac{S'(u)}{S(u)}\, du = -\log S(t) + \log S(0) = -\log S(t) \tag{1.12}$$

from which we obtain the relationship

$$S(t) = \exp\left(-\int_0^t h(u)\, du\right). \tag{1.13}$$

We leave the other relationships as an exercise. Table 1.1 shows most of the relationships.

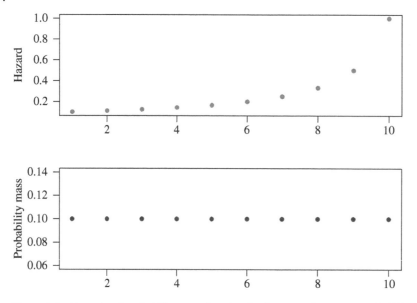

Figure 1.4 Hazard and probability mass function for the case of constant probability mass.

1.3 Censoring

A life testing experiment, where many units are operating simultaneously, may be terminated before all of the units have failed. This is a typical scenario, because many units will be highly reliable and will not fail during a test. This leads to the concept of censoring, where we cannot observe the lifetime of an item because of the design of the life testing experiment.

Definition 1.5 *When we cannot observe the exact failure time of an item, but rather we can only observe an interval in which the failure was observed, we say that the observation is* **censored**. *This interval could be an unbounded interval, such as* (c, ∞) *or a bounded interval such as* $(0, a)$.

The most common type of censoring occurs when the life test is stopped before all items have failed. Let τ denote the censoring time, that is, the time of termination of the test. In this case, we would know the exact failure times of all items that failed before time τ, but for those still operating at the end of the experiment, we know only that the failure would occur past time τ. For those item still operating, we would only know that the failure time was in the interval (τ, ∞). Actually, the censoring time need not be the same for all items. For example, if the items were placed into service at different times, then the censoring times would be different even if the test was terminated at the same (calendar) time. This type of lifetime censoring, where we observe a survival event (where we know that the failure must occur in the interval (τ, ∞)), is called **right censoring**.

Table 1.1 General relationships between PDF, CDF, hazard function, and cumulative hazard function.

	PDF $f(t)$	CDF $F(t)$	Hazard $h(t)$	Cumulative hazard $H(t)$
$f(t)$	—	$F(t) = \int_0^t f(u)\,du$	$h(t) = \dfrac{f(t)}{\int_t^\infty f(u)\,du}$	$H(t) = \exp\left(\dfrac{f(t)}{\int_t^\infty f(u)\,du}\right)$
$F(t)$	$f(t) = F'(t)$	—	$h(t) = \dfrac{F'(t)}{1 - F(t)}$	$H(t) = -\log\left(1 - F(t)\right)$
$h(t)$	$f(t) = h(t)\exp\left(-\int_0^t h(u)\,du\right)$	$F(t) = 1 - \exp\left(-\int_0^t h(u)\,du\right)$	—	$H(t) = \int_0^t h(u)\,du$
$H(t)$	$f(t) = H'(t)\exp(-H(t))$	$F(t) = 1 - \exp(-H(t))$	$h(t) = H'(t)$	—

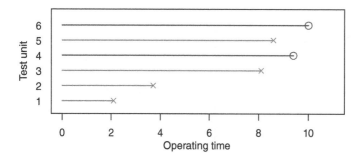

Figure 1.5 Times to failure for six items. The X indicates a failure time and the O indicates a censored time.

Figure 1.5 shows an illustration of right censoring. In this situation six items were placed on test at the same time. Items 1, 2, 3, and 5 failed during the test at times 2.1, 3.7, 8.1, and 8.6, respectively. The other two units were still operating at times 9.4 and 10.0. The failure times are denoted by an X and the censoring times are denoted by an O. The solid lines cover the times for which it is known that the items were operating.

It sometimes happens that observation of an item may begin well after the it was placed into service. This can occur when items are observed in the field. In some instances, it may be the case that the item is observed to have already failed when it is first inspected. For example, suppose an item is placed into service and then not inspected until an age of 100 days. If it was observed to be in a failed condition at that time, then we know only that it failed before time $t = 100$. In other words, the failure must have occurred sometime in the interval $(0, 100)$, but the exact failure time is unknown. This is called **left censoring**, because the failure is known to have occurred before time 100. Figure 1.6 illustrates the concept of left (and right) censoring. Here, items 1, 2, and 4 were placed on test and the failure times were observed to be 24, 37, and 77, respectively. Items 3 and 6 were observed at times 10 and 28, respectively, to be in a failed condition. A dashed line is used to indicate the possible times of actual failure. In general, we use a solid line to indicate times when the units were known to be operating. Finally, items 5 and 7 were still operating at time 99 when the test was terminated. Thus, items 3 and 6 were left censored and items 5 and 7 were right censored.

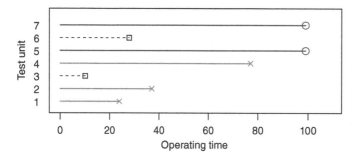

Figure 1.6 Times to failure for seven items. The X indicates a failure event and the O indicates a right censoring event, whereby the failure occurred to the right of the of the O. The dashed line that ends with a □ indicates a left censoring event, whereby the failure occurred to the left of the □.

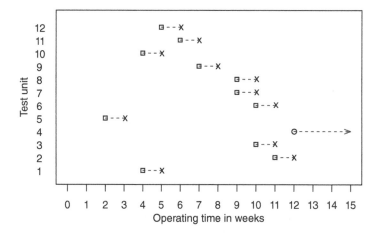

Figure 1.7 Intervals containing the failures for items that are inspected only once a week. The dashed lines indicate possible failure times. One item, the fourth one, was right censored at time $t = 12$. All other observations were interval censored. The □ on the left indicates that this is the left endpoint of the censoring interval. The X on the right indicates that the item had failed before the end of the interval.

Another type of censoring occurs when items are inspected only periodically and observed to be either operating or failed. For example, the items may be inspected once every week. If an item was observed at time $t = 2$ to be operating and at time $t = 3$ to be failed, then it is known that the failure must have occurred in the interval (2,3]. This is called **interval censoring**. The concept is illustrated in Figure 1.7.

Two special kinds of right censoring are called Type I and Type II censoring. For both we assume that all items were placed on test at the same time. Under **Type I** censoring, we terminate the test at a predetermined time T, whereas under **Type II** censoring, we terminate the test after a predetermined number r of failures. Thus, under Type I censoring, the testing time is fixed, but the number of failures is random. Conversely, for Type II censoring the number of failures is fixed, but the testing time is random. For planning reliability experiments it is important to note that time based censoring can result in a failed experiment if the time allowed is not sufficient for achieving the required number of failures for estimating the reliability distribution. Alternatively, Type II censoring guarantees a certain number of failures, but can lead to long experiment timelines.

Often, the censoring mechanism does not fit either of these categories. For example, items are placed on test at random points in time, and some are removed from test for reasons unrelated to the life test. This situation is called *arbitrary censoring*. We hypothesize two random variables T_i and C_i, where T_i is the failure time and C_i is the censoring time. We observe only the first of these to occur, that is, $U_i = \min(T_i, C_i)$. We define the indicator variable δ_i to be 1 if $T_i = \min(T_i, C_i)$ and 0 if $C_i = \min(T_i, C_i)$. Thus, $\delta_i = 1$ if we observe a failure, and $\delta = 0$ if the failure observation is right censored. In most reliability data sets, we will observe the event times $u_1, u_2, \ldots, u_n$ (i.e. the times of either failure or censoring) along with the corresponding indicators $\delta_1, \delta_2, \ldots, \delta_n$. These two sets of variables completely describe the observed failure process.

When the systems under test are repairable, we usually observe multiple failures per unit. These can be represented in a graph, similar to those shown earlier, but with multiple failures per line. In Figure 1.8, we see two repairable systems whose failure times are marked with an X on the time axis. System 1

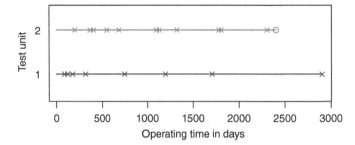

Figure 1.8 Failure times for two repairable systems. For the first system (bottom), the period of observation ended at the time of the last failure. For the second system, the period of observation ended *after* the last failure; this is indicated by the O at the end of the line segment.

(on the bottom part of the figure) is **failure truncated**, that is, the termination of the test coincided with the time of the last failure. System 2 (top) is **time truncated**, that is, the test terminated after the last failure. The fact that system 2 continued to operate failure free from time 1955 to time 2400 is a piece of information that cannot be ignored.

Note the terminology used to describe the termination of a life test. Nonrepairable systems whose (one and only) failure occur in some interval (not at some point in time) are said to be censored. A repairable system whose period of observation ends at some point (either at a failure time or extends past the last observed failure) is said to be truncated. Since we deal almost exclusively with nonrepairable systems, we will refer to censored systems often.

Problems

1.1 Suppose that there are only two possible lifetimes for an item. The lifetime will be either 0 or 1, with equal probability. Let T denote the lifetime. Find
(a) the probability mass function
(b) $E(T)$
(c) the hazard function.

1.2 Rework Problem 1.1, but assume that the probabilities of being 0 and 1 are p and $1 - p$.

1.3 Suppose that the lifetime is a discrete distribution with equal mass at $t = 0, 1, \ldots, A$. Find
(a) the probability mass function
(b) $E(T)$
(c) the hazard function.

1.4 Repeat Problem 1.3, except assume that the probability that the lifetime T equals i is p_i for $i = 1, 2, \ldots, A$ where $p_0 + p_1 + \cdots + p_A = 1$.

1.5 Suppose that an item is observed each day at the same time, and it is observed whether the item is or isn't working. During each day there is a constant probability p that a working item will fail. The support for the lifetime T is therefore $1, 2, \ldots$. Find

(a) the probability mass function

(b) $E(T)$

(c) the hazard function.

1.6 Your town announces that it will test the new severe weather warning system at noon on some day next week (Sunday through Saturday). They will pick a day at random with equal probability to conduct the test.

(a) What is the probability that the test is conducted on Saturday?

(b) If it is Saturday morning and the test has not yet been conducted, what is the probability that it occurs on Saturday?

(c) If it is Friday morning and the test has not yet been conducted, what is the probability that the test occurs on Friday?

1.7 Suppose that the distribution of a lifetime T has PDF

$$f(x) = \begin{cases} \dfrac{1}{(x+1)^2} & \text{if } x \geq 0, \\ 0 & \text{if } x < 0. \end{cases}$$

Find

(a) the CDF

(b) $E(T)$

(c) the hazard function.

1.8 Suppose that the distribution of a lifetime T has PDF

$$f(x) = \begin{cases} \dfrac{1}{2}\, e^{-x/2} & \text{if } x \geq 0, \\ 0 & \text{if } x < 0. \end{cases}$$

Find

(a) the CDF

(b) $E(T)$

(c) the hazard function.

1.9 Suppose that the hazard function for a lifetime distribution is

$$h(t) = t, \qquad t \geq 0.$$

Find

(a) the PDF

(b) the CDF

(c) $E(T)$.

1.10 Suppose that the hazard function for a lifetime distribution is

$$h(t) = \sqrt{t}, \qquad t \geq 0.$$

Find
(a) the PDF
(b) the CDF
(c) $E(T)$.

1.11 Suppose that n items are put on test at the same time. Let $T_1, T_2, \ldots, T_n$ denote the lifetimes, which we assume come from the distribution with the PDF given in Problem 1.7. The test will be terminated at time $t = 5$.
(a) What is the probability that one particular item will be censored?
(b) If we denote by X the number of items out of the n that fail before the censoring time $t = 5$, what is the distribution of X?
(c) Find $E(X)$.
(d) If we want the expected number of failed items to be at least 20, how large must n be?
(e) Describe the kind of censoring involved in this problem.

1.12 Repeat Problem 1.11 with the distribution from Problem 1.8.

1.13 Suppose that n items are put on test and then tested weekly, say, every Monday. Let T is the lifetime of these items, measured in hours. Describe the kind of censoring involved in this situation.

1.14 Suppose that $n = 10$ items are placed on test as described in the previous problem. The observed quantity X is the first week at which an item is not working. For example, if an item was working after week 3, but was not working when it was observed at week 4, then $x = 4$. The test is terminated after time 5 and the observations are

3	3	5	2	1	5+	3	1	4	5+

where a + indicates a right censored observation. Sketch a graph like the one in Figure 1.7 for this data.

1.15 In Eqs. (1.6), (1.5), (1.13), (1.12), and (1.11) we derived several of the relationships among the PDF, CDF, hazard, and cumulative hazard. For this problem, derive the other relationships:
(a) PDF → hazard
(b) hazard → PDF
(c) PDF → cumulative hazard
(d) cumulative hazard → PDF
(e) CDF → cumulative hazard
(f) cumulative hazard → CDF
(g) cumulative hazard → hazard.

2

Lifetime Distributions

While the normal distribution is a useful model for a number of circumstances, it is generally a poor model for lifetimes. Usually lifetimes have a distribution that is skewed to the right, that is, there are more observations far in the right tail of the distribution than in the left tail. Also, lifetimes have a natural lower boundary of zero (since lifetimes cannot ever be negative).

In this chapter, we present a number of lifetime distributions, including the exponential, Weibull, gamma, lognormal, and log-logistic. We also consider some distributions that are related to these, such as the extreme value distribution. The choice of a lifetime distribution is important, because these distributions have different properties, especially in the left and right tails. In reliability applications, making an inference about the behavior in the tails is often important.

2.1 The Exponential Distribution

The simplest model for lifetimes is the exponential distribution, which has the probability density function (PDF)

$$f(t|\theta) = \frac{1}{\theta} \exp(-t/\theta), \quad t > 0. \tag{2.1}$$

The survival function is

$$S(t|\theta) = \exp(-t/\theta), \quad t > 0 \tag{2.2}$$

and the hazard function is

$$h(t|\theta) = \frac{1}{\theta}, \quad t > 0. \tag{2.3}$$

The cumulative distribution function (CDF) is related to the survival function by

$$F(t|\theta) = 1 - S(t|\theta) = P(T \le t) = 1 - \exp(-t/\theta). \tag{2.4}$$

The exponential distribution is characterized by a single parameter, θ, which is equal to the mean (and also to the standard deviation) of the distribution. If T has an exponential distribution with mean θ, then we will write $T \sim \text{EXP}(\theta)$. Thus, if $T \sim \text{EXP}(\theta)$, then

$$E(T) = \theta \tag{2.5}$$

Design of Experiments for Reliability Achievement, First Edition.
Steven E. Rigdon, Rong Pan, Douglas C. Montgomery, and Laura J. Freeman.
© 2022 John Wiley & Sons, Inc. Published 2022 by John Wiley & Sons, Inc.
Companion website: www.wiley.com/go/rigdon/designexperiments

and

$$V(T) = \theta^2. \tag{2.6}$$

Figure 2.1 shows the PDFs, survival functions, and hazard functions for the exponential distributions with means $\theta = 10$, $\theta = 20$, and $\theta = 40$.

The simplicity of having just one parameter is also a drawback because the one parameter limits the flexibility of the model. The hazard function for the exponential distribution, given in (2.3), is independent of t. This implies that the hazard function, which is the (limit of) the probability of failure in a small interval divided by the length of the interval, is constant regardless of the age of the unit. Most mechanical systems have an increasing hazard function; that is, as the unit ages, the probability of an imminent failure increases. This is because the moving parts wear and often the connections among parts become less reliable. For example, a car engine that has run for 100,000 miles is in a weaker condition (i.e. is less reliable) than one that has run for 10,000 miles. Electrical systems with no moving parts are a different matter. Often, the wiring does not degrade, or degrades very little, so that a two-year-old part is essentially equivalent to a new part. Thus, the exponential distribution can sometimes be used as a model for electrical

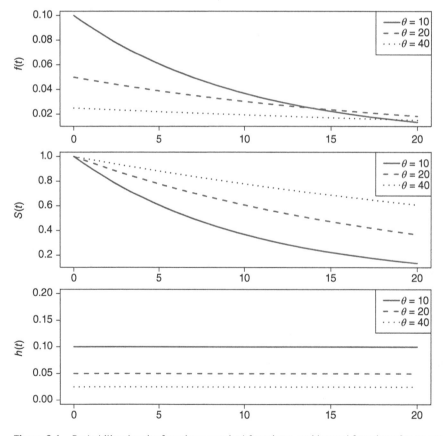

Figure 2.1 Probability density functions, survival functions, and hazard functions for several exponential distributions.

parts. Another aspect of reliability that cannot be modeled by the exponential distribution is that of infant mortality, followed by a gradually increasing hazard. Computer hard drives (which are both electrical and mechanical) provide a good example of this. Some drives have one or more manufacturing flaws that will cause failure at, or soon after, the time that they are first put into service. Most drives, however, do not have this type of flaw and will work without failure for a long time. Only after the mechanical parts begin to wear or there is a sudden electrical failure will the drive fail. This phenomenon leads to a "bathtub" shaped hazard function, as shown in Figure 2.2. Initially, the hazard is large, but it decreases rapidly. Then after being relatively flat for a while, the hazard increases as the parts wear out. The exponential distribution, having a perfectly flat hazard function, cannot model this situation. (Most two-parameter distributions, such as the Weibull or gamma, described in the next sections, cannot model this situation either.) For systems that have an inherently increasing (or decreasing) hazard function, the exponential distribution is too simple, and more complicated models, such as the Weibull or gamma distribution, are required.

In addition to having a constant hazard, the exponential distribution also has related property, called the *memoryless property*. That is, if $T \sim \text{EXP}(\theta)$, then

$$P(T > t + x | T > t) = P(T > x). \tag{2.7}$$

This means that conditioned on the lifetime T being greater than t, the probability that a unit survives an additional x time units is the same as the probability that a brand new unit survives for x time units. In fact, it is the same as saying that the hazard rate of the unit is a constant regardless of the age of the unit. To put this in perspective, suppose that the unit is an automobile and "time" is measured in miles. The memoryless property then says that a car with 200,000 miles has the same chance of failing as a brand new car. (Here, by "failing," we mean a catastrophic failure in a fixed interval, such as one week, that causes the automobile to be scrapped. We do not mean a condition that can be repaired, bringing the automobile to be returned to an operating condition.) Since a car with 200,000 miles has a much higher chance of failing than a new car (or a car with 20,000 miles), the memoryless assumption is not realistic for this example.

If the exponential distribution is assumed for the lifetime, then the memoryless property must be assumed as well. This limits the applicability of the exponential distribution for modeling real lifetimes.

The parameterizations of the exponential distribution given in (2.1), (2.2), and (2.3) provide one way to write the exponential distribution. Written in this way, the one parameter of the exponential

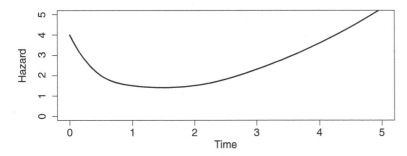

Figure 2.2 Example of a bathtub-shaped hazard function, which cannot be modeled by the exponential distribution.

distribution, θ, is equal to the mean. There is another parameterization that is sometimes used instead. Under this parameterization, the PDF, survival function, and hazard function are written as

$$f(t|\lambda) = \lambda \exp(-\lambda t), \; t > 0, \tag{2.8}$$

$$S(t|\lambda) = \exp(-\lambda t), \; t > 0, \tag{2.9}$$

and

$$h(t|\lambda) = \lambda, \; t > 0. \tag{2.10}$$

The parameter λ is called the **failure rate** and its units are in failures per time. The obvious relationship between these two parameterizations is $\lambda = 1/\theta$. Thus, with this parameterization, the mean of the distribution is $E(T) = 1/\lambda$ and the hazard function is its reciprocal, $h(t) = \lambda$. The choice of which parameterization to use may depend on which property is more important: the mean or the hazard. Of course, both parameterizations are essentially equivalent. Although not universal, the parameterization given in (2.1), (2.2), and (2.3) is usually used in the engineering literature, and the parameterization in (2.8), (2.9), and (2.10) is usually used in the biomedical literature.

Example 2.1 An example of a case where the exponential distribution may be applicable is given in Bartholomew (1957), also cited in Lawless (2003, pp. 4–5). The data are shown in Table 2.1. Ten pieces of equipment (of unspecified type) were put into service at different points in time and the days until failure were recorded. Bartholomew (1957), in order to illustrate the concept of censoring, artificially created a cutoff date in order to have censored observations. (Lawless (2003) used the censored values imposed by Bartholomew (1957), but here we present the complete uncensored data set.) We are using the full data set which has no censored values. A plot of the empirical CDF, together with the CDF estimated under the exponential distribution, are shown in Figure 2.3. (We will discuss in more detail the empirical CDF and estimation of the exponential parameters in Section 3.3.) The two functions in Figure 2.3 seem to agree very closely, indicating that the exponential distribution is a good fit. For small samples such as this, it is often difficult to detect a departure from any specific distribution such as the exponential. ■

It is a reasonable question to ask, especially with regard to human mortality, "what is the additional expected lifetime conditioned on survival to a certain point?" For example, one might ask what the expected additional life is given survival to age 58. To calculate such a property, we would need the conditional PDF of the lifetime T conditioned on survival to some time t. For the exponential distribution, the conditional survival function is simply

$$S(x|t) = P(T > x | T > t) = \exp(-(x - t)/\theta), \quad x > t. \tag{2.11}$$

Table 2.1 Failure times in days for equipment given in Bartholomew (1957).

i	1	2	3	4	5	6	7	8	9	10
T_i	2	119	51	77	33	27	14	24	4	37

Source: Modified from Bartholomew (1957).

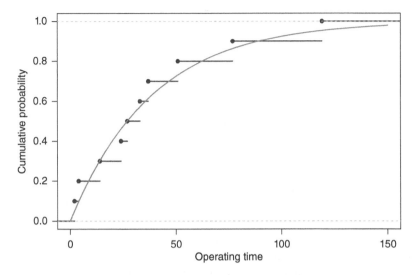

Figure 2.3 Empirical CDF (the step function) and the CDF estimated assuming an exponential distribution (solid curve).

The conditional PDF is then the negative of the derivative:

$$f(x|t) = -D_x S(x|t) = (1/\theta) \exp\left(-(x-t)/\theta\right), \; x > t. \tag{2.12}$$

The expected additional lifetime, called the *mean residual life* (MRL) and denoted MRL(), is then defined as

$$\text{MRL}(t) = E(T|T > t) - t \tag{2.13}$$

and is obtained by integrating the conditional PDF:

$$E(T|T > t) = \int_t^\infty (1/\theta) \exp\left(-(x-t)/\theta\right) - t = \theta. \tag{2.14}$$

Knowing that the exponential distribution has the memoryless property, this result should not be surprising: the MRL given survival to time t is just an additional θ time units, which is the same as the expected life for a brand new item. Thus, for the exponential distribution, $\text{MRL}(t) = \theta$ is independent of the age of the item. In other words, no matter how long an item has been operating, its expected additional lifetime is the same as its expected lifetime when it was brand new.

Percentiles for the exponential distribution are easily obtained from the CDF or the survival function. We define t_p, the pth percentile of the random lifetime variable, as that value of t_p for which $P(T \le t_p) = p$, for $0 < p < 1$. It is obtained by setting

$$F(t_p) = 1 - \exp(-t_p/\theta)$$

equal to p and solving for t_p; this yields

$$t_p = -\theta \log(1 - p). \tag{2.15}$$

For example, if the mean life is $\theta = 60$, then the 90th percentile is found to be

$$t_p = -60 \log(1 - 0.90) = 138.2,$$

meaning that 90% of all lifetimes will be 138.2 or less, or equivalently, only 10% of all lifetimes will exceed 138.2.

In the last example, we saw a small data set that seemed to fit the exponential distribution well. When the exponential distribution assumptions, such as the constant hazard, the memoryless property or the constant MRL, are not reasonable, or if the data do not seem to fit the exponential distribution, then other models must be considered. In Section 2.2, we look at the Weibull distribution, which is probably the most widely used distribution for lifetimes.

Data that fit the exponential distribution well are rather rare. One reason that the exponential distribution is given so much attention in reliability books is that its simplicity allows us to work through inference (point and interval estimates, hypothesis tests, prediction, goodness-of-fit tests, etc.) *exactly* rather than having to rely on approximations. Thus, the exponential distribution is a "stepping stone" to more complicated models for which approximations and numerical techniques must be used. While true, this pedagogical justification for studying the exponential distribution is rarely admitted in reliability textbooks.

2.2 The Weibull Distribution

The Weibull distribution has two parameters and is more flexible than the exponential distribution, although it cannot model the whole bathtub hazard curve described in Section 2.1. The Weibull distribution is usually parameterized so that the PDF is written as

$$f(t|\theta, \kappa) = \frac{\kappa}{\theta}\left(\frac{t}{\theta}\right)^{\kappa-1} \exp\left[-\left(\frac{t}{\theta}\right)^{\kappa}\right], \quad t > 0. \tag{2.16}$$

The parameters must satisfy $\kappa > 0$ and $\theta > 0$. The survival function and hazard function are then

$$S(t|\theta, \kappa) = \exp\left[-\left(\frac{t}{\theta}\right)^{\kappa}\right], \quad t > 0 \tag{2.17}$$

and

$$h(t|\theta, \kappa) = \frac{\kappa}{\theta}\left(\frac{t}{\theta}\right)^{\kappa-1}, \quad t > 0. \tag{2.18}$$

For the Weibull distribution, θ is a scale parameter and κ is a shape parameters. The parameter θ is sometimes called the **characteristic life** and it is the 0.632 quantile of the distribution regardless of the value of the shape parameter κ. (See Problem 2.16.) Often in other books, the parameters of the Weibull distribution are denoted θ and β; here we use κ rather than β so that we can use β to be a vector of coefficients in lifetime regression. We will cover this in more detail in Part III. If T has a Weibull distribution with parameters θ and κ, then we will write $T \sim \text{WEIB}(\theta, \kappa)$.

From (2.18) we see that the hazard function is proportional to t raised to the power $\kappa - 1$. Thus, if $\kappa > 1$, then the exponent $\kappa - 1$ will be positive and the Weibull hazard function will be increasing. If $\kappa < 1$, then the exponent $\kappa - 1$ will be negative and the Weibull hazard function will be decreasing. If $\kappa = 1$, then the

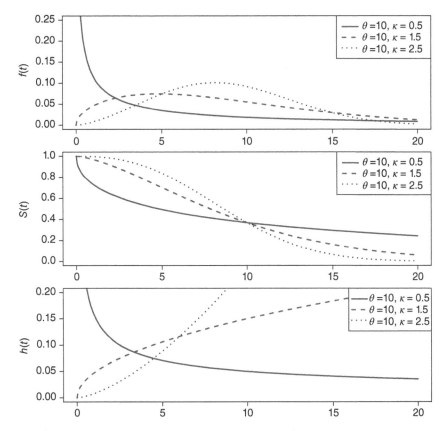

Figure 2.4 PDFs, survival functions, and hazard function s for three Weibull distributions. The three curves correspond to $\kappa = 0.5, 1.5$, and 2.5. $\theta = 10$ for all three graphs.

hazard function reduces to

$$h(t|\theta, 1) = \frac{1}{\theta}\left(\frac{t}{\theta}\right)^{1-1} = \frac{1}{\theta}, \quad t > 0,$$

which is constant. When $\kappa = 1$, the Weibull distribution becomes the exponential distribution. Since the Weibull hazard can be increasing ($\kappa > 1$), decreasing ($\kappa < 1$), or constant ($\kappa = 1$), the Weibull is a much more flexible model than the exponential. Figure 2.4 shows graphs of the PDFs, survival functions, and hazard functions for three different Weibull distributions. Since θ is a scale parameter, we held θ constant at $\theta = 10$ and varied κ ($\kappa = 0.5, 1.5$, and 2.5). For κ smaller than one, the PDF is asymptotic to the y-axis and has a very long right tail. For large κ, the PDF is nearly symmetric and looks much like a normal distribution.

The mean and variance of the Weibull distribution involve the gamma function, which is defined for $a > 0$ to be

$$\Gamma(a) = \int_0^\infty t^{a-1} e^{-t} \, dt.$$

It can be shown that $\Gamma(1) = 1$, and for $a > 0$, $\Gamma(a) = (a-1)\Gamma(a-1)$; from this, and from the fact that $\Gamma(1) = 1$, it follows that if a is a positive integer, then $\Gamma(a) = (a-1)!$ Thus, in some sense, the gamma function is a generalization of the factorial function; that is, it is defined for all positive a and not just for positive integers. If $T \sim \text{WEIB}(\theta, \kappa)$, then the mean and variance can be written as

$$E(T) = \theta\Gamma\left(1 + \frac{1}{\kappa}\right)$$

and

$$V(T) = \theta^2\left[\Gamma\left(1 + \frac{2}{\kappa}\right) - \Gamma^2\left(1 + \frac{1}{\kappa}\right)\right].$$

See Johnson et al. (1994) for a derivation of these results.

The MRL for the Weibull distribution is a bit more involved. The conditional survival function is

$$S(x|t) = P(T > x | T > t) = \exp\left(-(x/\theta)^\kappa + (t/\theta)^\kappa\right), \quad x > t.$$

Integrating the conditional PDF, which is found by differentiating $-S(x|t)$, gives the MRL. For the case of the Weibull distribution this can be written as

$$\text{MRL}(t) = E(T|T > t) = \theta\exp\left[\left(\frac{t}{\theta}\right)^\kappa\right]\Gamma\left(1 + \frac{1}{\kappa}, \left(\frac{t}{\theta}\right)^\kappa\right), \tag{2.19}$$

where $\Gamma(\cdot, \cdot)$ is the *incomplete gamma function* (sometimes called the upper incomplete gamma function) defined by

$$\Gamma(a, x) = \int_x^\infty t^{a-1}e^{-t}\, dt, \quad a > 0,\ x > 0. \tag{2.20}$$

Even though the MRL function for the Weibull distribution cannot be simplified any further, we can still compute the MRL using software such as *Mathematica* (Wolfram Research Inc., 2010), or R (R Core Team, 2020).

For example, if $\kappa = 1.2$ and $\theta = 1$, then some MRL values are

$$\text{MRL}(1) = 0.756, \quad \text{MRL}(2) = 0.688, \quad \text{MRL}(3) = 0.645, \quad \text{MRL}(4) = 0.614.$$

Thus, as an item whose lifetime has distribution WEIB(1, 1.2), can be expected to survive an additional 0.756 time units if it survived to time $t = 1$. If the item manages to survive to time $t = 2$, then it can be expected to survive an additional 0.688 time units, which is less than its expected additional lifetime when it was just one unit old. If the units were to survive to time $t = 3$ or $t = 4$, the MRL would be even smaller.

Consider now, a unit whose lifetime has a Weibull distribution with $\kappa = 0.8$ and $\theta = 1$. Calculations similar to the aforementioned would give

$$\text{MRL}(1) = 1.459, \quad \text{MRL}(2) = 1.592, \quad \text{MRL}(3) = 1.687, \quad \text{MRL}(4) = 1.763.$$

Thus, as the item ages, the MRL gets larger. An item that is four units old can then be expected to survive for a longer additional period than an unit that is only one unit old (or a brand new item, for that matter).

For the Weibull distribution, the shape of the MRL is related to the value of κ. If $\kappa > 1$, then the Weibull has an increasing hazard function and the MRL is decreasing. On the other hand, if $\kappa < 1$, then the hazard function is decreasing and the MRL is increasing. Of course, if $\kappa = 1$, then the hazard and the MRL are both constant, and are equal to $1/\theta$ and θ, respectively.

Because the CDF exists in closed form, we can easily solve for the percentiles of the Weibull distribution. Setting $F(t) = 1 - \exp(-(t_p/\theta)^\beta) = p$ and solving for t_p yields

$$t_p = \theta\left[-\log(1-p)\right]^{1/\kappa}. \tag{2.21}$$

Taking the logarithm of both sides of this expression yields

$$\log t_p = \log \theta + \frac{1}{\kappa} \log\left[-\log(1-p)\right]. \tag{2.22}$$

Just as the exponential distribution has two generally accepted parameterizations, the Weibull distribution also has a second parameterization. As an alternative to (2.16), the Weibull survival function is sometimes written as

$$S(t|\lambda, \gamma) = \exp\left(-\lambda t^\gamma\right), \quad t > 0. \tag{2.23}$$

The parameters must satisfy $\lambda > 0$ and $\gamma > 0$. From this expression for the survival function, the PDF and hazard function are found to be

$$f(t|\lambda, \gamma) = \gamma \lambda t^{\gamma-1} \exp\left(-\lambda t^\gamma\right), \quad t > 0 \tag{2.24}$$

and

$$h(t|\lambda, \gamma) = \gamma \lambda t^{\gamma-1}, \quad t > 0. \tag{2.25}$$

The parameters (θ, κ) and (λ, γ) are related by

$$\gamma = \kappa \quad \text{and} \quad \lambda = \theta^{-\kappa}.$$

If the Weibull distribution is written in this alternative form, which we denote as $\text{WEIB}(\lambda, \gamma)$, then $1/\lambda$ is a scale parameter (often called the **intrinsic failure rate**) and γ is a shape parameter. If software is used to estimate the parameters of the Weibull distribution, it is important to be aware of which parameterization the software uses. Usually the parameterization in (2.16) is used in engineering and reliability applications and the parameterization in (2.24) is used in biological and medical applications, although this distinction does not universally hold.

2.3 The Gamma Distribution

Another two-parameter alternative to the exponential distribution is the gamma, which has PDF

$$f(t|\theta, \alpha) = \frac{t^{\alpha-1}}{\theta^\alpha \Gamma(\alpha)} \exp\left(-\frac{t}{\theta}\right), \quad t > 0, \tag{2.26}$$

where the parameters must satisfy $\theta > 0$ and $\alpha > 0$. If T has a gamma distribution with parameters (θ, α), then we will write $T \sim \text{GAM}(\theta, \alpha)$. The survival function and the CDF cannot be expressed in terms of elementary functions, although they can be written in terms of the incomplete gamma function. The CDF of the gamma distribution is

$$F(t|\theta, \alpha) = 1 - \frac{\Gamma(\alpha, t/\theta)}{\Gamma(\alpha)}, \quad t > 0,$$

where $\Gamma(,)$ is the incomplete gamma function defined in (2.20). The hazard function can then be written as

$$h(t|\theta, \alpha) = \frac{t^{\alpha-1}}{\theta^\alpha \Gamma(\alpha, t/\theta)} \exp(-t/\theta), \quad t > 0.$$

If $\alpha < 1$, then the gamma distribution has a deceasing hazard function, and if $\alpha > 1$ then the gamma distribution has an increasing hazard function. The hazard function for the gamma distribution has the property that it approaches the value $1/\theta$ as $t \to \infty$, that is,

$$\lim_{t \to \infty} h(t|\theta, \alpha) = 1/\theta, \quad t > 0.$$

Plots of the PDFs, survival functions, and hazard functions for several different gamma distributions are shown in Figure 2.5. The parameter θ is a scale parameter and α is a shape parameter.

Like the Weibull distribution, the gamma distribution is a generalization of the exponential distribution, but in a different way. When $\alpha = 1$, in Eq. (2.26) the PDF becomes

$$f(t|\theta, 1) = \frac{t^{1-1}}{\theta^1 \Gamma(1)} \exp(-t/\theta) = \frac{1}{\theta} \exp(-t/\theta), \quad t > 0.$$

Thus, the GAM $(\theta, 1)$ becomes the EXP(θ) distribution.

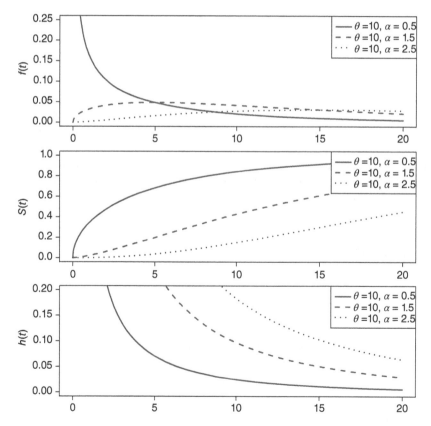

Figure 2.5 PDFs, survival functions, and hazard function s for three gamma distributions. The three curves correspond to $\alpha = 0.5, 1.5$, and 2.5. $\theta = 10$ for all three graphs.

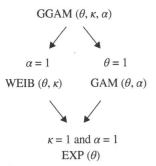

Figure 2.6 The generalized gamma distribution and special cases.

If $T \sim \text{GAM}(\theta, \alpha)$, then

$$E(T) = \theta\alpha$$

and

$$V(T) = \theta^2\alpha.$$

Higher moments can be obtained from the fact that

$$E(T^k) = \theta^k\alpha(\alpha + 1)\cdots(\alpha + k - 1).$$

Since the CDF of the gamma distribution does not have a closed form expression, there is no simple formula for the percentile t_p. (The CDF involves the incomplete gamma function.) Percentiles, can however be obtained by numerically solving the equation

$$F(t_p|\theta, \alpha) = 1 - \frac{\Gamma\left(\alpha, t_p/\theta\right)}{\Gamma(\alpha)} = p. \tag{2.27}$$

A special case of the gamma distribution occurs when the value of α is a positive integer. In this case, the gamma distribution is sometimes called the Erlang distribution. This distribution arises naturally when we take the sum of k i.i.d. $\text{EXP}(\theta)$ random variables. In this case the sum has a $\text{GAM}(\theta, k)$ distribution. This special case also leads to a closed form expression for the CDF:

$$F(t|\theta, k) = 1 - \sum_{i=0}^{k-1} \frac{(t/\theta)^i \exp(-t/\theta)}{i!}. \tag{2.28}$$

The terms in the summation are Poisson probabilities, so this expression leads to a relationship between the gamma and Poisson distributions. The fundamental connection can be traced to the homogeneous Poisson process, where the count of events by time t has a Poisson distribution and the time lap between two consecutive events has an exponential distribution.

Like the exponential and Weibull distributions, the gamma distribution has an alternative parameterization. The PDF is sometimes written as

$$f(t|\lambda, \alpha) = \frac{\lambda^\alpha t^{\alpha-1}}{\Gamma(\alpha)} \exp(-\lambda t), \quad t > 0. \tag{2.29}$$

In this parameterization, the parameter λ is often called a "rate" parameter since the hazard function, which is sometimes called the hazard rate, approaches the constant λ. The parameterization in (2.29) is used more often in the medical and biological literature.

Finally, there is a distribution, called the *generalized gamma distribution*, which is a generalization of both the gamma and the Weibull, and the log normal distribution is a limiting case. See Figure 2.6. It has PDF

$$f(t|\theta, \kappa, \alpha) = \frac{\kappa}{\theta^{\alpha\kappa}\Gamma(\alpha)} t^{\alpha\kappa} \exp\left(-\left(\frac{t}{\theta}\right)^{\kappa}\right), \qquad t > 0. \tag{2.30}$$

We denote this distribution by $\mathrm{GGAM}(\theta, \kappa, \alpha)$.

If $\kappa = 1$, then this PDF reduces to the PDF of the gamma distribution given in (2.26), and when $\alpha = 1$ this PDF reduces to the Weibull PDF given in (2.16). Of course, when $\kappa = \alpha = 1$ the generalized gamma distribution reduces to the exponential distribution with mean θ. Also, the half-normal is a special case of the generalized gamma distribution, obtained by setting $\kappa = 2$ and $\alpha = 1/2$. As α tends to infinity, the generalized gamma distribution approaches the log normal.

The mean and the variance of a random variable T having the generalized gamma distribution are

$$E(T) = \theta \, \frac{\Gamma\left(\alpha + \dfrac{1}{\kappa}\right)}{\Gamma(\alpha)}$$

and

$$V(T) = \theta^2 \left[\frac{\Gamma\left(\alpha + \dfrac{2}{\kappa}\right)}{\Gamma(\alpha)} - \left(\frac{\Gamma\left(\alpha + \dfrac{1}{\kappa}\right)}{\Gamma(\alpha)}\right)^2 \right].$$

Like the gamma distribution, the hazard function for the generalized gamma distribution does not have a closed form expression. Being a generalization of both the Weibull and gamma, the hazard can take on many different shapes. While the exponential hazard is always flat, and the Weibull and gamma hazard functions are monotonic (i.e. always increasing or always decreasing), the generalized gamma distribution need not be monotonic.

2.4 The Lognormal Distribution

If X has a normal distribution with mean μ and variance σ^2, (written $X \sim N(\mu, \sigma)$), then the random variable $T = \exp(X)$ has a lognormal distribution. We will denote this by writing $T \sim \mathrm{LOGN}(\mu, \sigma)$. The PDF of the lognormal distribution can be written as

$$f(t|\mu, \sigma) = \frac{1}{\sqrt{2\pi}\sigma t} \exp(-(\log t - \mu)^2 / 2\sigma^2), \qquad t > 0. \tag{2.31}$$

Conversely, if $T \sim \mathrm{LOGN}(\mu, \sigma)$, then $\log T \sim N(\mu, \sigma^2)$. (Here, log is the logarithm to the base e.) The survival function can be written as

$$S(t) = 1 - P(T \leq t) = 1 - P(X \leq \log t) = 1 - \Phi_{\mathrm{SN}}\left(\frac{\log t - \mu}{\sigma}\right),$$

where Φ_{SN} indicates the CDF of the standard normal distribution. The hazard function is

$$h(t|\mu,\sigma) = \frac{\dfrac{1}{\sqrt{2\pi}\sigma t}\exp(-(\log t - \mu)^2/2\sigma^2)}{1 - \Phi_{SN}\left(\dfrac{\log t - \mu}{\sigma}\right)}.$$

The mean and variance of the lognormal distribution are

$$E(T) = \exp(\mu + \sigma^2/2)$$

and

$$V(T) = \exp(2\mu + \sigma^2)\left(e^{\sigma^2} - 1\right).$$

Figure 2.7 shows graphs of the PDFs, survival functions and hazard functions for the lognormal distribution. We have chosen $\mu = 0$ for all curves, and varied the value of σ ($\sigma = 0.2, 0.5, 1.0$). The hazard function begins at zero when $t = 0$ and increases until it reaches its maximum, and then decreases toward zero as $t \to \infty$ (see Problem 2.15). The fact that the hazard goes to zero as $t \to \infty$ could make the log normal conceptually unattractive as a lifetime distribution.

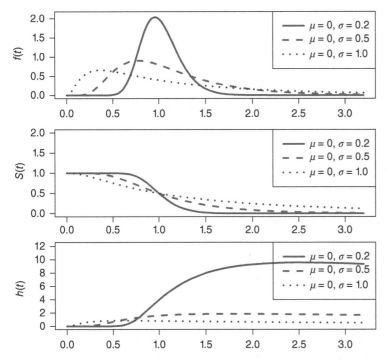

Figure 2.7 PDFs, survival functions, and hazard functions for three lognormal distributions. The three curves correspond to $\mu = 0$ and $\sigma = 0.2, \ 0.5,$ and 1.0.

2.5   Log Location and Scale Distributions

Practitioners often analyze the logarithm of the failure times, rather than the failure times themselves. Such a move transforms the support (the values where the PDF is nonzero) from $(0, \infty)$ to $(-\infty, \infty)$. On the unbounded interval $(-\infty, \infty)$, assumptions of a normal distribution, for example, might be more reasonable. Under some assumptions about the lifetime distribution, taking the logs of the failure times will lead to a distribution that depends on two parameters in a special way. This is why log-location-scale distributions play such an important role in reliability and survival analysis.

Suppose that a random variable X, whose distribution depends on two parameters, μ and σ, has the property that its CDF is a function of the quantity $(x - \mu)/\sigma$. In other words, there is some function Φ for which

$$P(X \leq x | \mu, \sigma) = \Phi\left(\frac{X - \mu}{\sigma}\right), \quad \infty < y < \infty. \tag{2.32}$$

If this is the case, we say that the distribution is a *location-scale distribution*, where μ is the *location parameter* and σ is the *scale parameter*. Although μ is often the mean of the distribution of X, as for example with the normal distribution, this need not be the case. Also, the scale parameter σ is not necessarily the standard deviation.

The PDF for a location-scale distribution can be found by differentiating (2.32) with respect to x. This yields

$$f(x | \mu, \sigma) = \frac{1}{\sigma}\, \phi\left(\frac{x - \mu}{\sigma}\right), \tag{2.33}$$

where $\phi(\cdot) = \Phi'(\cdot)$. We denote a location-scale distribution by writing $X \sim \text{LocScale}\,(\mu, \sigma)$.

The most obvious example of a location-scale distribution is the normal distribution $N(\mu, \sigma)$, since if $X \sim N(\mu, \sigma)$, then one of the fundamental properties of the normal distribution is that

$$Z = \frac{X - \mu}{\sigma} \sim N(0,1).$$

Thus, the CDF of X can be written as

$$P(X \leq x | \mu, \sigma) = F(x | \mu, \sigma)$$
$$= \int_{-\infty}^{x} \frac{1}{\sqrt{2\pi}} \exp\left(-(x - \mu)^2/(2\sigma)\right) dx$$
$$= \Phi_{\text{SN}}\left(\frac{x - \mu}{\sigma}\right),$$

where in this case the function Φ_{SN} is the CDF of the standard normal distribution:

$$\Phi_{\text{SN}}(z) = \int_{-\infty}^{z} \exp(-u^2/2)\, du.$$

Essentially this is saying that by standardizing (subtracting the mean and dividing by the standard deviation) we transform a normal random variable to the standard normal random variable. This property, which is taught in almost all introductory statistics courses, is why a single normal distribution table is sufficient.

In Section 2.4 we saw that if $T = \exp(X)$, where X is a normal random variable with the mean of μ and the standard deviation of σ, then T has a lognormal distribution, a commonly used distribution for modeling lifetimes. Thus, by taking the logarithms of survival times, we obtain the normal distribution, which is certainly a location-scale distribution.

2.5.1 The Smallest Extreme Value Distribution

Any choice of a CDF, Φ, so long as it has support $(-\infty, \infty)$ and doesn't depend on any other parameters, leads to a location-scale distribution family. Aside from the normal distribution, the most important location-scale distribution is the *smallest extreme value* (SEV) distribution. The standard smallest extreme value (SSEV) distribution has

$$\Phi_{\text{SSEV}}(z) = 1 - \exp(-\exp(z)), \quad -\infty < z < \infty, \tag{2.34}$$

and

$$\phi_{\text{SSEV}}(z) = \exp(z - \exp(z)), \quad -\infty < z < \infty. \tag{2.35}$$

The CDF for the SEV distribution with location parameter μ and scale parameter σ can then be written in terms of the CDF of the SSEV distribution as

$$\Phi_{\text{SEV}}(x|\mu, \sigma) = \Phi_{\text{SSEV}}\left(\frac{x - \mu}{\sigma}\right)$$

$$= 1 - \exp\left(-\exp\left(\frac{x - \mu}{\sigma}\right)\right), \quad -\infty < x < \infty, \tag{2.36}$$

where $\sigma > 0$. The PDF is then

$$\phi_{\text{SEV}}(x|\mu, \sigma) = \phi_{\text{SSEV}}\left(\frac{x - \mu}{\sigma}\right)$$

$$= \frac{1}{\sigma} \exp\left(\left(\frac{x - \mu}{\sigma}\right) - \exp\left(\frac{x - \mu}{\sigma}\right)\right), \quad -\infty < x < \infty. \tag{2.37}$$

If X is a random variable that has the SEV distribution with location and scale parameters μ and σ, respectively, then we write $X \sim \text{SEV}(\mu, \sigma)$.

The shape of the SEV distribution looks similar to the normal distribution but it is slightly asymmetric. The SEV distribution is slightly skewed to the left; that is, the left tail seems to extend further than the right tail. Figure 2.8 shows PDFs for some SEV distributions. The top shows the PDF of the SSEV distribution, that is the SEV$(0, 1)$ distribution. In the middle are plots of the PDFs of the SEV$(\mu, 1)$ distribution for $\mu = 0, 1, 2, 3$. Notice how the PDF is simply shifted to the right by the amount μ. The bottom plot in this figure shows the SEV$(0, \sigma)$ PDFs for $\sigma = 0.5, 1, 2$. Notice how increasing the scale parameter causes the spread of the distribution to increase.

Moments of the SEV distribution can be obtained by doing some straightforward, but messy, integrals. For the SSEV distribution, the mean is

$$E(Z) = \int_{-\infty}^{\infty} z \exp(z - \exp(z)) \, dz = -\gamma, \tag{2.38}$$

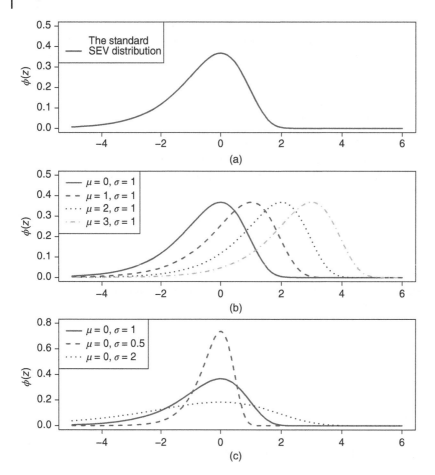

Figure 2.8 Plots of the PDFs for SEV distributions. (a) The standard SEV distribution, (b) SEV distributions with varying μ with constant $\sigma = 1$, (c) SEV distributions with constant $\mu = 0$ and varying σ.

where γ is Euler's constant, which is approximately $\gamma \approx 0.577216$. The variance of the SSEV is then

$$V(Z) = \frac{\pi^2}{6} \approx 1.645. \tag{2.39}$$

The required integrations can be performed on a computer algebra system such as *Mathematica* (Wolfram Research Inc., 2010). The mean and variance for the SEV(μ, σ) distribution are then

$$E(X) = \mu - \gamma\sigma \tag{2.40}$$

and

$$V(X) = \frac{\pi^2}{6}\sigma^2. \tag{2.41}$$

Thus, while μ and σ are the location and scale parameters, respectively, they are not the mean and standard deviation.

Percentiles of the SSEV distribution can be found by inverting the CDF found in (2.36). This leads to the following equation for the percentiles of the $SEV(\mu, \sigma)$ distribution:

$$x_p = \mu + \sigma \log\left(-\log(1 - p)\right). \tag{2.42}$$

The $SEV(\mu, \sigma)$ distribution is closely related to the Weibull distribution in the following way. If $T \sim WEIB(\theta, \kappa)$ then $Y = \log T$ has an SEV distribution. To see this, note that

$$
\begin{aligned}
F_T(t) &= P(\log T \leq y) \\
&= 1 - P(T > e^y) \\
&= 1 - \exp\left(-\left(\frac{e^y}{\theta}\right)^\kappa\right) \\
&= 1 - \exp\left(-\exp\left(\frac{y - \log\theta}{1/\kappa}\right)\right) \\
&= \Phi_{SEV}\left(\frac{y - \log\theta}{1/\kappa}\right).
\end{aligned}
$$

Thus $Y \sim SEV\left(\log\theta, 1/\kappa\right)$, and the parameters are related by

$$\mu = \log\theta, \tag{2.43}$$

$$\sigma = \frac{1}{\sigma}. \tag{2.44}$$

Another related distribution, the Gompertz distribution, can be derived from an SEV distribution truncated at $t = 0$ (since lifetime is a positive variable). Thus, the CDF of Gompertz distribution is given by

$$F(t|\mu, \sigma) = 1 - \left[\frac{1 - \Phi_{SEV}(t)}{1 - \Phi_{SEV}(0)}\right]. \tag{2.45}$$

2.5.2 The Logistic and Log-Logistic Distributions

Another example of a location-scale family is the logistic distribution. The CDF for the standard logistic distribution is

$$\Phi_{SL}(z) = \frac{e^z}{1 + e^z}.$$

The CDF for the general logistic distribution, which we will denote $Logistic(\mu, \sigma)$, is

$$F(x) = \Phi_{SL}\left(\frac{x - \mu}{\sigma}\right) = \frac{\exp\left(\frac{x - \mu}{\sigma}\right)}{1 + \exp\left(\frac{x - \mu}{\sigma}\right)}, \quad -\infty < x < \infty,$$

where $\sigma > 0$, and the PDF is

$$f(x) = F'(x) = \frac{1}{\sigma} \frac{\exp\left(\frac{x - \mu}{\sigma}\right)}{\left(1 + \exp\left(\frac{x - \mu}{\sigma}\right)\right)^2}, \quad -\infty < x < \infty.$$

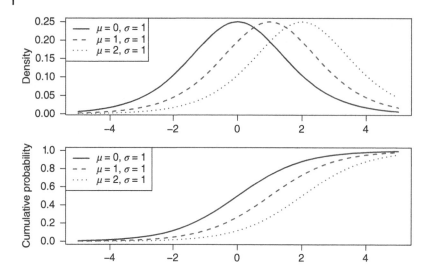

Figure 2.9 Plots of the PDFs and CDFs for Logistic distributions for varying μ for a constant σ.

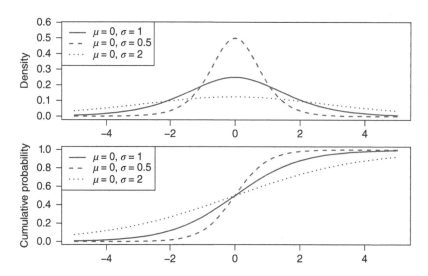

Figure 2.10 Plots of the PDFs and CDFs for Logistic distributions for varying σ for a constant μ.

Figures 2.9 and 2.10 show plots of the PDF and CDF for various choices of μ and σ. Although not readily apparent from its formula, the PDF of the standard logistic distribution, Logistic $(0, 1)$, is symmetric about 0. To see this, note that

$$F(-z) = \frac{e^{-z}}{1 + e^{-z}} = 1 - \frac{e^z}{e^z + 1} = 1 - F(z). \tag{2.46}$$

Differentiating both sides with respect to z yields

$$-f(-z) = 0 - f(x)$$

or equivalently,

$$f(-z) = f(z)$$

so the standard logistic distribution is symmetric about zero. In general, the logistic distribution is centered at μ. Similar to other location-scale distribution examples, if X has a logistic distribution, then $T = \exp(X)$ has a log-logistic distribution.

Problems

2.1 A company insures cell phones for screen breakage. Suppose that the time T until a user breaks the screen is exponentially distributed with mean $\theta = 80$ weeks.
  (a) Find the probability that the user will not have a broken screen over the two-year (104-week) life of the insurance policy.
  (b) Repeat part (a) for a 1.5-year (78-week) policy.
  (c) Comment on the reasonableness of the assumption that the lifetime follows an exponential distribution.

2.2 Suppose that the company from Problem 2.1 also sells a policy to protect the cell phone battery. Suppose that the battery lifetime has an EXP(60) distribution.
  (a) Find the probability that the user will not have a failed battery over the two-year (104-week) life of the insurance policy.
  (b) Repeat part (a) for a three-year (78-week) policy.
  (c) Comment on the reasonableness of the assumption that the lifetime follows an exponential distribution.

2.3 Show that for a continuous lifetime distribution, i.e. one for which $F(0) = 0$, the expected value can be found from

$$E(T) = \int_0^\infty S(t)\, dt.$$

  Hint: Use integration by parts.

2.4 The mean residual life is the expected lifetime given that the unit has been operating failure-free for t time units. That is,

$$m(t) = E(T - t \,|\, T \geq t).$$

Show that for the exponential distribution EXP(θ) the mean residual life is $1/\theta$. *Hint*: Use the result of Problem 2.3.

2.5 Consider the situation from Problem 2.1, and suppose that the user has a 52-week-old phone that has not failed (i.e. the screen hasn't broken). What is the mean residual life?

2.6 Repeat Problem 2.5 for the situation described in Problem 2.2, where the user's phone is 52 weeks old.

2.7 Suppose that headlights for a car have a lifetime that has an EXP(2000) distribution, where time is measured in hours of driving.

   (a) A customer buys a car with brand new headlights. The one on the left side burns out at 1000 hours and the user must replace it with a new one. Find the probability that this new headlight lasts 2000 hours or more.

   (b) The customer's right headlight is still working when the one on the left side fails at 1000 hours. Find the probability that the right headlight, which is already 1000 hours old, lasts an additional 2000 hours or more.

   (c) Suppose the manufacturer suggests replacing *both* headlights whenever one of them fails. Is this a good strategy for the customer? Explain.

2.8 Find the mean residual life for the WEIB(κ, θ) distribution.

2.9 For the Weibull distribution, what values of the parameters lead to a decreasing mean residual life? What values lead to an increasing mean residual life.

2.10 Consider the WEIB(60, 2) distribution.

   (a) Plot the PDF.

   (b) Plot the survival function.

   (c) Plot the hazard function.

   (d) Compute the mean lifetime.

   (e) Compute and plot the mean residual life function.

2.11 Suppose that the cell phone battery described in Problem 2.2 has a WEIB(60,2) distribution.

   (a) How likely is it that the battery fails before the 104-week policy expires?

   (b) Suppose that the battery survives two years. What is the probability that it survives another two years?

   (c) Suppose that after two years, the battery is replaced. What is the probability that this new battery survives two years.

2.12 Repeat Problem 2.7 assuming that the lifetimes have a WEIB(1800,2.2) distribution.

2.13 Verify that the mean and variance of the GAM(α, θ) distribution are $\alpha\theta$ and $\alpha\theta^2$ respectively. *Hint*: To evaluate the integrals, multiply (and divide) by the right constant so that when some of the constants are factored outside the integral, the resulting integrand is a PDF, which must integrate to one.

2.14   Verify Eq. (2.46).

2.15   Show that the hazard for the log normal distribution approaches 0 as $t \to \infty$. (*Hint*: Use L'Hopital's rule and the fundamental theorem of calculus.)

2.16   Show that the characteristic life of the Weibull distribution (the parameter θ) is the 0.632 quantile regardless of the value of the shape parameter κ.

3

Inference for Parameters of Life Distributions

When we use failure data to make inference about a lifetime distribution, we must make certain assumptions about the distribution. For example, if we assume that the lifetimes follow an exponential distribution, then we must estimate the distribution's one parameter θ. (The exponential distribution, having just one parameter, is too simple to model most situations, but its simplicity often gives us insight into more complicated models and thus serves as a stepping stone to understanding these models.) If we assume a Weibull distribution, then we must estimate its two parameters, which we have called θ and κ. The weakest assumption that we could make is that the lifetime is some continuous, but unspecified, distribution. In this case we usually use this data to estimate directly the survival function $S(t)$ or the cumulative distribution function (CDF) $F(t)$. This approach is often called *nonparametric* since it makes no parametric assumptions about the lifetime distribution.

In this chapter, we look at the problem of estimating characteristics of the lifetime distribution, beginning with nonparametric approaches, and then covering the exponential and Weibull distributions. In this chapter we assume that the distribution of lifetimes is the same for all units; that is, there are no predictor variables or covariates. Later, in Part III, we consider the case where one or more predictors can have an effect on the lifetime distribution.

3.1 Nonparametric Estimation of the Survival Function

When there are no censored observations, we can estimate the probability of surviving past time t by taking the fraction of units that did, in fact, survive past time t. Thus, for a fixed t, our estimate of $S(t)$ is

$$\hat{S}(t) = \frac{1}{n}\left(\# \text{ items surviving past time } t\right), \quad t > 0.$$

This step function begin at 1 when $t = 0$ (since all items survive past time $t = 0$, assuming a continuous lifetime distribution) and takes a step of $1/n$ downward at each failure time (assuming all failure times are distinct). For example, if the failure times are

3, 7, 12, 13, 20

Design of Experiments for Reliability Achievement, First Edition.
Steven E. Rigdon, Rong Pan, Douglas C. Montgomery, and Laura J. Freeman.
© 2022 John Wiley & Sons, Inc. Published 2022 by John Wiley & Sons, Inc.
Companion website: www.wiley.com/go/rigdon/designexperiments

then the estimated survival function is

$$\hat{S}(t) = \begin{cases} 1, & 0 \le t < 3, \\ 0.8, & 3 \le t < 7, \\ 0.6, & 7 \le t < 12, \\ 0.4, & 12 \le t < 13, \\ 0.2, & 13 \le t < 20, \\ 0, & t \ge 20. \end{cases}$$

The estimated survival function is shown in Figure 3.1.

The situation is, however, more complicated when there is censoring. If all units begin operating at the same time, and are censored at the same time (either through Type I or Type II censoring), then censoring does not create a problem. In this case the empirical survival function is

$$\hat{S}(t) = \frac{1}{n} \left(\# \text{ items surviving past time } t \right), \quad 0 < t < \text{ last censoring time.}$$

Beyond the last censoring time, the estimate of the survival function is undefined. All we know for sure is that the estimate should be less than or equal to the value just before the last censoring time.

For arbitrary censoring, we use the following notation. Let T_i^* be the true lifetime of the ith unit. This lifetime is not observable if system i is still operating when testing stops. Let C_i denote the censoring time, that is, the time that observation of system i ceases even though it is still working. We treat C_i as a random variable that is independent of the survival time T_i. What we observe are the following random variables:

$$T_i = \min(T_i^*, C_i)$$

and

$$\delta_i = \begin{cases} 1, & \text{if } T_i = T_i^*, \\ 0, & \text{if } T_i = C_i. \end{cases}$$

In other words, we observe whichever comes first, the actual failure time, or the censored time, and we observe the indicator variable that is 1 if we observed a failure, and 0 if we observed a censored value. Thus, the T_is are the *event* times, where the event could be either the observed failure time or the censoring time.

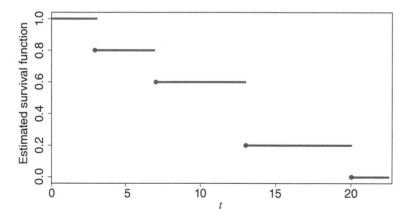

Figure 3.1 The empirical survival function when there is no censoring.

Let

$$t_{(1)} \leq t_{(2)} \leq \cdots \leq t_{(n)} \tag{3.1}$$

denote the *ordered* event times (either failure or censoring times). The notation t alone, with no subscript, indicates an arbitrary time, not necessarily one of the observed event times $t_{(1)}, t_{(2)}, \ldots, t_{(n)}$. Note that if we had a continuous distribution, then with probability one all of the failure times would be distinct and so the less than or equal to signs in (3.1) would be strictly less than signs. In practice, however, most event times are rounded to some extent, for example, to the nearest day (or hour or week); this can lead to simultaneous failure times, or even times when there are simultaneous failures and censoring times. We will keep the less than or equal sign to cover this possibility.

We will assume for now that the failure times have been rounded to the nearest integer, so that the possible failure times are $t = 0, 1, 2, \ldots$. Define

$$d_j = \text{the number of items that failed (or died) at time } t_{(j)} \tag{3.2}$$

and

$$r_j = \text{the number of items operating, i.e. at risk, just before time } t_{(j)}. \tag{3.3}$$

The estimated hazard function at time $t_{(j)}$ is

$$\hat{h}_j = \frac{d_j}{r_j}. \tag{3.4}$$

This is somewhat intuitive, since the hazard at time (the discrete time) t is equal to the probability of failure at time t conditioned on survival up to time t. This fraction represents the proportion of "at risk" units that fail at time t. In the appendix to this chapter we show that this is also the maximum likelihood estimator for h_j.

The probability that a unit survives past time t is then the probability that it survives past time $t = 1$, times the probability that it survives past time $t = 2$ given that it survives past time $t = 1$, times the probability that it survives past time $t = 3$ given that it survives past time $t = 2$, etc. Using estimates for these survival probabilities, the probability of all of these events happening is their product, so the estimated probability of surviving past time $t = j$, where j is an integer, is

$$\hat{S}(j) = (1 - \hat{h}_1)(1 - \hat{h}_2) \cdots (1 - \hat{h}_j).$$

Now, allowing t to be any real number, we obtain what is called the Kaplan–Meier estimate (Kaplan and Meier 1958) of the survival function:

$$\hat{S}(t) = \begin{cases} 1, & \text{if } t < t_{(1)}, \\ \prod_{i=1}^{j}(1 - \hat{h}_i), & \text{if } t_{(j)} \leq t < t_{(j+1)}. \end{cases} \tag{3.5}$$

The Kaplan–Meier estimate is often written in the following equivalent form

$$\hat{S}(t) = \prod_{t_j \leq t} \left(1 - \frac{d_j}{r_j}\right). \tag{3.6}$$

Note that if the argument t is less than the first event time t_1, then the product is a null product, which by definition is equal to 1. This agrees with the piecewise definition in (3.5). Also note that if there are no items surviving at the time of the last event time, then the estimated survival function is undefined for $t > \max(t_j)$; this is because past the last event time, both the number of items at risk is and the number of failures or deaths at that time are 0, and $0/0$ is undefined.

3.1.1 Confidence Bounds for the Survival Function

Approximate confidence bounds can be found for the estimated survival function at any event time t_k. Using the delta method theorem from statistical theory, we could say that

$$\hat{V}(\log \hat{S}(t)) \approx \sum_{j:\, t_{(j)} \leq t} \frac{d_j}{r_j(r_j - d_j)}$$

from which we could obtain the estimated variance of $\hat{S}(u)$:

$$\hat{V}(S(t)) \approx \left(\hat{S}(t)\right)^2 \sum_{j:\, t_{(j)} \leq t} \frac{d_j}{r_j(r_j - d_j)}.$$

This often provides confidence limits that exceed one or drop below zero. We are certain that the survival function remains in the interval $[0, 1]$, so this information is not helpful. Instead, we usually look at the log of the negative log of the estimated survival, and apply the delta method theorem again. This estimated variance is

$$\hat{V}\left[\log\left[-\log \hat{S}(t)\right]\right] \approx \left(\log \hat{S}(t)\right)^{-2} \sum_{j:\, t_{(j)} \leq t} \frac{d_j}{r_j(r_j - d_j)} \tag{3.7}$$

which we can use to get an approximate confidence interval for $\log(-\log S(t))$; we can easily convert this into a confidence interval for $S(t)$. For example, if (a, b) is a confidence interval for $\log\left(-\log S(t)\right)$, then

$$e^{-e^b} < S(t) < e^{-e^a} \tag{3.8}$$

is a confidence interval for $S(t)$. This method will always lead to a confidence interval that lies entirely in the interval $[0, 1]$. When converted back to the estimated variance for $\hat{S}(t)$, this method guarantees that the lower and upper limits are in the interval $[0, 1]$.

We give two examples of the Kaplan–Meier estimate: one simple one in order to understand the computation and another larger and more realistic one.

Example 3.1   Suppose that the ordered failure times are

$$2 \quad 2 \quad 4+ \quad 5 \quad 7+$$

Find the Kaplan-Meier estimate of the survival function.

Solution:
The computations required to obtain the Kaplan–Meier estimate are shown in Table 3.1 and the plot of the survival function is shown in Figure 3.2. The estimated survival function is the step function defined by

$$\hat{S}(t) = \begin{cases} 1, & 0 \leq t < 2, \\ \frac{3}{5}, & 2 \leq t < 4, \\ \frac{3}{10}, & 4 \leq t < 7, \\ \text{undefined}, & t \geq 7. \end{cases}$$

Table 3.1 Calculations for Kaplan–Meier estimate for the failure data 2, 2, 4+, 5, 7+.

j	$t_{(j)}$	d_j	r_j	$\hat{h}_j$	$\hat{S}_j$
1,2	2	2	5	$\frac{2}{5}$	$(1 - \frac{2}{5}) = \frac{3}{5}$
3	4	0	3	$\frac{0}{5}$	$(1 - \frac{2}{5})(1 - \frac{0}{5}) = \frac{3}{5}$
4	5	1	2	$\frac{1}{2}$	$(1 - \frac{2}{5})(1 - \frac{0}{5})(1 - \frac{1}{2}) = \frac{3}{10}$
5	7	0	1	$\frac{0}{1}$	$(1 - \frac{2}{5})(1 - \frac{0}{5})(1 - \frac{1}{2})(1 - \frac{0}{1}) = \frac{3}{10}$

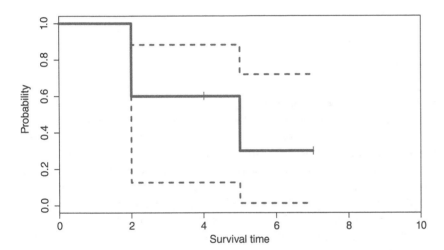

Figure 3.2 The Kaplan–Meier estimate of the survival function for the data in Example 3.1.

Technically, the estimate of the survival function is a step function that is right continuous, i.e. the value of the function at the jump is equal to that of values neighboring it on the right. The solid dot on the left of each "step" indicates this. Most software packages, however, connect all of the dots, resulting in vertical lines at the jumps. This, of course, violates the "vertical line test" for a function (i.e. a vertical line should intersect the graph of the function at most one point), but after this simple example we will show the output directly from software. ■

Example 3.2 This example involves a clinical trial with two arms for treating head and neck cancer. (An "arm" is one possible treatment that was used in the experiment.) The two arms of the experiment, as described in Efron and Hastie (2016), were "Chemotherapy" and "Chemotherapy plus Radiation." Clinical trials for serious or life-threatening diseases, like cancer, often involve a combination of medical interventions. It would be unethical to include a placebo in such a trial because this would subject a participant to an obviously inferior treatment. The experimental design for such a study usually involves the randomization of subjects to one of the treatments. The following data shown give the event time, where the event could be either death or censoring. A "+" indicates censoring:

```
Chemotherapy
  248    160    319+   277    440     91    241    594    157    140
  146    108    160    273     84     63    133    218    405    140
  523   1226+  1101      7     74+  1417     64   1116+  1412+  1146
 1349+   185+   297    139    149    583    420     83    133    176
  165    129    112     34    154    173    523+    42    417    225
  279+

Chemotherapy + Radiation
  179   1776     37   2146+  2297+   528+   130    817    169+   112
 1557    281   2023+   319    140    633   1771+  1897+   194    127
   94     84    432   1642+   159    155    146    173   1245+   469
 1331+  1092+   249    110    339    133     92    725    759+   209
  519    613+   547+   195    119
```

Find the Kaplan-Meier estimate of the survival function.

Solution:

The Kaplan–Meier estimate can be obtained and plotted using the `survival` package in R. To do this, first create a survival object consisting of the observed event times (both failure and censoring times) and the censoring indicator variable using the `Surv` command. Then apply the `survfit` command to the model with no covariates. The commands to do this are explained in the appendix to this chapter.

The Kaplan–Meier estimate of the survival function is a step function that changes only at the observed failure times, not the censoring times. The vertical "hash marks" along the curves indicates the points of censoring, since these are otherwise indistinguishable on the plot. The top plot in Figure 3.3 shows the estimated survival for both arms of the trial. Initially, the chemotherapy and chemotherapy plus radiation have similar survival probabilities. These curves begin to diverge after about 200 days after which the second arm has higher survival probabilities. ■

3.1.2 Estimating the Hazard Function

The hazard function measures how likely a failure is given survival up to some particular point in time. If a system is wearing out, it would seem like the hazard would be increasing in time. For example, an imminent failure is more likely for a unit that has experienced 10,000 cycles (or hours) without failure than a unit that has experienced only 1000 cycles (or hours) without failure. For this reason, the hazard function plays a central role in reliability and survival analysis. In Eq. (3.4) we saw that the raw estimate of the hazard is the number of failures at a particular time, divided by the number of individuals at risk at that time. This measure is usually quite variable and is often not helpful as an estimate. We will discuss two ways to make this estimate smoother and more stable.

The first approach is to consider divisions of the time variable, usually into equally spaced units. Let j denote the index for the intervals of time, $j = 1, 2, \ldots, J$. Also, let n_j denote the number of units at risk during that time, and d_j the number of failures or deaths in that interval. This binned estimate of the hazard across interval j is then

$$\hat{h}_j = \frac{d_j}{n_j}, \qquad j = 1, 2, \ldots, J. \tag{3.9}$$

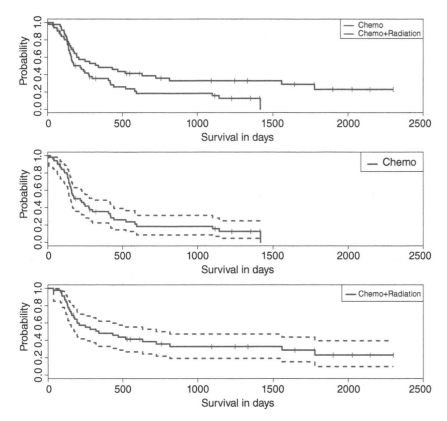

Figure 3.3 Kaplan–Meier estimates of the survival functions for the NCOG trial. The top plot shows the two estimates of the survival function. The middle and bottom plots show separately the Kaplan–Meier curves with confidence bands for the Chemo and the Chemo+Radiation treatments.

The estimated hazard function is then flat at $\hat{h}_j$ across the entirety of interval j. This estimate is necessarily a step function.

The Nelson–Aalen estimator[1] of the survival function is an alternative to the Kaplan–Meier estimator, which begins by estimating the cumulative hazard function. This is done by adding the hazard from each observed event time:

$$\hat{H}(t) = \sum_{t_i \le t} \frac{d_i}{n_i}.$$

The estimator for the survival function is then

$$\hat{S}_{NA}(t) = e^{-\hat{H}(t)}.$$

Since $\hat{H}(t)$ is a step function that takes jumps only at the observed failure times, the same is true for $\hat{S}_{NA}(t)$.

---

1 This is also called the Fleming–Harrington estimator, and in the surfit function from the survival package in R, the option to choose this estimator is type="fh".

The second approach to estimating the hazard function is to use a kernel density smoother that puts a kernel, usually a probability density function (PDF) with small variance, at each of the observed failure times. Specifically, let $K(z)$ denote a PDF such as the standard normal distribution, or the uniform distribution across $[-1, 1]$. Another popular kernel is the Epanechnikov kernel

$$h(u) = \begin{cases} 1 - u^2, & \text{if} -1 \leq u \leq 1, \\ 0, & \text{if otherwise.} \end{cases} \tag{3.10}$$

The smoothing estimate is then

$$\hat{h}_K(t) = \sum_{i \in \mathcal{U}} K\left(\frac{t - t_i}{\sigma}\right) \frac{d_i}{n_i}. \tag{3.11}$$

Here the parameter σ is a tuning parameter; larger values will make $K((t - t_i)/\sigma)$ more spread out, so $\hat{h}_K(t)$ will be smoother, and vice versa for smaller values of σ.

Example 3.3 Nelson (1984) presented lifetimes in cycles for two designs of a snubber; data are reproduced in Table 3.2. (A snubber is the mechanism on a toaster to soften the ejection force on a slice of toasted bread.) Estimate and compare the survival functions and the hazard functions.

Solution:
The Kaplan–Meier estimate of the survival function is given as the top graph in Figure 3.4. Light gray indicates the old design, and black indicates the new design. The survival probabilities are similar across time. Figure 3.4b,c shows the estimated hazard functions. The solid line in both graphs is the binned hazard estimate and the dashed line is the kernel density using the Epanechnikov kernel defined in (3.10). The graphs of the estimated hazard functions tell a slightly different story. For the new design, the hazard function is nearly constant across time, whereas the hazard for the old design seems to be increasing rapidly after about 800 cycles. ∎

3.2 Maximum Likelihood Estimation

Maximum likelihood methods are widely employed for estimating the parameters lifetime models when we assume a particular distribution for the lifetimes. Maximum likelihood methods are statistically optimum for large sample sizes, and they easily allow for non-normal data and censoring, both of which are common in reliability data. In addition to these benefits, likelihood based estimation methods provide a ready solution for statistical inference based on the information matrix derived from the log-likelihood.

3.2.1 Censoring Contributions to Likelihoods

The likelihood function is essentially the joint PDF of all random variables, viewed as a function of the parameters, with the observations being fixed. It measures the goodness-of-fit of a statistical model for a fixed set of parameters to a given sample of data. It is used to develop estimates of the parameters via maximization of the function (maximum likelihood estimation, or MLE) given a set of data. It is a

Table 3.2 Failure times two designs of a snubber.

			Old design		
t_i	δ_i	t_i	δ_i	t_i	δ_i
90	1	600	0	790	0
90	1	600	0	790	0
90	0	631	1	790	0
190	0	631	1	790	0
218	0	631	1	790	0
218	0	635	1	790	0
241	0	658	1	790	0
268	1	658	0	790	0
349	0	731	1	855	1
378	0	739	1	980	1
378	0	739	0	980	1
410	1	739	0	980	0
410	1	739	0	980	0
410	0	739	0	980	0
485	1	790	1	980	0
508	1	790	0	980	0
600	0	790	0		
600	0	790	0		

			New design		
t_i	δ_i	t_i	δ_i	t_i	δ_i
45	0	571	1	670	1
47	1	571	0	670	1
73	1	575	1	731	0
136	0	608	1	838	1
136	0	608	1	964	1
136	0	608	0	964	1
136	0	608	0	1164	0
136	0	608	0	1164	0
145	1	608	0	1164	0
190	0	608	0	1164	0
190	0	608	0	1164	0
281	0	608	0	1164	0
311	1	608	0	1164	0
417	0	608	0	1198	1
485	0	608	0	1198	0
485	0	608	0	1300	0
490	1	608	0	1300	0
569	0	630	1	1300	0

t_i is the event time, and δ_i is the failure indicator where 1 indicates a failure time and 0 indicates a censored time.

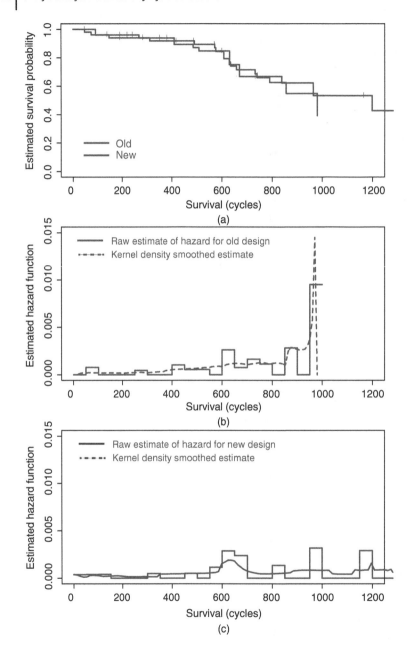

Figure 3.4 Kaplan–Meier estimates of the survival functions for the old and new designs for a snubber. Light gray indicates the old design and black indicates the new design. The other graphs show the binned estimate of the hazard (solid line) and the kernel density estimate (dashed) for the old design (middle) and new design (bottom).

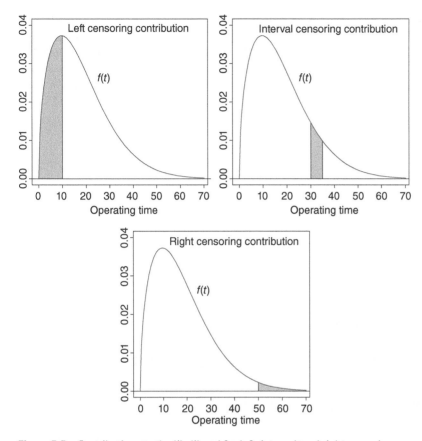

Figure 3.5 Contributions to the likelihood for left, interval, and right censoring.

powerful tool for incorporating censored data in that it provides a general framework for including the observed failures and the censored observations in estimating the parameters of any of the distributions that will be discussed in subsequent chapters. It can accommodate left, interval, and right censored data. Figure 3.5 shows shaded regions that corresponds to left, interval, and right censoring. The shaded areas represent the range in which a failure could have occurred. The likelihood contribution essentially integrates over that area to include the censored information in the estimation of model parameters. This general framework provides the methods for incorporating censoring in any analysis that leverages likelihood functions for assessing the fit of a model to a set of data.

Let $f(t_i|\theta)$ be the PDF of the ith lifetime given the value of the parameter θ, which is usually a vector. In the absence of censoring, the likelihood is the product of these marginal distributions:

$$L(\theta|t) = \prod_{i=1}^{n} f(t_i|\theta).$$

If observation i is right censored, then we observe only that the item lasted past the censoring time c_i, which occurs with probability $P(T_i > c_i|\theta) = 1 - F(c_i|\theta)$. If the item is left censored at time t_i, that is, the failure is known to have occurred before time t_i, then the contribution to the likelihood is therefore $P(T_i < t_i|\theta) = F(t_i|\theta)$. Let a_i be 1 if the ith observation is left-censored and 0 otherwise. Similarly, let r_i be 1 if the ith observation is right censored and 0 otherwise. Finally, define $\delta_i = 1 - a_i - r_i$; that is, δ_i is 1 if observation i is a failure time, and 0 if it is either left or right censored. Then, the total likelihood can be expressed as:

$$L(\theta|\boldsymbol{t}) = \prod_{i=1}^{n} \left[F(t_i|\theta)\right]^{a_i} \times \left[f(t_i|\theta)\right]^{\delta_i} \times \left[1 - F(t_i|\theta)\right]^{r_i}. \tag{3.12}$$

The logarithm of the likelihood function (called the log-likelihood) is then

$$
\begin{aligned}
\ell(\theta|\boldsymbol{t}) &= \sum_{i=1}^{n} a_i \log F(t_i|\theta) \sum_{i=1}^{n} \delta_i \log f(t_i|\theta) \\
&\quad + \sum_{i=1}^{n} r_i \log(1 - F(t_i|\theta)) \\
&= \sum_{i\in\text{left cens}} \log F(t_i|\theta) + \sum_{i\in\text{failures}} \log f(t_i|\theta) \\
&\quad + \sum_{i\in\text{right cens}} \log(1 - F(t_i|\theta)).
\end{aligned}
$$

Because it is often easier to maximize the log-likelihood, we usually work with $\ell(\theta|\boldsymbol{t})$.

3.3 Inference for the Exponential Distribution

The exponential distribution comes with some strong assumptions, such as a constant hazard function and the memoryless property. Therefore, the exponential distribution should be used only after appropriate consideration of the underlying assumptions. Model checking is certainly an appropriate step before using the exponential distribution. We discuss inference for the exponential distribution in detail, not because it is so widely applicable, but rather because we can perform most inference procedures simply and exactly. We look at the exponential distribution as a stepping stone to more complicated, but more realistic, distributions such as the Weibull or gamma.

3.3.1 Type II Censoring

In Section 1.3 we discussed censoring, the termination of a life test before all units have failed. Under Type II censoring, n items are placed on test and testing stops after the rth failure. Of course, if $r = n$, then there is no censoring at all, so that "no censoring" can be seen as a special case of Type II censoring. This is why we begin with Type II censoring.

If $T_1^*, T_2^*, \ldots, T_n^*$ denote the lifetimes of the n objects placed on test (some of which we might not observe because of censoring), then we define

$$T_{(1)}^* < T_{(2)}^* < \cdots < T_{(n)}^*$$

to be the order statistics, that is, the observations ordered from smallest to largest. With Type II censoring, we observe only the first r order statistics. Thus, we have the following observations

$$T_{(1)} = T^*_{(1)}$$

$$\vdots$$

$$T_{(r)} = T^*_{(1)}$$

$$T_{(r+1)} = T^*_{(r)}$$

$$\vdots$$

$$T_{(n)} = T^*_{(r)}.$$

Normally, when we refer to a data set with censoring, we will say "let $T_1, T_2, \dots, T_n$ denote the event times and $\delta_1, \delta_2, \dots, \delta_n$ the failure indicators." Here event times can be either failure times or censoring times. Although many books call δ_i the censoring indicator, the assignment of 1 to indicate a failure makes δ_i an indicator of a failure, so we call it a failure indicator. We will rarely have the need to refer to the unrealized failure time T^*_i.

The likelihood function for Type II censored data is then equal to the joint PDF of the first r order statistics (Bain and Engelhardt 1992), which for the exponential distribution is

$$f\left(t_{(1)}, t_{(2)}, \dots, t_{(r)} | \theta\right) = \frac{n!}{(n-r)!} \left(\prod_{i=1}^{r} f\left(t_{(i)} | \theta\right)\right) \left(S\left(t_{(r)} | \theta\right)\right)^{n-r}$$

$$= \frac{n!}{(n-r)!} \left(\prod_{i=1}^{r} \frac{1}{\theta} e^{-t_{(i)}/\theta}\right) \left(e^{-t_{(r)}/\theta}\right)^{n-r}$$

$$= \frac{n!}{(n-r)!} \frac{1}{\theta^r} \exp\left[-\frac{1}{\theta} \left(\sum_{i=1}^{r} t_{(i)} + (n-r) t_{(r)}\right)\right].$$

The first expression above has a nice interpretation: we have observed failures at times $t_{(1)} < t_{(2)} < \cdots < t_{(r)}$, so the PDF is evaluated at each of these values. For the censored observations, we have only observed their survival past time $t_{(r)}$; therefore, we include the factor $S\left(t_{(r)} | \theta\right)$ for each of the $n - r$ censored values. The constant $n!/(n-r)!$ in front counts the number of ways we can choose the r observations that are censored and the order in which they were observed.

The log likelihood is then the logarithm of the last expression previously. If we let $c = n!/(n-r)!$, then we can write

$$\ell\left(\theta | t_{(1)}, t_{(2)}, \dots, t_{(r)}\right) = \log c - r \log \theta - \left(\sum_{i=1}^{r} t_{(i)} + (n-r) t_{(r)}\right) \Big/ \theta$$

$$= \log c - r \log \theta - \frac{\text{TTT}}{\theta}, \tag{3.13}$$

where TTT is the total time on test, that is, the total amount of time that the units were operating. For the exponential distribution, the likelihood function depends on the data only through the TTT. For more complicated models, such as the Weibull, this is not true. If we differentiate (3.13) with respect to θ and set the result equal to zero, we find that the MLE is

$$\hat{\theta} = \frac{\text{TTT}}{r}.$$

Note that in the case of no censoring,

$$\text{TTT} = \sum_{i=1}^{n} t_{(i)} = \sum_{i=1}^{n} t_i$$

since adding the ordered failure times is the same as adding the unordered failure times. In this case the MLE is equal to

$$\hat{\theta} = \frac{1}{n}\sum_{i=1}^{n} t_i = \bar{t}$$

so the estimate of the mean of the distribution is equal to the sample mean.

For the exponential distribution, there are exact procedures for finding confidence intervals and constructing hypothesis tests. These methods are based on the fact that

$$\frac{2r\hat{\theta}}{\theta} \sim \chi^2(2r). \tag{3.14}$$

From this result, we can write

$$P\left(\chi^2_{2r,1-\alpha/2} < \frac{2r\hat{\theta}}{\theta} < \chi^2_{2r,\alpha/2}\right) = 1 - \alpha,$$

where $\chi^2_{v,\gamma}$ is that value for which the probability in the right tail of a $\chi^2(v)$ distribution is equal to γ. After some manipulation, this leads to a formula for a confidence interval for θ:

$$P\left(\frac{2r\hat{\theta}}{\chi^2_{2r,\alpha/2}} < \theta < \frac{2r\hat{\theta}}{\chi^2_{2r,1-\alpha/2}}\right) = 1 - \alpha. \tag{3.15}$$

Inside the probability, the random quantities are the endpoints $2r\hat{\theta}/\chi^2_{2r,\alpha/2}$ and $2r\hat{\theta}/\chi^2_{2r,1-\alpha/2}$; the parameter θ is considered fixed (i.e. not random) but unknown.

Since $r\hat{\theta} = \text{TTT}$, where TTT is the total time on test, we can write the confidence interval as

$$\frac{2\text{TTT}}{\chi^2_{2r,\alpha/2}} < \theta < \frac{2\text{TTT}}{\chi^2_{2r,1-\alpha/2}}. \tag{3.16}$$

One-sided confidence intervals, which might be of interest if we want to infer that the mean lifetime θ is *at least* some value, can be obtained similarly. In addition, confidence intervals can be obtained for any monotonic function of θ. For example, the reliability for a particular mission time t_m is equal to

$$S(t_m) = \exp(-t_m/\theta).$$

From (3.16) we can write successively

$$-\frac{2\text{TTT}}{\chi^2_{2r,1-\alpha/2}} < -\theta < -\frac{2\text{TTT}}{\chi^2_{2r,\alpha/2}}$$

$$-\frac{\chi^2_{2r,\alpha/2}}{2\text{TTT}} < -\frac{1}{\theta} < -\frac{\chi^2_{2r,1-\alpha/2}}{2\text{TTT}}.$$

$$-\frac{t_m \chi^2_{2r,\alpha/2}}{2\text{TTT}} < -\frac{t_m}{\theta} < -\frac{t_m \chi^2_{2r,1-\alpha/2}}{2\text{TTT}}$$

$$\exp\left(-\frac{t_m \chi^2_{2r,\alpha/2}}{2\text{TTT}}\right) < S\left(t_m\right) < \exp\left(-\frac{t_m \chi^2_{2r,1-\alpha/2}}{2\text{TTT}}\right).$$

Confidence intervals for other functions can be similarly obtained.

Example 3.4 Zelen (1959) gave one of the earliest examples of an experiment whose response variable was a lifetime. The unit was a glass capacitor and the two predictors were temperature (T, either 170 °C or 180 °C), or applied voltage (V, either 200, 250, 300, or 350 V). For this example, we will consider just one of the eight treatments ($T = 170$, $V = 200$). Eight units were put on test at each of the eight treatments and the test was terminated after the fourth failure. The lifetime data, where a "+" indicates a censored value, were

439, 904, 1092, 1105, 1105+, 1105+, 1105+, 1105 + .

Find point and interval estimates for θ and the reliability for a mission time of $t_m = 500$.

Solution:
The total time on test was

$$\text{TTT} = 439 + 904 + 1092 + 1105 + 1105 + 1105 + 1105 + 1105 = 7960.$$

Since $n = 8$, $r = 4$, we have

$$\hat{\theta} = \frac{\text{TTT}}{4} = \frac{7960}{4} = 1990.$$

We can obtain quantiles of the chi-square distribution using R. For example, the value that leaves 0.025 in the right tail of the $\chi^2(8)$ can be obtained by entering `qchisq(0.025,8,lower.tail=FALSE)`. This gives

$$\chi^2_{8,0.975} = 2.180 \qquad \text{and} \qquad \chi^2_{8,0.025} = 17.535.$$

The 95% confidence interval for θ is therefore

$$\frac{2 \times 7960}{17.535} < \theta < \frac{2 \times 7960}{2.18}$$
$$907.9 < \theta < 7302.8.$$

The point estimate for $S(500)$, which is the reliability for a mission time of $t_m = 500$, is

$$\hat{S}(500) = \exp(-500/1990) \approx 0.778$$

and the 95% confidence interval is

$$\exp\left(-\frac{500 \times 17.535}{2 \times 7960}\right) < S(500) < \exp\left(-\frac{500 \times 2.18}{2 \times 7960}\right)$$
$$0.577 < S(500) < 0.934.$$

3.3.2 Type I Censoring

Now suppose that n items are placed on test and testing stops at a predetermined time τ. We denote the failure times for the n items by

$$T_1^*, \ T_2^*, \ \ldots, \ T_n^*$$

although we are unable to observe those lifetimes that exceed the censoring time τ. What we observe is

$$T_i = \min \left(T_i^*, \tau \right)$$

and an indicator of whether the observed T_i was a failure time or a censoring time,

$$\delta_i = \begin{cases} 1, & \text{if } T_i = T_i^* \text{ (i.e. we observe a failure),} \\ 0, & \text{if } T_i = \tau \text{ (i.e. we observe a censored value.)} \end{cases}$$

Consider just unit i, for which we observe T_i and δ_i. If $\delta_i = 1$, then we observe a failure, which contributes $f(t_i)$ to the likelihood (here f is the PDF of any one of the times to failure). If $\delta_i = 0$, then we observe an event with probability $S(\tau)$ since we know that the item survived past time τ. Thus, unit i contributes

$$L_i = \left[f(t_i) \right]^{\delta_i} \left[S(t_i) \right]^{1-\delta_i} = \left[f(t_i) \right]^{\delta_i} [S(\tau)]^{1-\delta_i}$$

The likelihood function is therefore

$$L(\theta) = \prod_{i=1}^n L_i = \prod_{i=1}^n \left[f(t_i) \right]^{\delta_i} \left[S(t_i) \right]^{1-\delta_i}. \tag{3.17}$$

to the likelihood. For the exponential distribution, the likelihood is

$$L(\theta) = \prod_{i=1}^n \left[\frac{1}{\theta} e^{-t_i/\theta} \right]^{\delta_i} \left[e^{-t_i/\theta} \right]^{1-\delta_i}$$

$$= \theta^{-\sum_{i=1}^n \delta_i} \ e^{-\sum_{i=1}^n \delta_i t_i/\theta} \ e^{-\sum_{i=1}^n (1-\delta_i) t_i/\theta}$$

$$= \theta^{-r} \exp \left[-\sum_{i=1}^n t_i/\theta \right],$$

where $r = \sum_{i=1}^n \delta_i$ is the number of observed failures (which under Type I censoring is a random variable). The sum $\sum_{i=1}^n t_i$ is equal to the total time on test (TTT) as defined in Section 3.3.1. The log-likelihood function is therefore

$$\ell(\theta) = -r \log \theta - \frac{\text{TTT}}{\theta} \tag{3.18}$$

which looks nearly identical to the Type II censoring case. The MLE of θ is therefore

$$\hat{\theta} = \frac{r}{\text{TTT}}, \tag{3.19}$$

which again looks nearly identical to the case of Type II censoring. One difference is that under Type I censoring, r is a random variable, whereas in Type II censoring, r is fixed beforehand. On the other hand, in Type II censoring the time of termination of the test is random, whereas in Type I censoring it is fixed.

For Type I censoring, there is no exact distributional result analogous to (3.14). Instead, we must use asymptotic results. One method uses the asymptotic normal approximation to the distribution of the MLE $\hat{\theta}$. The second uses the delta method applied to the function $\log \hat{\theta}$. The former method always results in confidence intervals, which are symmetric about the MLE. Since the distribution of the MLE is usually skewed to the right, the former method often performs poorly unless the sample size is large. Usually, confidence intervals computed from the distribution of $\log \theta$ work better. The third approach is to define the *relative likelihood*

$$R(\theta, \hat{\theta}) = \frac{L(\theta)}{L(\hat{\theta})}.$$

The random variable

$$-2 \log R(\theta, \hat{\theta}) = -2 \left[\ell(\theta) - \ell(\hat{\theta}) \right]$$

has an approximate χ^2 distribution with one degree of freedom. Thus, the third method makes the confidence interval contain those values of θ for which

$$-2 \log R(\theta, \hat{\theta}) = -2 \left[\ell(\theta) - \ell(\hat{\theta}) \right] < \chi^2_{1-\alpha}(1)$$

which is equivalent to

$$R(\theta, \hat{\theta}) > \exp \left[-\chi^2_{1-\alpha}(1)/2 \right].$$

Details of the asymptotic methods are given in Section 3.3.4.

3.3.3 Arbitrary Censoring

In many cases, units are placed into service at random times or the termination of testing ends at a random time (or both). In this situation, the time to censoring is a random variable for each unit. We denote the censoring time by C_i and the (possibly unobserved) failure time by T_i^*. We therefore observe

$$T_i = \min(T_i^*, C_i)$$

and

$$\delta_i = \begin{cases} 1, & \text{if } T_i = T_i^*, \\ 0, & \text{if } T_i = C_i. \end{cases}$$

The likelihood is derived much like in the Section 3.3.2 and is found to be

$$L(\theta) = \prod_{i=1}^{n} [f(t_i)]^{\delta_i} [S(t_i)]^{1-\delta_i}$$

$$= \left(\prod_{i \in \mathcal{U}} f(t_i | \theta) \right) \left(\prod_{i \in \mathcal{C}} S(t_i | \theta) \right)$$

$$= \theta^{-r} \exp(-\text{TTT}/\theta),$$

where $r = \sum_{i=1}^{n} \delta_i$ is the number of observed failures (a random variable), and $\mathcal{U}$ and $\mathcal{C}$ are, respectively, the sets of uncensored and censored observations. The MLE of θ is then

$$\hat{\theta} = \frac{\text{TTT}}{r}. \tag{3.20}$$

Approximate confidence intervals for θ can be based on the asymptotic distributions as was the case for Type I censoring.

3.3.4 Large Sample Approximations

For the exponential distribution, exact methods are available. For example, closed form expressions for the MLE, exact formulas for confidence intervals, etc. exist for data from the exponential distribution. Thus, large sample approximations are generally not needed for the exponential distribution. One of the reasons we study the exponential distribution, despite its drawback of being too simple for most data, is that it serves as a stepping stone to more complicated models, such as the Weibull, where large sample methods *are* required to justify the approximations. We present some large sample approximations for the exponential distribution so that when we get to more complicated cases, the methodology will look familiar.

One approach to inference based on large samples is to use the approximation

$$Z = \frac{\hat{\theta} - \theta}{\left[I\left(\hat{\theta}\right)\right]^{-1/2}} \overset{\text{approx.}}{\sim} N(0,1),$$

where $I(\theta)$ is the Fisher information

$$I(\theta) = E\left[-\ell''(\theta)\right].$$

This is often called the *Wald approximation*. For the exponential distribution with a fixed number of failures r, the log likelihood function is

$$\ell(\theta) = \log c - r \log \theta - \frac{\text{TTT}}{\theta},$$

leading to

$$\ell'(\theta) = -\frac{r}{\theta} + \frac{\text{TTT}}{\theta^2}$$

and

$$\ell''(\theta) = \frac{r}{\theta^2} - \frac{2\text{TTT}}{\theta^3}.$$

Thus

$$\begin{aligned}
I(\theta) &= -E\left(\frac{r}{\theta^2} - \frac{2\text{TTT}}{\theta^3}\right) \\
&= -\frac{r}{\theta^2} + \frac{2}{\theta^3}E(\text{TTT}) \\
&= -\frac{r}{\theta^2} + \frac{2}{\theta^3}r\theta \\
&= \frac{r}{\theta^2}.
\end{aligned}$$

In this case, because of the simplicity of the exponential distribution, we can find an exact expression for $E\left[\ell''(\theta)\right]$, but this depends on the unknown parameter θ. For other models, an exact closed-form expression is not obtainable. (The derivation that $E(\text{TTT}) = r\theta$ is a bit involved, and uses Property 4.9 in Appendix A of Leemis (1995).) Since the value of θ is unknown, we estimate $E\left[\ell''(\theta)\right]$ by $\ell''\left(\hat{\theta}\right)$ where $\hat{\theta}$ is the MLE. The quantity

$$\hat{I}\left(\hat{\theta}\right) = -\ell''\left(\hat{\theta}\right) \tag{3.21}$$

is called the *observed Fisher information*. Using the observed Fisher information, we get

$$\hat{I}\left(\hat{\theta}\right) = -\ell''\left(\hat{\theta}\right) = \frac{r}{\hat{\theta}^2}.$$

The large sample result is therefore

$$\frac{\hat{\theta} - \theta}{\left(r/\hat{\theta}^2\right)^{-1/2}} = \frac{\hat{\theta} - \theta}{\hat{\theta}/\sqrt{r}} \overset{\text{approx.}}{\sim} N\left(0, 1\right). \tag{3.22}$$

An approximate $100\left(1 - \alpha\right)\%$ confidence interval for θ is then

$$\hat{\theta} - z_{\alpha/2}\frac{\hat{\theta}}{\sqrt{r}} < \theta < \hat{\theta} + z_{\alpha/2}\frac{\hat{\theta}}{\sqrt{r}}.$$

The approximation in (3.22) is usually not very good, however, unless the number of uncensored values r is large.

The distribution of $\log\hat{\theta}$ is often much closer to normal for small to moderate sample sizes. We can use the delta method theorem to show that the distribution of $g(\hat{\theta}) = \log\hat{\theta}$ is

$$g(\hat{\theta}) \overset{\text{approx.}}{\sim} N\left(g(\theta), [g'(\theta)]^2[I(\theta)]^{-1}\right)$$

which in the case of the log transformation is

$$\log\hat{\theta} \overset{\text{approx.}}{\sim} N\left(\log\theta, r^{-1}\right).$$

Thus,

$$\sqrt{r}\left(\log\hat{\theta} - \log\theta\right) \overset{\text{approx.}}{\sim} N(0,1).$$

An approximate confidence interval for $\log\theta$ is therefore

$$\log\hat{\theta} - \frac{z^{\alpha/2}}{\sqrt{r}} < \log\theta < \log\hat{\theta} + \frac{z^{\alpha/2}}{\sqrt{r}}.$$

This result can be transformed to obtain a confidence interval for θ, which is

$$\hat{\theta}\exp(-z_{\alpha/2}/\sqrt{r}) < \theta < \hat{\theta}\exp(z_{\alpha/2}/\sqrt{r}). \tag{3.23}$$

Another approach to finding confidence intervals for θ is to use the relative likelihood function

$$R(\theta, \hat{\theta}) = \frac{L(\theta)}{L(\hat{\theta})}$$

which has an approximate $\chi^2(1)$ distribution; that is

$$-2\log R(\theta, \hat{\theta}) = -2\log\frac{L(\theta)}{L(\hat{\theta})} \overset{\text{approx.}}{\sim} \chi^2(1). \tag{3.24}$$

The left side of (3.24) is equal to

$$-2\log\frac{L(\theta)}{L(\hat{\theta})} = 2\ell(\hat{\theta}) - 2\ell(\theta) = 2\left(-r\log\hat{\theta} - \frac{\text{TTT}}{\hat{\theta}} + r\log\theta + \frac{\text{TTT}}{\theta}\right).$$

A confidence interval for θ is found by finding those values of θ for which

$$2\left(-r\log\hat{\theta} - \frac{\text{TTT}}{\hat{\theta}} + r\log\theta + \frac{\text{TTT}}{\theta}\right) < \chi^2(1). \tag{3.25}$$

3.4 Inference for the Weibull

The exponential distribution, with a constant hazard function and the memoryless property, is too simple for most applications. The Weibull distribution, having two parameters, is much more flexible, making it probably the most widely used parametric distribution in reliability and life testing. We recall that the PDF and survival function for the Weibull distribution are

$$f(t|\theta,\kappa) = \frac{\kappa}{\theta}\left(\frac{t}{\theta}\right)^{\kappa-1}\exp\left(-\left(\frac{t}{\theta}\right)^{\kappa}\right), \qquad t > 0$$

and

$$S(t|\theta,\kappa) = \exp\left(-\left(\frac{t}{\theta}\right)^{\kappa}\right), \qquad t > 0,$$

respectively. Let $T_1, T_2, \ldots, T_n$ denote the event times and $C_1, C_2, \ldots, C_n$ the censoring times, that is the times to which the units would be observed. The likelihood for the censored sample $(T_1, \delta_1), (T_2, \delta_2), \ldots, (T_n, \delta_n)$, where T_i is either the failure time or the censoring time for item i, and δ_i is an indicator variable that is 1 if the ith item failed, and 0 if it is censored, is

$$L(\theta,\kappa) = \left(\prod_{i\in\mathcal{U}}f(t_i|\theta,\kappa)\right)\left(\prod_{i\in C}S(t_i|\theta,\kappa)\right),$$

where $\mathcal{U} = \{i|\delta_i = 1\}$ and $C = \{i|\delta_i = 0\}$; that is, $\mathcal{U}$ and C are the indices for the set of uncensored and censored observations, respectively. The likelihood function can be written using the censoring variable δ as

$$\begin{aligned}
L(\theta,\kappa) &= \prod_{i=1}^{n}\left[\left(f(t_i|\theta,\kappa)\right)^{\delta_i}\left(S(t_i|\theta,\kappa)\right)^{1-\delta_i}\right]\\[2mm]
&= \prod_{i=1}^{n}\left[\frac{\kappa}{\theta}\left(\frac{t_i}{\theta}\right)^{\kappa-1}\exp\left(-\left(\frac{t_i}{\theta}\right)^{\kappa}\right)\right]^{\delta_i}\left[\exp\left(-\left(\frac{t_i}{\theta}\right)^{\kappa}\right)\right]^{1-\delta_i}\\[2mm]
&= \frac{\kappa^{\sum\delta_i}}{\theta^{\kappa\sum\delta_i}}\left(\prod_{i=1}^{n}t_i^{(\kappa-1)\delta_i}\right)\exp\left[-\sum_{i=1}^{n}\delta_i\left(\frac{t_i}{\theta}\right)^{\kappa} - \sum_{i=1}^{n}(1-\delta_i)\left(\frac{t_i}{\theta}\right)^{\kappa}\right]\\[2mm]
&= \frac{\kappa^{r}}{\theta^{r\kappa}}\left(\prod_{i\in\mathcal{U}}t_i^{\kappa-1}\right)\exp\left[-\sum_{i=1}^{n}\left(\frac{t_i}{\theta}\right)^{\kappa}\right].
\end{aligned} \tag{3.26}$$

Here we have used r to represent the number of observed failures, that is, $r = \sum_{i=1}^{n} \delta_i$. Keep in mind, though, that unless r is fixed in advance, as in the case of Type II censoring, r is a random variable. There are therefore $n - r$ censored observations.

The log-likelihood is then

$$\ell(\theta, \kappa) = r \log \kappa - r\kappa \log \theta + (\kappa - 1) \sum_{i \in \mathcal{U}} \log t_i - \sum_{i=1}^{n} \left(\frac{t_i}{\theta} \right)^{\kappa}. \tag{3.27}$$

The maximum likelihood estimators do not have a closed form, but can be approximated using an iterative algorithm. If we differentiate (3.27) with respect to θ, we obtain

$$\frac{\partial \ell}{\partial \theta} = -\frac{r\kappa}{\theta} + \frac{\kappa}{\theta^{\kappa+1}} \sum_{i=1}^{n} t_i^{\kappa},$$

where r is the number of observed failures. We can set this result equal to zero and attempt to solve for θ in terms of κ to obtain

$$\hat{\theta} = \left(\frac{1}{r} \sum_{i=1}^{n} t_i^{\kappa} \right)^{1/\kappa}. \tag{3.28}$$

Note that we have not really *solved* for θ, because the expression on the right side still depends on the parameter κ, which is also unknown. The derivative with respect to κ is

$$\frac{\partial \ell}{\partial \kappa} = \frac{r}{\kappa} - r \log \theta + \sum_{i \in \mathcal{U}} \log t_i - \sum_{i=1}^{n} \left(\frac{t_i}{\theta} \right)^{\kappa} \log \frac{t_i}{\theta},$$

where $\mathcal{U}$ is the set of uncensored (failure) observations. We can substitute the expression in (3.28) for θ, and simplify to obtain

$$\kappa = \frac{r \sum_{i=1}^{n} t_i^{\kappa}}{r \sum_{i=1}^{n} t_i^{\kappa} \log t_i - \left(\sum_{i \in \mathcal{U}} \log t_i \right) \left(\sum_{i=1}^{n} t_i^{\kappa} \right)}.$$

Since κ appears on both sides, we have not really *solved* for κ, but at least this expression is independent of θ. This leaves us with one (nonlinear) equation in one unknown, which we can solved by an iterative method, such as fixed point iteration or Newton's method. Once we have approximated the solution of (3.28) for κ, which we call $\hat{\kappa}$, we can substitute the value into (3.28) in order to obtain the MLE of θ, which we call $\hat{\theta}$.

3.5 The SEV Distribution

Recall that if $T \sim \text{WEIB}(\theta, \kappa)$, then

$$Y = \log T \sim \text{SEV}(\mu, \sigma),$$

where the parameters are related by

$$\mu = \log \theta,$$

$$\sigma = \frac{1}{\kappa}.$$

As an alternative to computing the MLEs of the Weibull parameters using the Weibull likelihood, we could take the logarithms of the failure times (and censoring times) and apply the smallest extreme value (SEV) distribution. In this case the likelihood function is

$$L(\mu, \sigma) = \prod_{i=1}^{n} \left[\frac{1}{\sigma} \phi \left(\frac{y_i - \mu}{\sigma} \right) \right]^{\delta_i} \left[1 - \Phi \left(\frac{y_i - \mu}{\sigma} \right) \right]^{1 - \delta_i},$$

where

$$\phi(z) = \exp[z - \exp(z)]$$

is the PDF of the standard SEV distribution, and

$$\Phi(z) = 1 - \exp[-\exp(z)]$$

is the corresponding CDF. After some simplification the SEV likelihood becomes

$$\ell(\mu, \sigma) = -r \log \sigma + \sum_{i=1}^{n} \delta_i \frac{y_i - \mu}{\sigma} - \sum_{i=1}^{n} \exp \left(\frac{y_i - \mu}{\sigma} \right), \tag{3.29}$$

where $r = \sum_{i=1}^{n} \delta_i$ is the number of observed failures. Equation (3.29) can be differentiated with respect to μ and σ to obtain the likelihood equations. Numerical methods, which must be used to approximate the solution, tend to be more stable for the SEV(μ, σ) distribution compared with the Weibull. Large sample normal approximations that are used for confidence intervals and hypothesis tests also tend to be more accurate when the problem is posed in terms of the SEV. Because of the invariance property, the MLEs are related by

$$\hat{\mu} = \log \hat{\theta}$$

$$\hat{\sigma} = \frac{1}{\hat{\kappa}}$$

or

$$\hat{\theta} = \exp(\hat{\mu})$$

$$\hat{\kappa} = \frac{1}{\hat{\sigma}}.$$

3.6 Inference for Other Models

While the Weibull distribution is probably most widely applied distribution for lifetimes, there are other models that we should mention. Here we discuss the gamma distribution, the generalized gamma distribution, and the log normal distribution.

3.6.1 Inference for the GAM(θ, α) Distribution

Let $T_1, T_2, \ldots, T_n$ denote the event times and $\delta_1, \delta_2, \ldots, \delta_n$ the corresponding failure indicators. The likelihood for the GAM(θ, α) distribution for a sample with censoring is

$$
L(\theta, \alpha | t_1, \ldots, t_n, \delta_1, \ldots, \delta_n) = \left(\prod_{i \in \mathcal{U}} f(t_i | \theta, \alpha) \right) \left(\prod_{i \in C} S(t_i | \theta, \alpha) \right)
$$

$$
= \theta^{-r\alpha} [\Gamma(\alpha)]^{-n} \left(\prod_{i \in \mathcal{U}} t_i \right)
$$

$$
\times \exp\left(-(t_i/\theta) \sum_{i \in \mathcal{U}} \right) \prod_{i \in \mathcal{U}} \Gamma(\alpha, t_i/\theta),
$$

where r is the number of units that failed. This yields the rather messy log-likelihood function:

$$
\ell(\theta, \alpha | t_1, \ldots, t_n, \delta_1, \ldots, \delta_n) = -r\alpha \log \theta - n \log \Gamma(\alpha) + (\alpha - 1) \sum_{i \in \mathcal{U}} \log t_i
$$

$$
- (1/\theta) \sum_{i \in \mathcal{U}} t_i + \sum_{i \in C} \log \Gamma(\alpha, t_i/\theta).
$$

To obtain the MLEs of θ and α, we would first take the derivatives of ℓ with respect to θ and α, which are found to be

$$
\frac{\partial \ell}{\partial \theta} = -\frac{r\alpha}{\theta} + \frac{1}{\theta^2} \sum_{i \in \mathcal{U}} t_i + \sum_{i \in C} \frac{t_i^\alpha \exp(-t_i/\theta)}{\theta^{\alpha+1} \Gamma(\alpha, t_i/\theta)}
$$

$$
\frac{\partial \ell}{\partial \alpha} = -r \log \theta - n \frac{\Gamma(\alpha)}{\Gamma'(\alpha)} + \sum_{i \in \mathcal{U}} \log t_i
$$

$$
+ \sum_{i \in C} \frac{1}{\Gamma(\alpha, t_i/\theta)} \int_{t_i/\theta}^{\infty} u^{\alpha-1} (\log u) e^{-u} \, du.
$$

By setting these equal to zero, we obtain the likelihood equations, which are nonlinear in θ and α. Numerical methods can be used to approximate the solutions to these equations, thereby yielding the MLEs. The margins of error for $\hat{\theta}$ and $\hat{\alpha}$ can be approximated using the observed Fisher information matrix, as was done in Eq. (3.21).

3.6.2 Inference for the Log Normal Distribution

The log normal arises by assuming that the log of the failure time has a normal distribution with mean μ and variance σ^2. Thus, if $T \sim \text{LOGN}(\mu, \sigma)$, then $Y = \log T \sim N(\mu, \sigma^2)$. The PDF of the lognormal is

$$
f(t | \mu, \sigma) = \frac{1}{\sqrt{2\pi}\sigma t} \exp(-(\log t - \mu)^2 / 2\sigma^2), \qquad t > 0.
$$

If there is no censoring, then the MLEs of μ and σ are

$$
\hat{\mu} = \frac{1}{n} \sum_{i=1}^{n} \log t_i \tag{3.30}
$$

and

$$\hat{\sigma} = \sqrt{\frac{1}{n} \sum_{i=1}^{n} (\log t_i - \hat{\mu})^2} \ . \tag{3.31}$$

If there is censoring, the problem is much more complex, because the survival function for the log normal does not have a closed form expression. Fortunately, most software packages, such as JMP and R's `survival` package, can account for censoring and provide the MLEs as well as standard errors for these estimates. The log normal distribution should be used with caution because most real world systems will have an increasing hazard function for large t, whereas the hazard for the log normal approaches 0 as $t \to \infty$.

3.6.3 Inference for the GGAM(θ, κ, α) Distribution

The generalized gamma distribution GGAM(θ, κ, α) was introduced in Section 2.3. This distribution is a generalization of both the gamma and the Weibull distributions and has PDF

$$f(t|\theta, \kappa, \alpha) = \frac{\kappa}{\theta^{\alpha\kappa}\Gamma(\alpha)} t^{\alpha\kappa} \exp\left(-\left(\frac{t}{\theta}\right)^{\kappa}\right), \qquad t > 0. \tag{3.32}$$

Point estimates and estimated standard errors can be obtained using the R package `flexsurv`. JMP can handle the extended generalized gamma distribution (an extension of the generalized gamma) using the Reliability and Survival Methods set of functions.

What makes the generalized gamma distribution particularly useful is that we can use it for selecting a distributional model. We can assume a GGAM(θ, κ, α) distribution and test $H_0 : \kappa = 1$ versus $H_1 : \kappa \neq 1$ or we could test $H_0 : \alpha = 1$ versus $H_1 : \alpha \neq 1$. The results of these tests could suggest a distributional model. The tests can be carried out using the likelihood ratio test using the statistic

$$\Lambda = \frac{\max_{(\theta,\kappa)} L(\theta, \kappa, 1)}{\max_{(\theta,\kappa,\alpha)} L(\theta, \kappa, \alpha)} \tag{3.33}$$

or

$$\Lambda = \frac{\max_{(\theta,\alpha)} L(\theta, 1, \alpha)}{\max_{(\theta,\kappa,\alpha)} L(\theta, \kappa, \alpha)}. \tag{3.34}$$

In either case, $-2 \log \Lambda \sim \chi^2(1)$ is an approximation for large samples assuming the null hypothesis is true. If we were to test $H_0 : \kappa = \alpha = 1$ against the alternative that at least one of κ or α is not one, the likelihood ratio is

$$\Lambda = \frac{\max_{\theta} L(\theta, 1, 1)}{\max_{(\theta,\kappa,\alpha)} L(\theta, \kappa, \alpha)} \tag{3.35}$$

in which case $-2 \log \Lambda$ is approximately $\chi^2(2)$ when the null hypothesis is true.

Example 3.5 Consider the failure times of deep groove ball bearings, first studied by Lieblein and Zelen (1956). This data set gives the failure times of 23 bearings, measured in millions of cycles, and is shown in Table 3.5. There is no censoring.

Table 3.3 Failure times in millions of cycles for deep groove ball bearings.

i	1	2	3	4	5	6	7	8
T_i	17.88	28.92	33.00	41.52	42.12	45.60	48.40	51.84
i	9	10	11	12	13	14	15	16
T_i	51.96	54.12	55.56	67.80	68.64	68.64	68.88	84.12
i	17	18	19	20	21	22	23	
T_i	93.12	98.64	105.12	105.84	127.92	128.04	173.40	

Source: Modified from Lieblein and Zelen (1956).

Using this data, fit the exponential, Weibull, gamma, log normal, and generalized gamma distributions. Make comparisons among the models.

Solution:
Figure 3.6 shows Q-Q plots for the five distributions. The upper left plot suggests that the exponential distribution is not a good fit to the data. This isn't surprising since the exponential distribution's hazard function is a constant, which is an unrealistic assumption for most data sets. The other plots all seem to provide a reasonably good fit.

To fit the Weibull distribution, R's `survreg` function in the `survival` package can be used. The `survreg` function is ordinarily used for regression with lifetime data, and when there are no predictor variables, we simply use 1 as the independent variable. The following code accomplishes this.

```
library( survival )
model.weibull = survreg( Surv(t) ~ 1, dist="weibull" )
summary( model.weibull )
```

We do not need to say `dist="weibull"` because the Weibull distribution is the default. The response variable (the lifetime) must be a survival object in R; this can be accomplished using the `Surv` function (notice the capitalization). If there were a censoring variable, called `Cens`, the function would be called with two arguments: `Surv(t, Cens)`. The censoring variable `Cens` must be coded as 1 for a failure time and 0 for a censored time. The output from this is

```
Call:
survreg(formula = Surv(t) ~ 1, dist = "weibull")
                Value Std. Error     z       p
(Intercept)     4.405      0.105 41.93  <2e-16
Log(scale)     -0.743      0.156 -4.75  2e-06

Scale= 0.476

Weibull distribution
Loglik(model)= -113.7   Loglik(intercept only)= -113.7
Number of Newton-Raphson Iterations: 6
n= 23
```

In the `survival` package, R reports estimates based on the distribution of log T which has a smallest extreme value distribution with location parameter μ and scale parameter σ. These are related to the

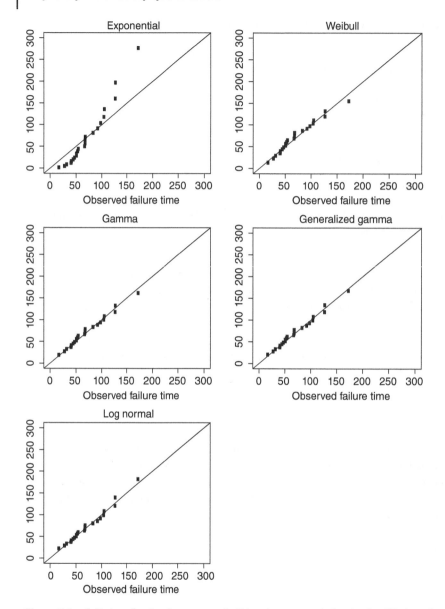

Figure 3.6 Q-Q plots for the deep groove ball bearings example for the five lifetime distributions: exponential, Weibull, gamma, generalized gamma, and log normal.

parameters of the Weibull in this way:

$$\mu = \log \theta$$
$$\sigma = \frac{1}{\kappa}$$

or equivalently

$$\theta = \exp \mu$$
$$\kappa = \frac{1}{\sigma}.$$

We can extract estimates of μ and σ from the code:

```
muhat = model.weibull$icoef[1]
sigmahat = exp( model.weibull$icoef[2] )
estimates.SEV = c( muhat , sigmahat )
names( estimates.SEV ) = c("muhat","sigmahat")
print( estimates.SEV )

thetahat = exp( muhat )
kappahat = 1/sigmahat
estimates.weibull = c( thetahat , kappahat )
names( estimates.weibull ) = c("thetahat","kappahat" )
print( estimates.weibull )
```

This yields the output

```
    muhat  sigmahat
4.4051883 0.4757721
```

and

```
 thetahat   kappahat
81.874559  2.101847
```

Thus, we see that the estimates of the Weibull parameters are

$$\hat{\theta} = \exp(4.4051833) = 81.874559,$$
$$\hat{\kappa} = \frac{1}{0.4757721} = 2.101847.$$

The survival package does not allow the gamma distribution, so we use the flexsurv package instead. The code

```
library( flexsurv )
model.gamma = flexsurvreg( Surv(t) ~ 1, dist="gamma" )
print( model.gamma )
```

produces the output

```
Call:
flexsurvreg(formula = Surv(t) ~ 1, dist = "gamma")

Estimates:
       est     L95%    U95%    se
shape  4.0247  2.3089  7.0155  1.1410
rate   0.0557  0.0308  0.1007  0.0168

N = 23,  Events: 23,  Censored: 0
Total time at risk: 1661.08
Log-likelihood = -113.0298, df = 2
AIC = 230.0596
```

Note that `flexsurvreg` uses the distribution of log T in a manner similar to that of `survreg`. Finally, we can apply the generalized gamma distribution by using `dist= "gengamma.orig"`. The output is

```
Call:
flexsurvreg(formula = tSurv ~ 1, dist = "gengamma.orig")

Estimates:
       est       L95%      U95%      se
shape  6.13e-01  1.86e-02  2.02e+01  1.09e+00
scale  1.55e+00  3.33e-11  7.26e+10  1.95e+01
k      1.02e+01  1.15e-02  9.10e+03  3.54e+01

N = 23,  Events: 23,  Censored: 0
Total time at risk: 1661.08
Log-likelihood = -112.9693, df = 3
AIC = 231.9386
```

The parameter estimates are summarized in Table 3.4. The most general distribution is the three-parameter GGAM(θ, κ, α). The Weibull ($\alpha = 1$) and gamma ($\kappa = 1$) are special cases of the GGAM distribution, and the log normal is a limiting case ($\alpha \to \infty$). The values of the log likelihood at the MLE are shown, allowing us to perform the likelihood ratio tests in Eqs. (3.33), (3.34), and (3.35). We can test $H_0 : \alpha = \kappa = 1$ since

$$-2 \log \Lambda = -2 \left(\log \frac{\max_{(\theta,\alpha)} L(\theta, 1, 1)}{\max_{(\theta,\kappa,\alpha)} L(\theta, \kappa, \alpha)} \right) = 2 \left(\log L(\theta, \kappa, \alpha) - \log L(\theta, 1, 1) \right).$$

Using the values from Table 3.4, we find that

$$-2 \log \Lambda = 2(-112.9693 + 121.4338) = 16.929.$$

The difference in the dimension of the parameter space between the numerator and denominator is 2, since there are two more parameters in the denominator. We compare this result to $\chi^2_{0.05}(2) = 5.99$. Since the test statistic, 16.929, exceeds 5.99 we have evidence that the exponential distribution is not an adequate model for the failure times. If we use the values of the log likelihood for the Weibull distribution, we obtain

$$-2 \log \Lambda = 2(-112.9693 + 113.6920) = 1.4448.$$

Table 3.4 Summary of point estimates, log-likelihood function evaluated at the MLE, and the AIC for various model applied to the data of Lieblein and Zelen (1956).

Distribution	$\hat{\theta}$	$\hat{\kappa}$	$\hat{\alpha}$	Log-like	AIC
Exponential	72.221			−121.4338	244.8675
Weibull	81.875	2.1018		−113.6920	231.3839
Gamma	17.944		4.0247	−113.0298	230.0596
Log normal	$\hat{\mu} = 4.1504$	$\hat{\sigma} = 0.5217$		−113.1286	230.2571
Generalized gamma	1.5673	0.6141	10.2006	−112.9693	231.9386

This time the difference in the dimension of the parameter space is 1, so we compare this test statistic with $\chi^2_{0.05}(2) = 3.84$. In this case there is no evidence that the generalized gamma distribution fits substantially better than the Weibull distribution. The test for the gamma distribution leads to a similar conclusion. The test statistic,

$$-2 \log \Lambda = 2(-112.9693 + 113.0298) = 0.1210,$$

is less than $\chi^2_{0.05}(2) = 3.84$, so there is no evidence that the generalized gamma fits the data better than a gamma distribution. The argument for the log normal distribution is similar, but more subtle because the log normal is a limiting case, not a special case. The same computation can be done for the log normal, yielding

$$-2 \log \Lambda = 2(-112.9693 + 113.1286) = 0.3186.$$

Table 3.4 also shows the Akaike information criterion (AIC), a measure of model fit that assesses model predictions for out-of-sample data, that is, data that were not used to fit the model. The AIC penalizes overfitting, thereby accounting for the number of parameters. Smaller values of AIC indicate a better fit. The last column in Table 3.4 shows the AIC. By this criterion, the best fit is the gamma distribution, although the log normal, Weibull, and generalized gamma distributions fit nearly as well. The exponential distribution is a poor fit.

Thus, using either the χ^2 tests on the log likelihood function or the AIC, we conclude that any of the Weibull, gamma, or log normal distributions fits the data as well as the generalized gamma distribution, and the exponential does not fit the data well. ∎

When selecting a model, our belief about the shape of the hazard function often plays a key role in selecting a model. Among the five distributions:

1. The exponential distribution has a constant hazard function.
2. The Weibull distribution has a hazard function that is either always increasing or always decreasing. In most situations the hazard is increasing. Moreover, if the Weibull's hazard function is increasing, it increases without bound.
3. The gamma distribution also has a hazard function that is either always increasing or always decreasing, but it approaches the fixed value α/θ.
4. The log normal distribution's hazard function increases initially, reaches a maximum, and then decreases toward 0.
5. The generalized gamma distribution has a very flexible hazard function that can be always increasing, always decreasing, or increasing then decreasing.

Figure 3.7 shows the estimated hazard functions for the ball bearing data from Table 3.3 for each of the five distributions.

3.7 Bayesian Inference

The difference between the classical (or frequentist) approach and the Bayesian approach to statistic lies in our definition and interpretation of probability. Frequentists (i.e. those who believe in the classical

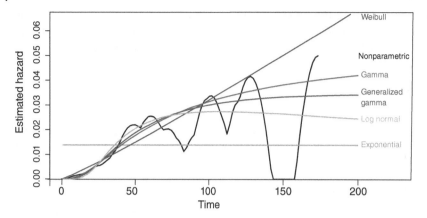

Figure 3.7 Estimates of the hazard function for the five lifetime distributions.

approach) define probability as the relative frequency that an event would occur in a hypothetically infinite sequence of trials. As such, probability is an innate property of an event, and is something we may assume, or we may agree that it is unknown, to be estimated from data. For example, we may assume that a coin is fair and the probability of getting a head is 0.5. On the other hand, we may develop new product whose probability of lasting 1000 hours, call it p, is unknown, but which can be estimated from the data from a life test. Even though p is unknown, frequentists would agree that it is an intrinsic property of the product design and the type of conditions under which it will be operated. Thus, p is an unknown *constant*. Just because p is unknown, this doesn't change the fact that it is really a fixed number; we just don't know what it is.

By contrast, Bayesians define probability in a more subjective sense. They, in effect, treat uncertainty and randomness as equivalent. Thus, In the Bayesian paradigm, unknown parameters are treated as *uncertain*, and Bayesians treat uncertainty in probabilistic terms. For example, if we don't know the value of the survival probability p described in the previous paragraph, we express our knowledge (or ignorance) about that value in a probability distribution.

Of course, Bayes' theorem plays the key role in the Bayesian approach, so we now describe it for a finite collection of events. Later, we'll generalize how to apply this to the case of a continuous outcome.

Theorem 3.1 (Law of Total Probability) *Suppose $B_1, B_2, \ldots, B_k$ are mutually exclusive and whose union consists of the set of all possible outcomes. (In other words, one and only one of the B_i will occur.) If A is any event, the probability of A can be computed from*

$$P(A) = \sum_{j=1}^{k} P(B_j)P(A|B_j).$$

We omit the proof here, but the idea is simple. The event A occurs if and only if exactly one of the events $A \cap B_1$, or $A \cap B_2$, or $\ldots A \cap B_k$ occurs. A straightforward application of the rule for conditional probability yields the result.

Given the Law of Total Probability, Bayes' theorem follows directly.

Theorem 3.2 (Bayes' Theorem) *Suppose $B_1, B_2, \ldots, B_k$ are mutually exclusive and whose union consists of the set of all possible outcomes, and let A be some event. Then*

$$P(B_i|A) = \frac{P(B_i|A)}{\sum_{j=1}^{k} P(B_j)P(A|B_j)}.$$

Often, the events B_j represent the outcome of some random variable. If the variable is discrete, then there is no problem with the version of Bayes' theorem that we stated earlier. (Even if the number of possible outcomes is countably infinite, the theorem still applies if we make the denominator an infinite series (that converges)). If the underlying random variable is continuous, then the denominator must be modified so that it *integrates*, rather than sums, over all possible values. Let Z denote the random variable that takes on values in the set S and has PDF $p(z)$. Then

$$P(A|Z = z) = \frac{p(z)P(A|z)}{\int_S p(z)P(A|z)\,dz}. \tag{3.36}$$

Suppose θ is an unknown parameter and x is the collection of all the available data whose distribution depends on the parameters, then the Bayesian approach to statistics can be described in the following two steps:

- Assign a distribution to the parameter θ that encapsulates what we believe about θ before observing any data. The PDF $p(\theta)$ of this distribution is called the *prior distribution*.
- Apply Bayes' theorem to determine the conditional PDF of θ given the data x; this is called the *posterior distribution*.

Once the prior distribution is determined, we combine the prior distribution and the likelihood function using Bayes' theorem to obtain the posterior distribution as in (3.36). Note that the denominator of Bayes' theorem is a function of the data only (since we integrate over the entire parameter space). Also, if there are p components to the parameter θ, then the dimension of the integral in the denominator will be p. (Even for a higher dimensional integral, we will write a single integral, with the understanding that the dimension of the integral is equal to the number of components in the parameter θ.) If p is greater than three or four, an analytic or numerical approximation to the integral will be difficult. Monte Carlo methods have been developed to simulate any number of values from the posterior distribution. These methods involve Markov chains and are usually called Markov chain Monte Carlo (MCMC) methods. If the formula for the posterior distribution is messy, it will likely be the case that 10,000 simulated values from the posterior distribution will provide more information than inspection of a messy formula that is the posterior distribution.

The posterior distribution reflects what we believe about θ given the observed data. Notice that the likelihood function $f(x|\theta)$ is the conditional PDF of x given θ, whereas the posterior is the conditional PDF of θ given x. Recall that Bayes' theorem tells us how to swap the events in a conditional probability.

We could take some measure of the center of the posterior distribution if we wanted a point estimate of θ. The posterior mean

$$\hat{\theta} = E(\theta|x) = \frac{\int \theta \pi(\theta) f(x|\theta)\,d\theta}{\int \pi(\theta) f(x|\theta)\,d\theta} \tag{3.37}$$

is often used as a point estimate. Sometimes the posterior median or mode is used as a point estimate instead. If we want an interval estimate of a parameter, we could determine an interval I for the case of a scalar parameter, or a region R for the case of a vector parameter, with the property that

$$P(\theta \in I|\boldsymbol{x}) = 1 - \alpha. \tag{3.38}$$

Such an interval is called a *credible interval*.

By contrast the classical approach determines an interval $I(\boldsymbol{x})$ with the property that

$$P(\theta \in I(\boldsymbol{X})) = 1 - \alpha, \tag{3.39}$$

where $\boldsymbol{X}$ is the data. While these two formulations look similar, they differ in an important and fundamental way. In (3.38), θ is the random variable while $\boldsymbol{x}$ is considered fixed; in (3.39) the endpoints of the interval $I(\boldsymbol{X})$ are random, depending on the random vector X, and θ is considered fixed.

The next example shows the steps involved in a Bayesian analysis in a life testing framework.

Example 3.6 Many new cars today are equipped with a mechanism to turn off the engine when the driver comes to a complete stop and then restart when the driver moves from the brake to the accelerator pedal. This puts a greater stress on the car's starter, requiring it to have a higher reliability. Suppose that design changes have been made to a starter that previously had a an average lifetime of about 40,000 starts. What assumptions are reasonable for a lifetime distribution and a prior distribution for that distribution's parameters?

Solution:
First, if the lifetime is measured in successful "starts" then the outcome is really a discrete distribution. However, with the responses measuring in the thousands, a continuous distribution is likely to be a reasonable approximation. Even in the case of a continuous life time measured, say, in hours, it is likely to be rounded to the nearest hour, giving about the same amount of discreteness we expect to see in this example. We will define the lifetime in terms of the number of starts in thousands.

Since the starter is a mechanical device, it is likely to fail due to wear out of at least one component. Because of the wear out, the exponential distribution is probably a bad choice since it has the memoryless property and it has a constant hazard function. Here we expect the hazard to be increasing. Further, we expect the hazard to increase without bound as time increases. The gamma distribution does have an increasing hazard function, but it approaches a constant for large t. Thus, the gamma distribution is not a good choice either. The Weibull distribution is probably the best choice.

If we assume a WEIB(θ, κ) distribution, we would have to place a joint prior distribution on (θ, κ). The κ parameter for mechanical parts like this is often greater than 1 and less than 3 or 4. If $\kappa = 1$ we would have a constant hazard (implying the exponential distribution). If $\kappa < 1$ the hazard would be decreasing, which we don't expect. Our prior should have most probability between 1 and 3, with some small chance of being less than 1 or greater than 3. In any case, the prior distribution must have zero probability for $\kappa < 0$ because a negative κ is not possible. The gamma distribution, with parameters α and γ, is a reasonable model for a parameter that must be positive. The mean and variance of the GAM(α, γ) distribution are

$$\mu = \gamma \alpha \quad \text{and} \quad \sigma^2 = \gamma^2 \alpha.$$

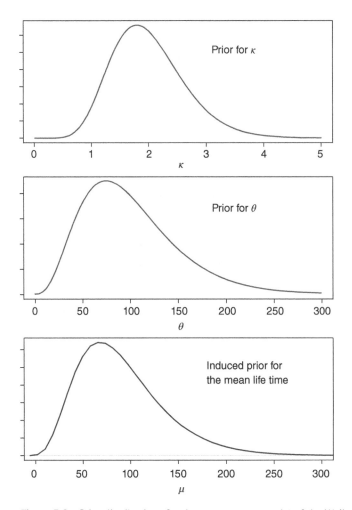

Figure 3.8 Prior distributions for the parameters κ and θ of the Weibull distribution. The prior for the mean life, measured in thousands of starts, that results from these priors is shown in the bottom figure; values for the bottom graph were obtained by simulation

Often, a little trial and error is required to select the prior parameters to reflect what we believe about the parameter. If we are looking for a prior for the shape parameter κ we might use information that we expect κ to be greater than one, implying an increasing hazard function, and probably no larger than $\kappa = 3$. The top graph in Figure 3.8 shows the GAM(10, 1/5) distribution, which seems to agree with the stated assumptions about κ.

The parameter θ is related to the mean life through the function

$$\mu = \theta\Gamma\left(1 + \frac{1}{\kappa}\right)$$

which is a quantity we might have a better intuitive feel for. It is possible to use the method of transformation of variables to find the prior PDF of μ given the priors for γ and κ, but this is a messy process.

The method of simulation works just as well. We can choose a GAM prior distribution for θ and simulate thousands of draws from it, along with thousands of draws from the prior for κ. We would then plug these simulations into the function for the mean μ given earlier. We found that a GAM(4, 25,000) distribution gives the frequency distribution for mean lifetime shown at the bottom of Figure 3.8. If this is judged to be a good reflection about what we believe about the parameters, we can specify our model as follows:

$$\kappa \sim \text{GAM}(10, 0.2) \qquad [\text{dgamma}(10, 5) \text{ in WinBUGS}]$$

$$\theta \sim \text{GAM}(4, 25) \qquad [\text{dgamma}(4, 0.04) \text{ in WinBUGS}]$$

$$T|(\theta, \kappa) \sim \text{WEIB}(\theta, \kappa).$$

∎

Here WinBUGS (Lunn et al., 2000) is a software package for performing MCMC. The abbreviations given in brackets show how the gamma distribution is referred to in WinBUGS. Note that the second parameter is the reciprocal of the value in our parameterization.

There is no single correct answer to the question posed in the previous example. Those who would prefer to inject less prior information about the parameters could choose more diffuse priors, that is, priors that are much more spread out. Fortunately, for even moderately large samples the effect of the prior is usually small.

Let's now address the problem of making inference for the parameters θ and κ from the Weibull distribution. Assume that 30 starters were tested until either they failed or the total operating time reached 120,000 starts. Suppose we observe the data shown in Table 3.5.

In principle, the problem is simple. Elicit the prior distributions, based on what is believed about the parameters. This is what the last example did. Then take the prior and the data, and apply Bayes' theorem to get the posterior distribution of all parameters. The likelihood function is obtained from Eq. (3.26)

$$L(\theta, \kappa) = \frac{\kappa^r}{\theta^{r\kappa}} \left(\prod_{i \in \mathcal{U}}^{n} t_i^{\kappa-1} \right) \exp \left[-\sum_{i=1}^{n} \left(\frac{t_i}{\theta} \right)^{\kappa} \right].$$

Table 3.5 Lifetimes of engine starters, measured in thousands of starts.

Unit	Lifetime	Unit	Lifetime	Unit	Lifetime
1	31	11	92	21	120+
2	42	12	98	22	120+
3	57	13	104	23	120+
4	58	14	118	24	120+
5	68	15	120+	25	120+
6	70	16	120+	26	120+
7	73	17	120+	27	120+
8	78	18	120+	28	120+
9	89	19	120+	29	120+
10	90	20	120+	30	120+

Observations were truncated at time 120.

Assuming the priors for κ and θ are independent, we find that the joint prior is

$$p(\kappa, \theta) = c\,\kappa^{10-1} \exp(-\kappa/0.2)\,\theta^{4-1} \exp(-\theta/25), \qquad \kappa > 0,\ \theta > 0.$$

Note that measuring the lifetimes in thousands of starts means the scale parameter in the prior for θ is reduced by a factor of $1/1000$, yielding 25, not 25,000 as before. Also, the constant c does not depend on either κ or θ. Applying Bayes' theorem yields the joint posterior distribution

$$p(\kappa, \theta | \text{data}) = \frac{p(\kappa, \theta)\,L(\kappa, \theta | \text{data})}{\displaystyle\int_0^\infty \int_0^\infty p(\kappa, \theta)\,L(\kappa, \theta | \text{data})\,d\theta\,d\kappa}. \tag{3.40}$$

You can see that this gets messy, and if the model had even more than two parameters, it would be even messier. For just two parameters, the double integral in the denominator can be evaluated, but for models that involve even just a few more parameters become intractable. A model with p parameters will involve a p-dimensional integral in the denominator. For p greater than 4 or 5, evaluating the integral becomes more difficult.

The joint posterior given in Eq. (3.40) gives us the distribution of the parameters given the data. In a sense, Bayes' theorem tells us how to change our minds, from the prior to the posterior, as we take into account the observed data.

There are several ways we could summarize the information contained in the posterior. We could create a contour plot of the joint posterior distribution. We would, in addition, evaluate the posterior means of each parameter:

$$E(\kappa | \text{data}) = \frac{\displaystyle\int_0^\infty \int_0^\infty \kappa\,p(\kappa, \theta)\,L(\kappa, \theta | \text{data})\,d\theta\,d\kappa}{\displaystyle\int_0^\infty \int_0^\infty p(\kappa, \theta)\,L(\kappa, \theta | \text{data})\,d\theta\,d\kappa}.$$

$$E(\theta | \text{data}) = \frac{\displaystyle\int_0^\infty \int_0^\infty \theta\,p(\kappa, \theta)\,L(\kappa, \theta | \text{data})\,d\theta\,d\kappa}{\displaystyle\int_0^\infty \int_0^\infty p(\kappa, \theta)\,L(\kappa, \theta | \text{data})\,d\theta\,d\kappa}.$$

In a similar manner, we could get the posterior variance and standard deviation, which would serve as a measure of our uncertainty in each parameter after observing the data. We could even determine an interval (L, U) with the property that

$$P(L < \kappa < U | \text{data}) = 1 - \alpha,$$

and similarly for θ. Such an interval is called a *credible interval* and is the Bayesian analogue of the classical confidence interval.

The computations described here are messy, but doable for a model with just two parameters. For more complex models, other approaches must be taken. We will describe two of these.

The first involves making approximations to the posterior distribution. For large samples, the joint posterior distribution for an arbitrary parameter ϕ should be approximately multivariate normal with mean equal to the posterior mode $\hat{\phi}$ and covariance matrix $[nI(\hat{\phi})]^{-1}$ where $I(\hat{\phi})$ is the Fisher information matrix

$$I(\hat{\phi}) = -E\left[\frac{\partial^2 \log L}{\partial \phi_i\,\partial \phi_j}\right].$$

Since the posterior mode depends on the data, the data enter the picture through $\hat{\phi}$. There are still some numerical difficulties here, because we must maximize a function of p variables, assuming $\hat{\phi}$ is a vector parameter of length p, and we must be able to evaluate the partial derivatives of the log-likelihood function and take the expected values (although there are ways to approximate this too). These are easier problems than evaluating the p-dimensional integral in the denominator of Bayes' theorem.

The second method involves simulating values from the posterior distribution. The key idea here is that having a large number, say, 10,000, of simulations from the posterior distribution is often better than having an exact formula for the posterior that we are unable to evaluate. Such simulation algorithms date back to 1953 when Metropolis et al. (1953) developed an algorithm that produced a Markov chain whose steady state distribution is the distribution we'd like to simulate from. This algorithm was extended by Hastings (1970), but most statisticians were unaware of the algorithm until Geman and Geman (1984) applied it to a Bayesian analysis of image restoration. This new approach to Bayesian statistics opened up the door to allow statisticians to analyze much more complicated models than ever before. In the appendix to this chapter we describe what is called the Metropolis–Hastings algorithm, although some recent enhancements have improved on the method.

The theory behind Markov chain Monte Carlo says that the steady state distribution of the Markov chain is the posterior distribution. We can learn about the posterior distribution by this by first applying MCMC for enough steps that we are convinced, for all practical purposes, that the steady state has been reached. This is called the burn-in. Second, we simulate additional iterations through the Markov chain. We take these as observations from the posterior distribution. With a large number of these simulations, say, 10,000 or more, we can learn what the posterior distribution of any parameter looks like. These simulated values are, however, not independent. Thus, the information in 10,000 iterations is less that it would be if we had 10,000 independent simulations. The stronger the autocorrelation in the simulated values, the less information we have in those values.

Example 3.7   Suppose that 30 starters were produced and tested until failure or until they reached 120,000 starts, whichever came first. Data are shown in Table 3.5. Use the priors developed in Example 3.6 to obtain the posterior distributions of both parameters using MCMC.

Solution:

We used R and WinBUGS to produce 1,000,000 simulations from the joint posterior distribution for (κ, θ), after running a burn-in run of 50,000. Because the successive values of the Markov chain showed strong autocorrelation, we retained only every 100th simulation, yielding 10,000 simulations from the posterior. This is called *thinning*.

The following WinBUGS (Lunn et al. 2000) code can be sent to WinBUGS in order to perform the MCMC:

```
model {
   shape ~ dgamma( 10 , 5 )
   scale ~ dgamma( 4 , 0.04 )
   rate <- 1/pow( scale , shape )
   for (j in 1:n)
   {
     t.obs[j]   ~   dweib( shape , rate )I(t.cen[j],)
   }
}
```

See Sturtz et al. (2005) for a description of the R2WinBUGS package in R and how to run the aforementioned code in the R environment.

The data must be in the following format. Notice how t.obs is equal to the failure time for those items that failed and is NA for the censored values. Also, t.cen is equal to 0 for each failure and is equal to the censoring time (120 in this case) for the censored values. The data frame in R is therefore:

```
         t  cen  t.obs  t.cen
 [1,]   31    0     31      0
 [2,]   42    0     42      0
 [3,]   57    0     57      0
 [4,]   58    0     58      0
 [5,]   68    0     68      0
 [6,]   70    0     70      0
 [7,]   73    0     73      0
 [8,]   78    0     78      0
 [9,]   89    0     89      0
[10,]   90    0     90      0
[11,]   92    0     92      0
[12,]   98    0     98      0
[13,]  104    0    104      0
[14,]  118    0    118      0
[15,]  120    1     NA    120
[16,]  120    1     NA    120
[17,]  120    1     NA    120
[18,]  120    1     NA    120
[19,]  120    1     NA    120
[20,]  120    1     NA    120
[21,]  120    1     NA    120
[22,]  120    1     NA    120
[23,]  120    1     NA    120
[24,]  120    1     NA    120
[25,]  120    1     NA    120
[26,]  120    1     NA    120
[27,]  120    1     NA    120
[28,]  120    1     NA    120
[29,]  120    1     NA    120
[30,]  120    1     NA    120
```

The trace plots are shown on the left sides of Figure 3.9. These plots exhibit what appears to be random noise, indicating that we have probably reached the steady state. The right sides of Figure 3.9 show the posterior distribution estimated from the 10,000 simulations. This is a kernel density estimate of the PDF obtained using the density command in R. From this plot, we can see that the most likely values for κ and θ are just above 2 and slightly below 150, respectively. The posterior means can be obtained by averaging the 10,000 simulations that were kept; this yields

$$\hat{\kappa} = 2.1657,$$
$$\hat{\theta} = 147.75.$$

It is often helpful to compare the prior and posterior distributions. This can tell us how much information the data provided to us. Both posterior distributions have less variability indicating the extent to which we have more information after the data than before.

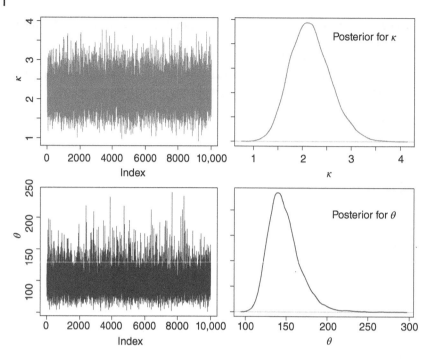

Figure 3.9 Trace plots (on left side) for parameters κ and θ. The posterior distributions based on 10,000 simulations are shown on the right side.

The joint posterior distribution for (κ, θ) can be shown in the scatter plot of Figure 3.10. Here we see that there is a slight dependence between κ and θ. If κ is large, then θ tends to be small and vice versa. The 10,000 simulated ordered pairs from the joint posterior distribution can be used to obtain the posterior distribution of any function of these two parameters. Simply substitute each ordered pair from the 10,000 simulations into the function and we obtain a simulation of the desired function. For example, if we are looking for the posterior distribution of the mean lifetime $\mu = \theta \, \Gamma(1 + 1/\kappa)$, we could compute

$$\mu^{(i)} = \theta^{(i)} \, \Gamma\left(1 + \frac{1}{\kappa^{(i)}}\right),$$

where $(\kappa^{(i)}, \theta^{(i)})$ is the ith simulation from the Markov chain. The posterior distribution for the mean lifetime is shown in Figure 3.11. This figure represents what we believe about the mean lifetime given the observed data from Table 3.7. This figure is often misread. It shows our belief about the mean lifetime, not what individual lifetimes might be. For example, since 50 is in the far left tail of this distribution, we can be quite certain that the mean lifetime exceeds 50; it does not say that almost all starters will have a lifetime that exceeds 50.

The phenomenon described in the last paragraph requires the *posterior predictive distribution* for a new observation. This is defined as the distribution of one additional observation T given the observed data.

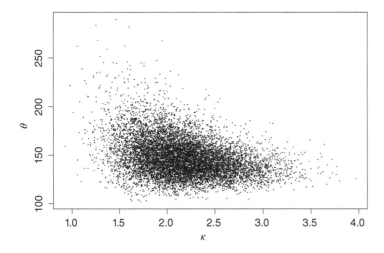

Figure 3.10 Joint posterior distribution for κ and θ based on 10,000 MCMC simulations.

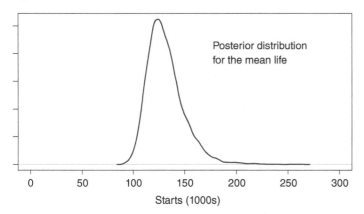

Figure 3.11 Posterior mean lifetime for starter measured in 1000s.

Once we've obtained the simulations from the joint posterior, getting the posterior predictive distribution involves a simple two-step algorithm:

1. Simulate one ordered pair $(\kappa^{(i)}, \theta^{(i)})$ from the posterior.
2. Given this value of $(\kappa^{(i)}, \theta^{(i)})$, simulate one observation from the WEIB$(\kappa^{(i)}, \theta^{(i)})$ distribution and call it $t^{(i)}$.

With this two-step algorithm we can easily simulate 10,000 values from the predictive distribution. The PDF of the predictive density, estimated from these 10,000 simulations, is shown in Figure 3.12.[2]

---

2 You might notice that the left tail of the estimated PDF yields positive probability for negative values of the lifetime, which is of course impossible for a lifetime distribution. This is an artifact of the method of kernel density estimation where the kernels, symmetric PDFs centered at each data point, for the smallest values lap over to negative values.

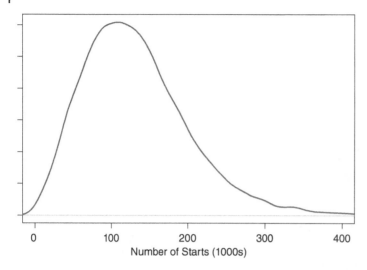

Figure 3.12 Predictive distribution for starter lifetime measured in 1000s of starts.

Given the data from this experiment, how reliable is the new starter design? The answer depends on how many times the starter will be needed to start the engine. For an engine that shuts down and restarts every time the car comes to a stop, we might require 20 starts per day. Multiply this by 365.25 days per year and the normal life of 8 years for a car, this comes up to 58,440 starts. Compare this value of 58.44 (1000) starts to the PDFs in Figures 3.11 and 3.12. We are quite certain that the mean life exceeds 58.44, but there is about a 13% chance that a single observation will fall below 58.44. The reliability for a mission time of $T_{mission}$ is, assuming a WEIB(θ, κ),

$$R(T_{mission}) = \exp\left(-\left(\frac{T}{\theta}\right)^{\kappa}\right) = \exp\left(-\left(\frac{58.44}{\theta}\right)^{\kappa}\right).$$

Once again, this is a function of unknown parameters, so given 10,000 simulations from these parameters' posterior distributions, we can apply the algorithm from earlier to simulate from the posterior of $R(T_{mission})$. The resulting density (really the kernel density estimate of the PDF) is shown in Figure 3.13. Here we assess the uncertainty in our estimate of the reliability. The reliability seems to be just below 0.90, which is consistent with our 13% estimate of the unreliability computed earlier.

It is instructive to compare Figures 3.8 and 3.12, which give the posterior for the mean lifetime and the predictive distribution for one additional lifetime. The former conveys what we believe about the mean lifetime after observing the data. For a large amount of data, the posterior for the mean lifetime can be quite narrow, indicating that we have strong confidence about what the mean lifetime is. The latter conveys what we believe about one additional observation will be. Alternatively, we could think of the predictive distribution as being the population from which we sampled. Even if we knew the exact values of the parameters θ and κ (which we never do) we would still not be able to give an exact prediction for a new observation. The posterior for the mean life involves our uncertainty in the parameters θ and κ only, while the predictive distribution involves the uncertainty of θ and κ as well as the randomness of the additional observation. ∎

Figure 3.13 Posterior distribution of reliability for a mission of 58,440 starts.

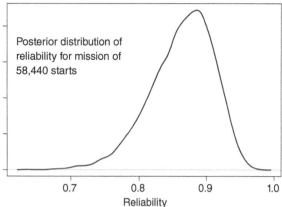

Posterior distribution of reliability for mission of 58,440 starts

Reliability

The Bayesian approach involves a different way of thinking about uncertainty. Bayesians treat randomness and uncertainty as the same, so uncertainty in parameters is quantified in terms of probability. Thus, the Bayesian approach treats parameters as random variables, something the classical approach never does.

There are a number of advantages of the Bayesian approach. Here are some:

1. We are able to incorporate prior knowledge. The shape parameter κ of the Weibull is a good example of a parameter for which we usually have some strong prior belief. Although the only requirement for κ is that be a positive number, we often have a much better idea about what it might be. Mechanical parts will likely wear out, resulting in $\kappa > 1$. A value of $\kappa = 3$ would result in rapid wear out, much more rapid than is usually observed. For many problems it is reasonable that κ will be in the interval $(1, 3)$, much narrower than the $(0, \infty)$ that is required for κ. Thus selecting a prior with a small probability of being below 1 or above 3 seems reasonable for most problems.
2. In principle, the Bayesian method is straightforward, in the sense that we select the prior, determine the likelihood and compute the posterior using Bayes' theorem. If the integral in the denominator of Bayes' theorem is not tractable then apply MCMC to estimate the posterior distribution. This straightforward nature of Bayesian statistics is sometimes referred to as "turning the Bayesian crank."
3. Bayesian inference usually does not rely on normal approximations. With MCMC, the posteriors can be determined with the error being due to just the random simulation. This Monte Carlo error is usually very small if the number of simulations is large. By contrast, classical confidence intervals and hypothesis tests usually rely on the large-sample approximate normality of the estimators. This can be determined from the expected or observed Fisher information matrix. These are sometimes poor approximations and result in confidence intervals whose coverage is not close to the nominal; for example, a desired 90% confidence interval may cover the true parameter 80% of the time (or 96% of the time). Bayesian posterior distributions, even estimated with MCMC, are nearly exact reflections of our belief after observing the data.
4. Classical statistics makes inference in an indirect way. For example, the correct interpretation of a 95% confidence interval is "95% of the time an interval calculated in this way will include the true value of the parameter." It says something about a future sequence of experiments, none of which will ever be

conducted. A Bayesian 95% credible interval, by contrast, says that the probability that true parameter will be in the interval, conditioned on the observed data, is 0.95.

5. With the Bayesian approach, making inference for functions of parameters is easy. We saw in the last example that the posterior for the mean lifetime can be gotten by simulating from the posterior and then computing the quantity of interest. We did the same thing for the mission reliability. Classical methods usually rely on linearizing the function and applying an approximation method such as the delta method theorem.

6. Many problems in statistics are sequential. We collect data, make an inference, then collect more data, make inference again, and so on. This problem of making sequential inference is often called *filtering*, and Bayesian statistics in particularly suited for this. The posterior distribution at one stage becomes the prior distribution at the next stage. The old saying is "Yesterday's posterior is today's prior."

Two of the most commonly mentioned objections to Bayesian statistics are these:

1. Users must select a prior distribution. This is often perceived as injecting subjectivity into statistics. They say that statistics should be a purely objective science.

2. If different statisticians take different priors, they will make different inferences based on the same data.

Both of these are true, but require some explanation. First, it is usually possible to select priors with large variances. These can be thought of as being noninformative priors. Also, for large samples, the posterior does not depend much on the prior. It's only for small samples that the prior plays a large role in the posterior. Regarding the second objection, it is true that two people selecting different priors will get different posteriors. This is not so much a contradiction as it is a difference of opinion. If my priors reflect my belief going in to an experiment, then the posterior reflects my belief afterward. The same applies to your priors and posteriors. If we end up with different posteriors it's because we disagreed on our priors, which is not a contradiction.

3.A Kaplan–Meier Estimate of the Survival Function

Here we look at the derivation of the Kaplan–Meier estimate of the survival function for the case of discrete failure times. The generalization to continuous random variables is straightforward, but subtle. Suppose that the possible survival times are

$$1, \ 2, \ 3, \ \dots.$$

Think of the lifetime as measured to the nearest time unit, such as hours. The "parameters" to be estimated are the values of the hazard function at each of these integer values; that is, we would like to estimate each of

$$h_1, \ h_2, \ h_3, \ \dots.$$

The likelihood function is the product over all the failure times of the probability mass, times the product over all the censored times of the survival function. Letting $\pi_k = P(T = k)$ and $\boldsymbol{h} = [h_1, h_2, \dots]$, we see that

$$L(\boldsymbol{h}) = \left(\prod_{k \in U} \pi_k \right) \left(\prod_{k \in C} S_k \right). \tag{3.A.1}$$

Note that in order for the failure to occur at time k, the item must survive past time 1 (with probability $1 - h_1$), and it must survive past time 2 given that it survives past time 1 (with probability $1 - h_2$), etc.; finally, it must fail at time k given that it survived past time $k - 1$ (with probability h_k). The probability of all this occurring is

$$\pi_k = P(\text{failure occurs at time } k) \tag{3.A.2}$$
$$= (1 - h_1)(1 - h_2) \cdots (1 - h_{k-1})h_k$$
$$= \left(\prod_{j=1}^{k-1}(1 - h_j) \right) h_k.$$

Similarly, the survival function is

$$S_k = P(T > 1 \text{ and } T > 2 \text{ and } \cdots \text{ and } T > k) \tag{3.A.3}$$
$$= P(T > 1)P(T > 2 | T > 1) \cdots P(T > k | T > k - 1)$$
$$= (1 - h_1)(1 - h_2) \cdots (1 - h_k).$$

To see the general form of the likelihood, it is instructive to evaluate (3.A.1) for the data described in Section 3.A.1. For the data set

2, 2, 4+, 5, 7+

we can apply (3.A.2) and (3.A.3). The likelihood function from (3.A.1) is then

$$L(\boldsymbol{h}) = \left(\pi_2\pi_2\pi_5 \right) \left(S_4 S_7 \right)$$
$$= \left[(1 - h_1)h_2(1 - h_1)h_2(1 - h_1)(1 - h_2)(1 - h_3)(1 - h_4)h_5 \right]$$
$$\left[(1 - h_1)(1 - h_2)(1 - h_3)(1 - h_4)\ (1 - h_1)(1 - h_2)(1 - h_3)(1 - h_4) \right.$$
$$\left. (1 - h_5)(1 - h_6)(1 - h_7) \right]$$
$$= h_2^2 h_5 (1 - h_1)^5 (1 - h_2)^3 (1 - h_3)^3 (1 - h_4)^3 (1 - h_5)^2 (1 - h_6)(1 - h_7).$$

With this particular case worked out, we can see that in general, the likelihood equation is

$$L(\boldsymbol{h}) = \prod_{j=1}^{\max(\boldsymbol{t})} h_j^{d_j} (1 - h_j)^{r_j - d_j} \tag{3.A.4}$$

which yields a log-likelihood function of

$$\ell(\boldsymbol{h}) = \sum_{j=1}^{\max(\boldsymbol{t})} d_j \log h_j + \sum_{j=1}^{\max(\boldsymbol{t})} (r_j - d_j) \log(1 - h_j). \tag{3.A.5}$$

To get the MLE for each h, we differentiate this with respect to h_J, $J = 1, 2, \ldots, \max(\boldsymbol{t})$ to obtain

$$\frac{\partial \ell}{\partial h_J} = \frac{d_J}{h_J} - \frac{r_J - d_J}{1 - h_J} = 0.$$

A bit of algebra yields

$$h_J = \frac{d_J}{r_J}.$$

This has a clear interpretation: the estimate of the hazard at time J is equal to the number of failures (deaths) at time J divided by the number of units that were at risk at that time.

The `survival` package in R is used to perform many of the inference procedures for lifetime data. The package can be installed using the command

```
install.packages( "survival" )
```

and once installed, it can be loaded using the command `library( survival )`. Given arrays for the event (failure or censoring) times, and the indicator for whether each event was a failure or a censored time, a survival object can be created using the `Surv` command. For example, if the event time and indicator arrays are, respectively, `t0` and `d0` then we can run the `survfit` function to fit a model with no covariates using

```
survfit( Surv(t0,d0) ~ 1 )
```

The `Surv(t0,d0) ~ 1` tells R to fit the model with just a constant term, and no covariates. The aforementioned command creates an object with class `survfit` that can then be plotted using R's `plot` command.

The following code will create the Kaplan–Meier plot; the options given will add confidence intervals using the log-log transformation described in Eq. (3.A.8); it will mark the censoring times with a vertical line segment; the x-axis will go from 0 to 2500, and the estimated survival function and confidence limits will be plotted in light gray with a line width of 2.

```
KM0 = survfit( Surv(t0,d0) ~ 1 , conf.type="log-log" )
plot( KM0 , conf.int=TRUE , mark="|" , col="light gray" ,
    xlim=c(0,2500) , xlab="Survival in Days" ,
    ylab="Probability" , lwd=2 )
```

3.A.1 The Metropolis–Hastings Algorithm

Suppose we want to simulate from a posterior distribution $p(\theta|\boldsymbol{x})$.

Step 1: Starting value. Begin with an initial guess for a value of $\boldsymbol{\phi}$ that is near the posterior mode, and call it $\boldsymbol{\phi}^{(0)}$. This must be a value for which $p(\boldsymbol{\phi}^{(0)}|\text{data}) > 0$. Set $i = 1$.

Step 2a: Propose a move. Let $g(\boldsymbol{\phi}|\boldsymbol{\phi}^{(i-1)})$ be a conditional PDF having the same support as $p(\theta|\boldsymbol{x})$. (This PDF may depend on the previous value $\boldsymbol{\phi}^{(i-1)}$.) The PDF g is called the *proposal density*. Simulate a single value from $g(\boldsymbol{\phi}|\boldsymbol{\phi}^{(i-1)})$ and call it $\boldsymbol{\phi}^{(\text{prop})}$.

Step 2b: Compute the acceptance probability. Compute

$$\alpha = \frac{p(\boldsymbol{\phi}^{(\text{prop})}|\text{data})\, g(\boldsymbol{\phi}^{(i-1)}|\boldsymbol{\phi}^{(\text{prop})})}{p(\boldsymbol{\phi}^{(i-1)}|\text{data})\, g(\boldsymbol{\phi}^{(\text{prop})}|\boldsymbol{\phi}^{(i-1)})}. \tag{3.A.6}$$

Step 2c: Accept or reject the proposed move to $\boldsymbol{\phi}^{(\text{prop})}$.
If $\alpha \geq 1$, then accept the move to $\theta^{(\text{prop})}$. If $\alpha < 1$ then accept the move to $\theta^{(\text{prop})}$ with probability α. If we accept the move then set $\boldsymbol{\phi}^{(i)} = \boldsymbol{\phi}^{(\text{prop})}$. Otherwise we remain at $\boldsymbol{\phi}^{(i-1)}$, in which case we set $\boldsymbol{\phi}^{(i)} = \boldsymbol{\phi}^{(i-1)}$.

Thus, we set

$$\phi^{(i)} = \begin{cases} \phi^{(\text{prop})}, & \text{if we accept the move,} \\ \phi^{(i-1)}, & \text{if we reject the move.} \end{cases} \tag{3.A.7}$$

Step 3: Branch. If i reaches the required number of simulations, then stop. Otherwise, set $i \leftarrow i + 1$ and go to Step 2a.

The successive values generated by this process create a Markov chain, in the sense that the conditional distribution of $\phi^{(i)}$ given every value of $\phi^{(i)}$ that came before it is the same as the conditional distribution of $\phi^{(i)}$ given $\phi^{(i-1)}$. In other words, the distribution of the current $\phi^{(i)}$ depends on the previous value, $\phi^{(i-1)}$, but not the ones that came before. The values generated by the Metropolis–Hastings algorithm are not independent, partly because consecutive values might be the same, which happens when we reject the proposed move. The Metropolis–Hastings algorithm is a realization of a Markov chain. The algorithm therefore produces a sequence

$$\phi^{(0)} \to \phi^{(1)} \to \phi^{(2)} \to \cdots \to \phi^{(i-1)} \to \phi^{(i)} \to \phi^{(i+1)} \to \cdots$$

The Markov chain generated in this way converges to a steady state distribution, that is, a distribution where the unconditional distribution of the random variable at time i is independent of i. (There are some conditions required here, and this is a deep result in probability theory. Interested readers should consult a book on Bayesian statistics such as Gelman et al. (2013).)

The Metropolis algorithm is a special case of the Metropolis–Hastings algorithm that occurs when the proposal density is symmetric about the current value $\phi^{(i-1)}$. In this case

$$g(\phi^{(\text{prop})}|\theta^{(i-1)}) = g(\phi^{(i-1)}|\theta^{(\text{prop})})$$

so the g functions cancel in the calculation of α. For data $\boldsymbol{x}$, this leads to

$$\alpha = \frac{p(\theta^{(\text{prop})}|\boldsymbol{x})}{p(\theta^{(i-1)}|\boldsymbol{x})} = \frac{p(\theta^{(\text{prop})}) f(\boldsymbol{x}|\theta^{(\text{prop})})}{p(\theta^{(i-1)}) f(\boldsymbol{x}|\theta^{(i-1)})}. \tag{3.A.8}$$

Since the theory says that the steady state distribution of the Markov chain defined by the Metropolis–Hastings algorithm is the posterior distribution, we can learn about the posterior distribution by this two-step algorithm:

Burn-in. Apply the Metropolis–Hastings algorithm for enough steps that we are convinced, for all practical purposes, that the steady state has been reached.

Simulate from the posterior. Simulate a number of additional iterations from the Markov chain and use these as simulations from the desired posterior. Remember, though, that these are not independent observations, so the information in, say, 10,000 iterations is less that it would be if we had 10,000 independent simulations.

Problems

3.1 Suppose that the following failure times were observed, with no censoring. Fit the exponential distribution. Find a point and interval estimate for the mean θ and estimate $S(400)$.

6   10   11   41   46   56   99 133 160 278

3.2 Suppose that the following failure times were observed, where the + indicates censoring. Fit the exponential distribution. Find a point and interval estimate for the mean θ and estimate $S(500)$.

      0.1 1 4 4 5 6 6 10 14 15 20 20+ 20+ 20+ 20+

3.3 Suppose that twenty items are placed on test until time 3000. Only 10 items failed, and the total time that all units were tested was 16,335. Assume an exponential distributions. Is this enough information to estimate the mean lifetime? If so, find point and interval estimates for θ. If not, explain what additional information is needed.

3.4 Consider the data shown in Table 3.6 on the failure times of diesel generator fans. This data set is from Nelson (1969) and the data frame is available in R's `survival` function. The censoring

Table 3.6 Data for Problem 3.4.

i	Hours	Status	i	Hours	Status	i	Hours	Status
1	4.5	1	25	32	0	49	63	0
2	4.6	0	26	34.5	1	50	64.5	0
3	11.5	1	27	37.5	0	51	64.5	0
4	11.5	1	28	37.5	0	52	67	0
5	15.6	0	29	41.5	0	53	74.5	0
6	16	1	30	41.5	0	54	78	0
7	16.6	0	31	41.5	0	55	78	0
8	18.5	0	32	41.5	0	56	81	0
9	18.5	0	33	43	0	57	81	0
10	18.5	0	34	43	0	58	82	0
11	18.5	0	35	43	0	59	85	0
12	18.5	0	36	43	0	60	85	0
13	20.3	0	37	46	1	61	85	0
14	20.3	0	38	48.5	0	62	87.5	0
15	20.3	0	39	48.5	0	63	87.5	1
16	20.7	1	40	48.5	0	64	87.5	0
17	20.7	1	41	48.5	0	65	94	0
18	20.8	1	42	50	0	66	99	0
19	22	0	43	50	0	67	101	0
20	30	0	44	50	0	68	101	0
21	30	0	45	61	0	69	101	0
22	30	0	46	61	1	70	115	0
23	30	0	47	61	0			
24	31	1	48	61	0			

variable is `Status`, which is 1 for a failure and 0 for a censored time. The values in Table 3.6 are measured in hundreds of hours; in R's `survival` function, the `hours` variables gives times measured in hours. Assume an exponential distribution for the lifetimes and find a point and interval estimate for the mean lifetime θ.

3.5 Consider the data in Table 3.7 given by Wang (2000) on the failure times of an electrical component. There was no censoring. Assume an exponential distribution. Find point and interval estimates for the mean θ. Generate a probability plot to assess the fit.

Table 3.7 Data for Problem 3.5.

i	t_i	i	t_i
1	5	10	165
2	11	11	196
3	21	12	224
4	31	13	245
5	46	14	293
6	75	15	321
7	98	16	330
8	122	17	350
9	145	18	420

3.6 Consider the data given by Aarset (1987) and shown in Table 3.8 on the failure times of an unspecified device. There was no censoring. Assume an exponential distribution. Find point and interval estimates for the mean θ. Plot a probability plot to assess the fit.

3.7 Assume a WEIB(θ, κ) distribution for the data in Problem 3.1. Find point and interval estimates for θ and κ.

3.8 Assume a WEIB(θ, κ) distribution for the data in Problem 3.2. Find point and interval estimates for θ and κ.

3.9 Assume a WEIB(θ, κ) distribution for the data in Problem 3.3. Is this enough information to estimate θ and κ? If so, find point and interval estimates for both parameters. If not, explain what additional information is needed.

3.10 Assume a WEIB(θ, κ) distribution for the data in Problem 3.4. Find point and interval estimates for θ and κ.

Table 3.8 Data for Problem 3.6.

i	t_i	i	t_i	i	t_i
1	0.1	18	18	35	72
2	0.2	19	21	36	75
3	1	20	32	37	79
4	1	21	36	38	82
5	1	22	40	39	82
6	1	23	45	40	83
7	1	24	46	41	84
8	2	25	47	42	84
9	3	26	50	43	84
10	6	27	55	44	85
11	7	28	60	45	85
12	11	29	63	46	85
13	12	30	63	47	85
14	18	31	67	48	85
15	18	32	67	49	86
16	18	33	67	50	86
17	18	34	67		

3.11 Assume a WEIB(θ, κ) distribution for the data in Problem 3.5. Find point and interval estimates for θ and κ.

3.12 Assume a WEIB(θ, κ) distribution for the data in Problem 3.6. Find point and interval estimates for θ and κ.

3.13 For the data in Problem 3.4 fit a log normal distribution and a gamma distribution. Compute the AIC for all four models: exponential, Weibull, log normal, and gamma. Which model seems to fit best?

3.14 For the data in Problem 3.5 fit a log normal distribution and a gamma distribution. Compute the AIC for all four models: exponential, Weibull, log normal, and gamma. Which model seems to fit best?

3.15 For the data in Problem 3.6 fit a log normal distribution and a gamma distribution. Compute the AIC for all four models: exponential, Weibull, log normal, and gamma. Which model seems to fit best?

3.16 Use each of the four fitted models from Problems 3.4 and 3.13 to estimate $S(10)$, $S(40)$, and $S(120)$.

3.17 Use each of the four fitted models from Problems 3.5 and 3.14 to estimate $S(5)$, $S(150)$, and $S(500)$.

3.18 Use each of the four fitted models from Problems 3.6 and 3.15 to estimate $S(1)$, $S(60)$, and $S(100)$.

3.19 Verify that the MLEs of the parameters μ and σ for the log normal with no censoring are given by Eqs. (3.30) and (3.31).

3.20 The Burr XII distribution was proposed by Wang (2000) as a distribution with a (possibly) bathtub-shaped hazard function. The hazard function is

$$h(t) = \frac{k_1 c_1 (t/s_1)^{c_1-1}}{s_1[1 + (t/s_1)^{c_1}]} + \frac{k_2 c_2 (t/s_2)^{c_2-1}}{s_2[1 + (t/s_2)^{c_2}]}, \qquad t > 0.$$

Derive the CDF and the PDF.

3.21 Plot the hazard function of the Burr XII distribution for the following sets of parameters:
(a) $c_1 = 0.75$, $k_1 = 3$, $s_1 = 200$, $c_2 = 40$, $k_2 = 4$, and $s_2 = 70$.
(b) $c_1 = 0.25$, $k_1 = 2$, $s_1 = 8,000,000$, $c_2 = 40$, $k_2 = 2$, and $s_2 = 70$.

3.22 Plot the PDF of the Burr XII distribution for the following sets of parameters:
(a) $c_1 = 0.75$, $k_1 = 3$, $s_1 = 200$, $c_2 = 40$, $k_2 = 4$, and $s_2 = 70$.
(b) $c_1 = 0.25$, $k_1 = 2$, $s_1 = 8,000,000$, $c_2 = 40$, $k_2 = 2$, and $s_2 = 70$.

3.23 Find the likelihood for an uncensored sample from the Burr XII distribution. Use it along with R's `optim` function to find the MLEs for the data set from Problem 3.5.

3.24 Consider the data from Problem 3.4 on the failure times of diesel generator fans. The censoring variable is `Status`, which is 1 for a failure and 0 for a censored time. Suppose that prior to the life test, the investigators had found that similar fans had Weibull-distributed lifetimes, but κ was near, but slightly larger than 1, and θ was approximately 400. Discuss whether the priors

$$\kappa \sim \text{GAM}(2, 0.5)$$

$$\theta \sim \text{GAM}(200, 0.5)$$

are reasonable. Run the MCMC with these priors in order to find Bayes estimates for θ and κ.

3.25 Repeat Problem 3.24, but with *more* informative priors.

3.26 Repeat Problem 3.24, but with *less* informative priors.

Part II

Design of Experiments

4

Fundamentals of Experimental Design

4.1 Introduction to Experimental Design

Investigators perform experiments in virtually all fields of inquiry, usually to discover something about a particular process or system or to confirm previous experience or theory. Each experimental run is a test. More formally, we can define an *experiment* as a series of tests or runs in which purposeful changes are made to the input variables of a process or system so that we may observe and identify the reasons for changes that may be observed in the output response. We may want to determine which input variables are responsible for the observed changes in the response, develop a model relating the response to the important input variables, and use this model for process or system improvement or other decision-making.

This book is about planning and conducting experiments and about analyzing the resulting data so that valid and objective conclusions are obtained. Our focus is on experiments that deal with the reliability of components, products, and systems. This is an important aspect of *technology commercialization* and *product realization* activities. Designed experiments have extensive applications in many other areas, such as new product design and formulation, manufacturing process development, and process improvement. There are also many applications of designed experiments in a nonmanufacturing or non-product-development setting, such as marketing, service operations, and general business operations. Designed experiments are a key technology for innovation. Both *breakthrough innovation* and *incremental innovation* activities can benefit from the effective use of designed experiments. While this book focuses on reliability applications, Montgomery (2020) provides a more general introduction with applications to many of these other areas.

As an example of an experiment, suppose that a team of engineers wants to investigate how to increase the shear strength of solder joints for components on a printed circuit board. This printed circuit board is used in a product that is used in an application where it often experiences excessive vibration and is occasionally dropped, which can result in failure of one or more of the solder joints. Failure analysis has identified broken solder joints as a major failure mode for the product. When the printed circuit boards are manufactured the components are mechanically and electrically attached through the board by passing them through a flow solder machine. To prevent oxidization of the solder and potentially weakening the solder joints, an atmosphere of 100% hydrogen is maintained in the flow solder machine.

Design of Experiments for Reliability Achievement, First Edition.
Steven E. Rigdon, Rong Pan, Douglas C. Montgomery, and Laura J. Freeman.
© 2022 John Wiley & Sons, Inc. Published 2022 by John Wiley & Sons, Inc.
Companion website: www.wiley.com/go/rigdon/designexperiments

The engineering teams suspect that a 100% nitrogen atmosphere would be a better choice. Here the objective of the experimenters (the engineers) is to determine which atmosphere produces the maximum shear strength for this particular product. The engineers decide to produce a number of printed circuit boards using each atmosphere and measure the shear strength of the solder joints. The average shear strength of the solder joints on the boards manufactured in each atmosphere will be used to determine which atmosphere is best.

This is an example of a *single-factor experiment*. In the language of experimental design it is a single-factor experiment with two levels of the factor. The *response* or *outcome variable*, is the solder joint shear strength. Many experiments involve several factors and the design of these experiments requires an understanding of experimental strategy. For example, other potential factors that could influence solder joint shear strength would include solder temperature, conveyor speed, solder wave height, and type of solder (there are different formulations of lead-free solders, which have become widely used in many applications). One approach would be to select some arbitrary combination of the levels of these factors, test them, and see what happens. If the results are satisfactory, the experimenters may decide to stop. Otherwise, another combination of factors and levels could be selected and another text performed. This approach could be continued almost indefinitely, switching the levels of one or two (or perhaps several) factors for the next test, based on the outcome of the current test. This strategy of experimentation, which we call the *best-guess approach*, is frequently used in practice by engineers and scientists. It often works reasonably well, too, because the experimenters often have a great deal of technical or theoretical knowledge of the system they are studying, as well as considerable practical experience. The best-guess approach has at least two disadvantages. First, suppose the initial best-guess does not produce the desired results. Now the experimenter has to take another guess at the correct combination of factor levels. This could continue for a long time, without any guarantee of success. Second, suppose the initial best-guess produces an acceptable result (as we mentioned earlier). Now the experimenter is tempted to stop testing, although there is no guarantee that the best solution has been found.

Another strategy of experimentation that is used extensively in practice is the *one factor-at-a-time* (OFAT) approach. The OFAT method consists of selecting a starting point, or baseline set of levels, for each factor, and then successively varying each factor over its range with the other factors held constant at the baseline level. After all tests are performed, a series of graphs are usually constructed showing how the response variable is affected by varying each factor with all other factors held constant. Using these OFAT graphs, we would select the optimal combination of factor that should be used or the factors that are most influential on the outcome of the experiment.

The major disadvantage of the OFAT strategy is that it fails to consider any possible *interaction* between the factors. An interaction is the failure of one factor to produce the same effect on the response at different levels of another factor. Interactions between factors are very common, and if they occur, the OFAT strategy will usually produce poor results. Many people do not recognize this, and, consequently, OFAT experiments are run frequently in practice. (Some individuals actually think that this strategy is related to the scientific method or that it is a "sound" engineering principle.) OFAT experiments are always less efficient than other methods based on a statistical approach to design.

The correct approach to dealing with several factors is to conduct a *factorial* experiment. This is an experimental strategy in which factors are varied together, instead of one at a time. For example, suppose that the experimenters are interesting in investigating two factors, atmosphere and solder temperature.

There are two levels of the atmosphere and suppose that they are interested in two levels of the solder temperature. A factorial experiment would test all four combinations of the two different levels of these two factors. The factorial experimental design concept is extremely important, and examples of its use and a number of useful variations and special cases will be presented.

4.2 A Brief History of Experimental Design

There have been four eras in the modern development of statistical experimental design. The agricultural era was led by the pioneering work of Sir Ronald A. Fisher in the 1920s and early 1930s. During that time, Fisher was responsible for statistics and data analysis at the Rothamsted Agricultural Experimental Station near London, England. Fisher recognized that flaws in the way the experiment that generated the data had been performed often hampered the analysis of data from systems (in this case, agricultural systems). By interacting with scientists and researchers in many fields, he developed the insights that led to the three basic principles of experimental design: *randomization*, *replication*, and *blocking*. Randomization is the act of running the individual trials in an experiment in random order to eliminate the effect of unknown factors that may vary over the course of the experiment and invalidate the conclusions. Replication involves running some trials more than one so that an independent estimate of experimental error can be obtained, and blocking is a technique for minimizing the effect of known nuisance variables on the outcome of the experiment.

Fisher systematically introduced statistical thinking and principles into designing experimental investigations, including the factorial design concept and the analysis of variance (ANOVA), which remains today the primary method for analyzing data from designed experiments. His two books (the most recent editions are Fisher (1958, 1966)) had profound influence on the use of statistics, particularly in agricultural and related life sciences. For an excellent biography of Fisher, see Box (1978).

Although applications of statistical design in industrial settings certainly began in the 1930s, the second, or industrial, era was catalyzed by the development of *response surface methodology* (RSM) by Box and Wilson (1951). They recognized and exploited the fact that many industrial experiments are fundamentally different from their agricultural counterparts in two ways: (i) the response variable can usually be observed (nearly) immediately and (ii) the experimenter can quickly learn crucial information from a small group of runs that can be used to plan the next experiment. Box (1999) calls these two features of industrial experiments *immediacy* and *sequentiality*. Over the next 30 years, RSM and other design techniques spread throughout the chemical and the process industries, mostly in research and development work. George Box was the intellectual leader of this movement. However, the application of statistical design at the plant or manufacturing process level was still not extremely widespread. Some of the reasons for this include an inadequate training in basic statistical concepts and methods for engineers and other process specialists and the lack of computing resources and user-friendly statistical software to support the application of statistically designed experiments.

It was during this second or industrial era that work on *optimal design* of experiments began. Kiefer (1959, 1961) and Kiefer and Wolfowitz (1959) proposed a formal approach to selecting a design based on specific objective optimality criteria. Their initial approach was to select a design that would result in the model parameters being estimated with the best possible precision. This approach did not find much application because of the lack of computer tools for its implementation. However, there have been great advances in both algorithms for generating optimal designs and computing capability over the last 25 years. Optimal designs have great application to reliability experiments and are discussed at several places in the book.

The increasing interest of Western industry in quality improvement that began in the late 1970s ushered in the third era of statistical design. The work of Genichi Taguchi (Taguchi and Wu 1980; Kackar 1985; Taguchi 1987, 1991) had a significant impact on expanding the interest in and use of designed experiments. Taguchi advocated using designed experiments for what he termed *robust parameter design*, or

- making processes insensitive to environmental factors or other factors that are difficult to control,
- making products insensitive to variation transmitted from components, and
- finding levels of the process variables that force the mean to a desired value while simultaneously reducing variability around this value.

Taguchi suggested highly fractionated factorial designs and other orthogonal arrays along with some novel statistical methods to solve these problems. The resulting methodology generated much discussion and controversy. Part of the controversy arose because Taguchi's methodology was advocated in the West initially (and primarily) by entrepreneurs, and the underlying statistical science had not been adequately peer reviewed. By the late 1980s, the results of peer review indicated that although Taguchi's engineering concepts and objectives were well founded, there were substantial problems with his experimental strategy and methods of data analysis. For specific details of these issues, see Box (1988), Box et al. (1988), Hunter (1985, 1989), Myers et al. (2016), and Pignatiello and Ramberg (1991). Many of these concerns are also summarized in the extensive panel discussion in the May 1992 issue of *Technometrics* (see (Nair 1992)).

There were several positive outcomes of the Taguchi controversy. First, designed experiments became more widely used in the discrete parts industries, including automotive and aerospace manufacturing, electronics and semiconductors, and many other industries that had previously made little use of the technique. Second, the fourth or modern era of statistical design began. This era has included a renewed general interest in statistical design by both researchers and practitioners and the development of many new and useful approaches to experimental problems in the industrial world, including alternatives to Taguchi's technical methods that allow his engineering concepts to be carried into practice efficiently and effectively. There has been extensive growth in the use of designed experiments to evaluate and improve the reliability of components and products. Third, computer software for construction and evaluation of designs has improved greatly with many new features and capability. Fourth, formal education in statistical experimental design is becoming part of many engineering programs in universities, at both undergraduate and graduate levels. The successful integration of good experimental design practice into engineering and science is a key factor in future industrial competitiveness.

Applications of designed experiments have grown far beyond their agricultural origins. There is not a single area of science and engineering that has not successfully employed statistically designed experiments. In recent years, there has been a considerable utilization of designed experiments in many other areas, including the service sector of business, financial services, government operations, and many nonprofit business sectors. An article appeared in *Forbes* magazine on 11 March 1996, entitled "The New Mantra: MVT," where MVT stands for "multivariable testing," a term authors use to describe factorial designs. The article notes the many successes that a diverse group of companies have had through their use of statistically designed experiments.

4.3   Guidelines for Designing Experiments

Even experimental problems that sound relatively straightforward usually require that everyone involved have a clear idea about what the questions are that are going to be addressed in the experiment, exactly how the data will be collected, and at least a general idea of the methods that will be used to analyze the data. For example, in the simple problem that involved the effect of the two soldering atmospheres on the strength of solder joints, some possible items that need to be addressed include:

1. Are these two atmospheres the only ones of potential interest?
2. Are there any other factors that might affect strength that should be investigated or controlled in this experiment? (We suggested some of these previously, such as solder temperature, solder wave height, and conveyor speed. There may be others.)
3. How many printed circuit boards (solder joints) should be tested in each atmosphere?
4. How should the printed circuit boards be assigned to the two different atmospheres, and in what order should the data be collected?
5. What method of data analysis should be used?
6. What difference in average observed shear strength between the two atmospheres will be considered important?

For the experiment to be successful, questions like these must be considered and addressed carefully. We think of designing an experiment as a process, and all processes should have a rigorously defined set of tasks to be accomplished. Table 4.1 gives an outline of the seven-step process that we recommend.

Steps 2 and 3 are often done simultaneously, or in reverse order. We refer to steps 1 through 3 as *pre-experimental planning*, and in our experience this is a critical part of the process. If this is poorly done, the results of the experiment may be disappointing. We now give a brief discussion of the seven steps. For more details, see Coleman and Montgomery (1993), and the references therein.

1. Recognition of and statement of the problem. This may seem to be a rather obvious point, but in practice often it is not simple to realize that a problem requiring experimentation exists, nor is it simple to develop a clear and generally accepted statement of this problem. It is necessary to develop all ideas about the objectives of the experiment. Usually, it is important to solicit input from all concerned parties: engineering, quality assurance, manufacturing, marketing, management, customer, and operating personnel (who usually have much insight and who are too often ignored). For this reason, a *team approach*

Table 4.1   Guidelines for designing and experiment.

1.   Recognition of and statement of the problem
2.   Selection of the response variable(s)
3.   Choice of factors, levels, and ranges
4.   Choice of experimental design
5.   Performing the experiment
6.   Statistical analysis of the data
7.   Conclusions and recommendations

to designing experiments is recommended. A really bad approach that often leads to failure is to let the one person who is an "expert" in the area design the experiment alone. Often this person will design an experiment that is too narrow and intended to reinforce his or her preconceived ideas about the answer.

It is usually helpful to prepare a list of specific problems or questions that are to be addressed by the experiment. A clear statement of the problem often contributes substantially to better understanding of the phenomenon being studied and the final solution of the problem.

It is also important to keep the overall objectives of the experiment in mind. There are several broad reasons for running experiments and each type of experiment will generate its own list of specific questions that need to be addressed. Some (but by no means all) of the reasons for running experiments include:

(a) **Factor screening or characterization.** When a system or process is new, it is usually important to learn which factors have the most influence on the response(s) of interest. Often there are a lot of factors. This usually indicates that the experimenters do not know much about the system so screening is essential if we are to efficiently get the desired performance from the system. Screening experiments are extremely important when working with new systems or technologies so that valuable resources will not be wasted using best guess and OFAT approaches.

(b) **Optimization.** After the system has been characterized and we are reasonably certain that the important factors have been identified, the next objective is usually optimization, that is, find the settings or levels of the important factors that result in desirable values of the response. For example, if a screening experiment on a chemical process results in the identification of time and temperature as the two most important factors, the optimization experiment may have as its objective finding the levels of time and temperature that maximize yield, or perhaps maximize yield while keeping some product property that is critical to the customer within specifications. An optimization experiment is often a follow-up to a screening experiment, although there has been successful work in recent years to develop experimental designs that can combine screening and optimization in a single step. See Montgomery (2020) and Myers et al. (2016) for details.

(c) **Confirmation.** In a confirmation experiment, the experimenter is usually trying to verify that the system operates or behaves in a manner that is consistent with some theory or past experience. For example, if theory or experience indicates that a particular new material is equivalent to the one currently in use and the new material is desirable (perhaps less expensive, or easier to work with in some way), then a confirmation experiment would be conducted to verify that substituting the new material results in no change in product characteristics that impact its use.

Moving a new manufacturing process to full-scale production based on results found during experimentation at a pilot plant or development site is another situation that often results in confirmation experiments – that is, are the same factors and settings that were determined during development work appropriate for the full-scale process?

(d) **Discovery.** In discovery experiments, the experimenters are usually trying to determine what happens when we explore new materials, or new factors, or new ranges for factors. Discovery experiments often involve screening of several (perhaps many) factors. In the pharmaceutical industry, scientists are constantly conducting discovery experiments to find new materials or combinations of materials that will be effective in treating disease.

(e) **Robustness.** These experiments often address questions such as under what conditions do the response variables of interest seriously degrade? Degradation studies are a special type of reliability experiment. Or what conditions would lead to unacceptable variability in the response variables, or reliability of the product? A variation of this is determining how we can set the factors in the system that we can control to minimize the variability transmitted into the response from factors that we cannot control very well.

Obviously, the specific questions to be addressed in the experiment relate directly to the overall objectives. An important aspect of problem formulation is the recognition that one large comprehensive experiment is unlikely to answer the key questions satisfactorily. A single comprehensive experiment requires the experimenters to know the answers to a lot of questions, and if they are wrong, the results will be disappointing. This leads to wasting time, materials, and other resources and may result in never answering the original research questions satisfactorily.

A *sequential* approach employing a series of smaller experiments, each with a specific objective, such as factor screening, is a widely recommended strategy.

2. Selection of the response variable. In selecting the response variable, the experimenter should be certain that this variable really provides useful information about the process under study. Most often, the average or standard deviation (or both) of the measured characteristic will be the response variable. Multiple responses are not unusual. In many reliability experiments the response can be censored. That is, the experiment is terminated at some point before all runs have gone to completion. There are several types of censoring and analyzing censored data is often a fairly complicated problem. The experimenters must decide how each response will be measured, and address issues such as how will any measurement system be calibrated and how this calibration will be maintained during the experiment. The gauge or measurement system capability (or measurement error) is also an important factor. If gauge capability is inadequate, only relatively large factor effects will be detected by the experiment or perhaps additional replication will be required. In some situations where gauge capability is poor, the experimenter may decide to measure each experimental unit several times and use the average of the repeated measurements as the observed response. It is often critically important to identify issues related to defining the responses of interest and how they are to be measured *before* conducting the experiment. Sometimes designed experiments are employed to study and improve the performance of measurement systems.

3. Choice of factors, levels, and ranges. (As noted previously, steps 2 and 3 in Table 4.1 are often done simultaneously or in the reverse order.) When considering the factors that may influence the performance of a process or system, the experimenter usually discovers that these factors can be classified as either *potential design factors* or *nuisance factors*. The potential design factors are those factors that the experimenter may wish to vary in the experiment. Often we find that there are a lot of potential design factors, and some further classification of them is helpful. Some useful classifications are: *design factors*, *held-constant factors*, and *allowed-to-vary factors*. The design factors are the factors actually selected for study in the experiment. Held-constant factors are variables that may exert some effect on the response, but for purposes of the present experiment these factors are not of interest, so they will be held at a specific level. For example, in an etching experiment in the semiconductor industry, there may be an effect that is unique to the specific plasma etch tool used in the experiment. However, this factor would be very difficult to vary in an experiment, so the experimenter may decide to perform all experimental runs on one particular (ideally "typical") etcher. Thus, this factor has been held constant. As an example of allowed-to-vary factors, the experimental units or the "materials" to which the design factors are applied are usually non-homogeneous, yet we often ignore this unit-to-unit variability and rely on randomization to balance out any material or experimental unit effect. We often assume that the effects of held-constant factors and allowed-to-vary factors are relatively small.

Nuisance factors, on the other hand, may have large effects that must be accounted for, yet we may not be interested in them in the context of the present experiment. Nuisance factors are often classified as *controllable*, *uncontrollable*, or *noise factors*. A controllable nuisance factor is one whose levels may be set by the experimenter. For example, the experimenter can select different samples from different batches of components for a life test. The blocking principle, mentioned previously, is often useful in dealing with controllable nuisance factors. If a nuisance factor is uncontrollable in the experiment, but it can be measured, an analysis procedure called the *analysis of covariance* can often be used to compensate for its effect. For example, the relative humidity in the process environment may affect process performance, and if the humidity cannot be controlled, it probably can be measured and treated as a covariate. When a factor that varies naturally and uncontrollably in the process can be controlled for purposes of an experiment, we often call it a noise factor. In such situations, our objective is usually to find the settings of the controllable design factors that minimize the variability transmitted from the noise factors. This is sometimes called a process robustness study or a robust design problem.

Once the experimenter has selected the design factors, he or she must choose the ranges over which these factors will be varied and the specific levels at which runs will be made. Thought must also be given to how these factors are to be controlled at the desired values and how they are to be measured. In many reliability experiments the test conditions are accelerated; that is, they are more severe than the conditions under which the component or product will actually be used. For example, components may be tested at higher levels of temperature than the component will actually be exposed to. This is done to accelerate the failure mechanisms of the component. Process or system knowledge is required to do this. This process knowledge is usually a combination of practical experience and theoretical understanding. It is important to investigate all factors that may be of importance and to be not overly influenced by past experience, particularly when we are in the early stages of experimentation or when the process is not very mature.

When the objective of the experiment is factor screening or process characterization, it is usually best to keep the number of factor levels low. Generally, two levels work very well in factor screening studies. Choosing the region of interest is also important. In factor screening, the region of interest should be relatively large – that is, the range over which the factors are varied should be broad. As we learn more about which variables are important and which levels produce the best results, the region of interest in subsequent experiments will usually become narrower.

3. Choice of experimental design. If the aforementioned pre-experimental planning activities are done correctly, this step is usually relatively easy. Choice of design involves consideration of sample size (number of replicates), selection of a suitable run order for the experimental trials, and determination of whether blocking or other randomization restrictions are involved. This book discusses some of the more important types of experimental designs, and it can ultimately be used as a guide for selecting an appropriate experimental design for a wide variety of problems. There are also several interactive statistical software packages that support this phase of experimental design. The experimenter can enter information about the number of factors, levels, and ranges, and these programs will either present a selection of designs for consideration or recommend a particular design. (We usually prefer to see several alternatives instead of relying entirely on a computer recommendation in most cases.) Most software packages also provide some diagnostic information about how each design will perform. This is useful in evaluation of different design alternatives for the experiment. These programs will usually also provide a worksheet (with the order of the runs randomized) for use in conducting the experiment. Design selection also involves thinking about and selecting a tentative empirical model to describe the results. The model is just a quantitative relationship (equation) between the response and the important design factors. In many cases, a low-order polynomial model will be appropriate. A *first-order* model in two variables is

$$y = \beta_0 + \beta_1 x_1 + \beta_2 x_2 + \epsilon,$$

where y is the response, the xs are the design factors, the βs are unknown parameters that will be estimated from the data in the experiment, and ϵ is a random error term that accounts for the experimental error and noise in the system that is being studied. The first-order model is also sometimes called a main effects model. First-order models are used extensively in screening or characterization experiments. A common extension of the first-order model is to add an *interaction* term, say,

$$y = \beta_0 + \beta_1 x_1 + \beta_2 x_2 + \beta_{12} x_1 x_2 + \epsilon,$$

where the cross-product term $x_1 x_2$ is just the product of the two terms and represents the two-factor interaction between the design factors. An interaction allows the effect of one variable to be different for different values of the other variable. For example, in a drug interaction, drug A may be beneficial if taken alone, and similarly for drug B, but if both are taken together, the result is harmful. In this case drug B's effect is positive if drug A is not being taken, and it is negative if drug A is being taken. Because interactions between factors is relatively common, the first-order model with interaction is widely used. Higher-order interactions can also be included in experiments with more than two factors if necessary. Another widely used model is the *second-order* model

$$y = \beta_0 + \beta_1 x_1 + \beta_2 x_2 + \beta_{12} x_1 x_2 + \beta_{11} x_1^2 + \beta_{22} x_2^2 + \epsilon.$$

Second-order models are often used in optimization experiments.

In selecting the design, it is important to keep the experimental objectives in mind. In many engineering experiments, we already know at the outset that some of the factor levels will result in different values for the response. Consequently, we are interested in identifying which factors cause this difference and in estimating the magnitude of the response change. In other situations, we may be more interested in verifying uniformity. For example, two production conditions A and B may be compared, A being the standard and B being a more cost-effective alternative. The experimenter will then be interested in demonstrating that, say, there is no difference in yield between the two conditions.

5. Performing the experiment. When running the experiment, it is vital to monitor the process carefully to ensure that everything is being done according to plan. Errors in experimental procedure at this stage will usually destroy experimental validity. One of the most common mistakes is that the people conducting the experiment failed to set the variables to the proper levels on some runs. Someone should be assigned to check factor settings before each run. Up-front planning to prevent mistakes like this is crucial to success. It is easy to underestimate the logistical and planning aspects of running a designed experiment in a complex manufacturing or research and development environment.

Coleman and Montgomery (1993) suggest that prior to conducting the experiment a few trial runs or pilot runs are often helpful. These runs provide information about consistency of experimental material, a check on the measurement system, a rough idea of experimental error, and a chance to practice the overall experimental technique. This also provides an opportunity to revisit the decisions made in steps 1–4, if necessary.

6. Statistical analysis of the data. Statistical methods should be used to analyze the data so that results and conclusions are *objective* rather than judgmental in nature. If the experiment was designed correctly and performed according to the design, the statistical methods required are not elaborate. There are many excellent software packages that can assist in data analysis, and many of the programs used in step 4 to select the design provide a seamless, direct interface to the statistical analysis. Often we find that simple graphical methods play an important role in data analysis and interpretation. Because many of the questions that the experimenter wants to answer can be cast into an hypothesis-testing framework, hypothesis testing and confidence interval estimation procedures are very useful in analyzing data from a designed experiment.

It is also usually very helpful to present the results of many experiments in terms of an *empirical model*, that is, an equation derived from the data that express the relationship between the response and the important design factors. Residual analysis and model adequacy checking are also important analysis techniques. We will discuss these issues in detail later. Remember that statistical methods cannot prove that a factor (or factors) has a particular effect. They only provide guidelines as to the reliability and validity of results. When properly applied, statistical methods do not allow anything to be proved experimentally, but they do allow us to measure the likely error in a conclusion or to attach a level of confidence to a statement. The primary advantage of statistical methods is that they add objectivity to the decision-making process. Statistical techniques coupled with good engineering or process knowledge and common sense will usually lead to sound conclusions.

7. Conclusions and recommendations. Once the data are analyzed, the experimenter must draw *practical* conclusions about the results and recommend a course of action. Graphical methods are often useful in this stage, particularly in presenting the results to others. *Follow-up runs* and *confirmation testing* should also be performed to validate the conclusions from the experiment. Throughout this entire process, it is important to keep in mind that experimentation is an important part of the learning process, where we tentatively formulate hypotheses about a system, perform experiments to investigate these hypotheses, and based on these results formulate new hypotheses, and so on. This suggests that experimentation is iterative. It is usually a major mistake to design a single, large, comprehensive experiment at the start of a study. A successful experiment requires knowledge of the important factors, the ranges over which these factors should be varied, the appropriate number of levels to use, and the proper units of measurement for these variables. Generally, we do not perfectly know the answers to these questions, but we learn about them as we go along. As an experimental program progresses, we often drop some input variables, add others, change the region of exploration for some factors, or add new response variables. Consequently, we usually experiment *sequentially*, and not too many of the available resources should be invested in the first experiment. Sufficient resources must available to perform follow-up experiments and confirmation runs to ultimate accomplish the final objective of the experiment.

Finally, it is important to recognize that *all* experiments are designed experiments. The important issue is whether they are well-designed or not. Good pre-experimental planning will usually lead to a good, successful experiment. Failure to do such planning usually leads to wasted time, money, and other resources and often poor or disappointing results.

4.4 Introduction to Factorial Experiments

Many experiments involve the study of the effects of two or more factors. In general, *factorial designs* are most efficient for this type of experiment. By a factorial experimental design, we mean that in each complete trial or replicate of the experiment all possible combinations of the levels of the factors are investigated. For example, if there are a levels of factor A and b levels of factor B, each replicate contains all ab treatment combinations. When factors are arranged in a factorial design, they are often said to be *crossed*.

The effect of a factor is defined to be the change in response produced by a change in the level of the factor. This is frequently called a *main effect* because it refers to the primary factors of interest in the experiment. For example, consider the simple experiment in Figure 4.1. This is a two-factor factorial experiment with both design factors at two levels. We have called these levels "low" and "high" and denoted them "−" and "+," respectively. The main effect of factor A in this two-level design can be thought of as the difference between the average response at the low level of A and the average response at the high level of A. Numerically, this is

$$A = \frac{40 + 52}{2} - \frac{20 + 30}{2} = 21.$$

That is, increasing factor A from the low level to the high level causes an average response increase of 21 units. Similarly, the main effect of B is

$$B = \frac{30 + 52}{2} - \frac{20 + 40}{2} = 11.$$

If the factors appear at more than two levels, the aforementioned procedure must be modified because there are other ways to define the effect of a factor. The definition of effects in a factorial experiment when the factors have more than two levels is discussed more completely later.

Figure 4.1 A two-factor factorial experiment without interaction.

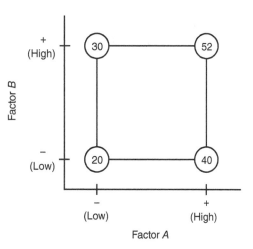

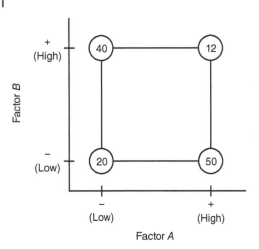

Figure 4.2 A two-factor factorial experiment with interaction.

In some experiments, the difference in response between the levels of one factor is not the same at all levels of the other factors. When this occurs, we say that the factors interact in their effect on the response. We call this an *interaction effect*. For example, consider the two-factor factorial experiment shown in Figure 4.2. At the low level of factor B (or $B-$), the A effect is

$$A = 50 - 20 = 30$$

and at the high level of factor B (or $B+$), the A effect is

$$A = 12 - 40 = -28.$$

Because the effect of A depends on the level chosen for factor B, we see that there is interaction between A and B. The magnitude of the interaction effect is the average difference in these two effects, or

$$AB = \frac{-28 - 30}{2} = -29.$$

Clearly, the interaction is large in this experiment.

These ideas may be illustrated graphically. Figure 4.3 plots the response data in Figure 4.3 against factor A for both levels of factor B. Note that the $B-$ and $B+$ lines are approximately parallel, indicating

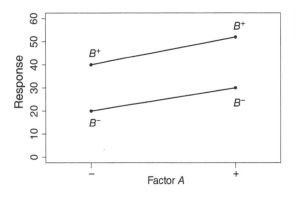

Figure 4.3 A factorial experiment with no significant interaction.

Figure 4.4 A factorial experiment with an interaction.

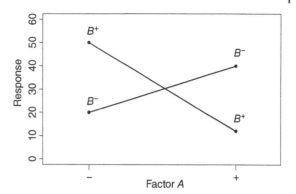

that the effect of A does not depend on the level of factor B, or a lack of interaction between factors A and B. Similarly, Figure 4.4 plots the response data in Figure 4.2. Here we see that the $B-$ and $B+$ lines are not parallel. This indicates an interaction between factors A and B. Two-factor interaction graphs such as these are frequently very useful in interpreting significant interactions and in reporting results to non-statistically trained personnel. However, they should not be utilized as the primary technique of data analysis because their interpretation is subjective and their appearance can be misleading.

This chapter focuses on the design and analysis of experiments with two factors. We consider situations where the factors can be either categorical or continuous. Both situations arise often in reliability experiments.

4.4.1 An Example

As an example of an experiment involving two factors, consider a situation where an engineering team is designing a battery for use in a device that will be subjected to some extreme variations in temperature. The only design parameter that can be selected at this point in the study is the plate material for the battery, and there are three possible choices. When the device is manufactured and is shipped to the field, the designers have no control over the temperature extremes that the device will encounter, and from experience it is known that temperature will probably affect the effective battery life. However, temperature can be controlled in the product development laboratory for the purposes of a test. The response variable is the battery life time measured in hours. In this example there was no censoring since all batteries were tested until failure.

The engineering team suspects that there may be an interaction between material type and temperature; that is, some materials may react differently to temperature and exhibit different lifetimes over the range of temperatures likely to be encountered in the field. They are also reluctant to assume that the effect of temperature is linear over its range, and that a second-order (quadratic) effect is more likely. This means that temperature should have at least three levels. Based on past experience, they select the temperature levels of 15, 70, and 125 °F and decide to replicate the design four times. The 36 runs are made in random order, producing the completely randomized design shown in Table 4.2.

The battery life experiment is a specific example of the general case of a two-factor factorial design. In general, let y_{ijk} denote the kth response when factor A is at level i ($i = 1, \ldots, a$), and factor B is at level j ($j = 1, \ldots, b$). In general, if there are n replicates at each treatment (i.e. each combination of factor levels),

Table 4.2 Data for the battery life experiment.

Material Type	Temperature (°F)					
	15		70		125	
1	130	155	34	40	20	70
	74	180	80	75	82	58
2	150	188	136	122	25	70
	159	126	106	115	58	45
3	138	110	174	120	96	104
	138	160	150	139	82	60

There were three material types, coded as 1, 2, and 3, which were run under three temperature conditions (15, 70, and 125 °F).

Table 4.3 General arrangement of responses for a two-factor factorial experiment.

		Factor B			
		1	2	$\cdots$	b
Factor A	1	$y_{111}, y_{112}, \ldots, y_{11n}$	$y_{121}, y_{122}, \ldots, y_{12n}$		$y_{1b1}, y_{1b2}, \ldots, y_{1bn}$
	2	$y_{211}, y_{212}, \ldots, y_{21n}$	$y_{221}, y_{222}, \ldots, y_{22n}$		$y_{2b1}, y_{2b2}, \ldots, y_{2bn}$
	$\vdots$				
	a	$y_{a11}, y_{a12}, \ldots, y_{a1n}$	$y_{a21}, y_{a22}, \ldots, y_{a2n}$		$y_{ab1}, y_{ab2}, \ldots, y_{abn}$

then the experiment will have *abn* runs, which are assumed to have been performed in random order. The notation for this *completely randomized design* is shown in Table 4.3.

The observations in a factorial experiment can be described in the following model:

$$y_{ijk} = \mu + \tau_i + \beta_j + (\tau\beta)_{ij} + \epsilon_{ijk} \qquad \begin{cases} i = 1, 2, \ldots, a, \\ j = 1, 2, \ldots, b, \\ k = 1, 2, \ldots, n. \end{cases} \qquad (4.1)$$

Here μ is the overall mean effect, τ_i is the effect of the *i*th level of factor A, β_j is the effect of the *j*th level of B, and $(\tau)_{ij}$ is the interaction effect between factors A and B. (Think of $(\tau\beta)_{ij}$ as a single entity, not a product.) The random error component ϵ_{ijk} is assumed to have a mean of zero and a variance of σ^2. Here, we assume that both factors are *fixed* and are defined as *deviations* from the overall mean. Because of this, we assume that

$$\sum_{i=1}^{a} \tau_i = 0 \quad \text{and} \quad \sum_{j=1}^{b} \beta_j = 0.$$

We also require that sums of the interaction term $(\tau\beta_{ij})$ are zero for both indices; that is,

$$\sum_{i=1}^{a}(\tau\beta)_{ij} = \sum_{j=1}^{b}(\tau\beta)_{ij} = 0.$$

Because there are exactly n replicates for every treatment, there are abn total observations.

By a *fixed factor* we mean that the levels of the factor were chosen by the experimenter because they are of specific interest, and that conclusions will be drawn about those specific levels. There are also experiments where the factor levels are chosen randomly from a large population of levels and the conclusions extend to the population. Effects due to these factors are called *random effects*.

For the two-factor factorial design described in Table 4.3, both row and column factors (factors A and B) are of equal interest. We are often interested in testing whether factor A has *any* effect at all on the response. This is done by testing the null hypothesis that all of the factor A effects are equal to zero. The alternative hypothesis is that at least one of the effects is nonzero. This leads to the hypothesis test

$$H_0 : \tau_1 = \tau_2 = \cdots = \tau_a = 0 \text{ versus } H_1 : \text{ at least one } \tau_i \neq 0.$$

Similarly, we can test whether all of the factor B effects are zero by testing

$$H_0 : \beta_1 = \beta_2 = \cdots = \beta_a = 0 \text{ versus } H_1 : \text{ at least one } \beta_i \neq 0.$$

Finally, we can test whether there is an AB interaction by testing

$$H_0 : (\tau\beta)_{ij} = 0 \text{ for all } i, j \text{ versus } H_1 : \text{ at least one } (\tau\beta)_{ij} \neq 0.$$

4.4.2   The Analysis of Variance for a Two-Factor Factorial

Let $y_{i..}$ denote the total of all observations under the ith level of factor A; also let $\bar{y}_{i..}$ denote the average of all observations under the ith level of factor A. Expressed mathematically,

$$y_{i..} = \sum_{j=1}^{b}\sum_{k=1}^{n} y_{ijk}$$

and

$$\bar{y}_{i..} = \frac{y_{i..}}{bn}.$$

In other words, the dot indicates summing across all indices indicated by the dots, and dots used with $\bar{y}$ indicate averaging across the indices indicated by the dots. The set of sums and means that will be of interest are shown in Table 4.4.

In creating the ANOVA table, the sums of squares about some mean are important quantities. The *corrected total sum of squares* is the sum of squared deviations between each observation and the grand mean $\bar{y}_{...}$; that is,

$$SST = \sum_{i=1}^{a}\sum_{j=1}^{b}\sum_{k=1}^{n} (y_{ijk} - \bar{y}_{...})^2.$$

The corrected total sum of squares can be decomposed into other sums of squares, each representing the variability of the observations around some mean. By cleverly adding and subtracting certain means

Table 4.4 Summary of dot notation.

$$y_{i..} = \sum_{j=1}^{b}\sum_{k=1}^{n} y_{ijk} \qquad \bar{y}_{i..} = \frac{y_{i..}}{bn} \qquad i = 1, 2, \ldots, a$$

$$y_{.j.} = \sum_{i=1}^{a}\sum_{k=1}^{n} y_{ijk} \qquad \bar{y}_{.j.} = \frac{y_{.j.}}{an} \qquad i = 1, 2, \ldots, b$$

$$y_{ij.} = \sum_{k=1}^{n} y_{ijk} \qquad \bar{y}_{ij.} = \frac{y_{ij.}}{bn} \qquad i = 1, 2, \ldots, a; \ j = 1, 2, \ldots, b$$

$$y_{...} = \sum_{i=1}^{a}\sum_{j=1}^{b}\sum_{k=1}^{n} y_{ijk} \qquad \bar{y}_{...} = \frac{y_{...}}{abn}$$

(so that the net effect is adding zero), we can obtain

$$SS_T = \sum_{i=1}^{a}\sum_{j=1}^{b}\sum_{k=1}^{n}[(\bar{y}_{i..} - \bar{y}_{...}) + (\bar{y}_{.j.} - \bar{y}_{...}) + (\bar{y}_{ij.} - \bar{y}_{i..} - \bar{y}_{.j.} + \bar{y}_{...}) + (y_{ijk} - \bar{y}_{ij.})]^2$$

$$= bn\sum_{i=1}^{a}(\bar{y}_{i..} - \bar{y}_{...})^2 + an\sum_{j=1}^{b}(\bar{y}_{.j.} - \bar{y}_{...})^2$$

$$+ n\sum_{i=1}^{a}\sum_{j=1}^{b}(\bar{y}_{ij.} - \bar{y}_{i..} - \bar{y}_{.j.} + \bar{y}_{...})^2 + \sum_{i=1}^{a}\sum_{j=1}^{b}\sum_{k=1}^{n}(y_{ijk} - \bar{y}_{ij.})^2 \qquad (4.2)$$

because when all this is multiplied out the six cross products on the right side of the first line are zero. Notice that the corrected total sum of squares had been partitioned into a sum of squares due to factor A, a sum of squares due to factor B, a sum of squares due the interaction between A and B, and a sum of squares due to error; we denote these respectively as SS_A, SS_B, SS_{AB}, and SS_E. This partitioning is the fundamental ANOVA equation for the two-factor factorial experimental design. From the last component on the right side of Eq. (4.2) we see that there must be at least two replicates ($n \geq 2$) to obtain an error sum of squares.

We can write Eq. (4.2) symbolically as

$$SS_T = SS_A + SS_B + SS_{AB} + SS_E.$$

The degrees of freedom associated with each sum of squares is shown below and in Table 4.5.

Effect	Degrees of freedom
A	$a - 1$
B	$b - 1$
AB	$(a - 1)(b - 1)$
Error	$ab(n - 1)$
Total	$abn - 1$

Table 4.5 ANOVA table for two-factor experiment with interaction.

Source of variation	Sum of squares	Degrees of freedom	Mean square	F statistic
Factor A	SS_A	$a - 1$	$MS_A = \dfrac{SS_A}{a - 1}$	$F = \dfrac{MS_A}{MS_E}$
Factor B	SS_B	$b - 1$	$MS_B = \dfrac{SS_B}{b - 1}$	$F = \dfrac{MS_B}{MS_E}$
Interaction	SS_{AB}	$(a - 1)(b - 1)$	$MS_{AB} = \dfrac{SS_{AB}}{(a - 1)(b - 1)}$	$F = \dfrac{MS_{AB}}{MS_E}$
Error	SS_E	$ab(n - 1)$	$MS_E = \dfrac{SS_E}{ab(n - 1)}$	
Total	SS_T	$abn - 1$		

We may justify this allocation of the $abn - 1$ total degrees of freedom to the sums of squares as follows. The main effects A and B have a and b levels, respectively, so they have $a - 1$ and $b - 1$ degrees of freedom as shown. The interaction degrees of freedom are simply the number of degrees of freedom for cells (which is $ab - 1$) minus the number of degrees of freedom for the two main effects A and B; that is

$$ab - 1 - (a - 1) - (b - 1) = (a - 1)(b - 1).$$

Within each of the ab cells, there are $n - 1$ degrees of freedom among the n replicates; thus there are $ab(n - 1)$ degrees of freedom for error. Note that the number of freedom on the right side of (4.2) adds to the total number of degrees of freedom (i.e. the degrees of freedom for the corrected total sum of squares, which is $abn - 1$).

Each sum of squares divided by its degrees of freedom is called the *mean square*. The expected values of the mean squares are

$$E(MS_A) = E\left(\frac{SS_A}{a - 1}\right) = \sigma^2 + \frac{bn\sum\limits_{i=1}^{a}\tau_i^2}{a - 1},$$

$$E(MS_B) = E\left(\frac{SS_B}{b - 1}\right) = \sigma^2 + \frac{an\sum\limits_{j=1}^{b}\beta_i^2}{a - 1},$$

$$E(MS_{AB}) = E\left(\frac{SS_{AB}}{(a - 1)(b - 1)}\right) = \sigma^2 + \frac{b\sum\limits_{i=1}^{a}\sum\limits_{j=1}^{b}(\tau\beta)_{ij}^2}{(a - 1)(b - 1)},$$

and

$$E(MS_E) = E\left(\frac{SS_E}{(a - 1)(b - 1)}\right) = \sigma^2.$$

Note that if the null hypotheses of no factor A or factor B effects and no interaction effects are all true, then MS_A, MS_B, MS_{AB}, and MS_E are all unbiased estimators of σ^2. However, if there are nonzero factor A effects, then $E(MS_A)$ will be larger than σ^2; thus, MS_A will tend to be larger than MS_E. To test the significance of the factor A main effect, we should consider the ratio MS_A/MS_B. Larger values of this ratio would indicate that the data do not support the null hypothesis. Similar statements apply to MS_B and MS_{AB} and their relationship to MS_E.

If we assume that the model in (4.1) is adequate and that the error terms ε_{ijk} are normally and independently distributed with constant variance σ^2, then each of the rations of mean squares has an F distribution whose numerator degrees of freedom is equal to the degrees of freedom from Table 4.5, and whose denominator degrees of freedom is $ab(n-1)$. Thus,

$$\frac{MS_A}{MS_E} \sim F(a-1, ab(n-1)),$$

$$\frac{MS_A}{MS_E} \sim F(b-1, ab(n-1)),$$

$$\frac{MS_A}{MS_E} \sim F((a-1)(b-1), ab(n-1)).$$

The critical region for each test is the upper tail of the F distribution; that is, we would reject for sufficiently large values of the F statistic. The test procedure is usually summarized in the ANOVA table as shown in Table 4.5.

Computationally, we almost always use a software package to perform the ANOVA calculations. However, it is occasionally useful to have manual calculation formulas available. The total sum of squares is computed as

$$SS_T = \sum_{i=1}^{a} \sum_{j=1}^{b} \sum_{k=1}^{n} y_{ijk}^2 - \frac{y_{\cdots}^2}{abn}$$

and the sums of squares for the factors are

$$SS_A = \frac{1}{bn} \sum_{i=1}^{a} y_{i\cdot\cdot}^2 - \frac{y_{\cdots}^2}{abn}$$

and

$$SS_B = \frac{1}{an} \sum_{j=1}^{b} y_{\cdot j\cdot}^2 - \frac{y_{\cdots}^2}{abn}.$$

It is convenient to obtain SS_E in two stages. First, we compute the sum of squares between the ab cell totals, which is called the sum of squares due to "subtotals:"

$$SS_{\text{Subtotals}} = \frac{1}{n} \sum_{i=1}^{a} \sum_{j=1}^{b} y_{ij\cdot}^2 - \frac{y_{\cdots}^2}{abn}.$$

The sum of squares due to the interaction can then be obtained as

$$SS_{AB} = SS_{\text{Subtotals}} - SS_A - SS_B.$$

The sum of squares due to error can be obtained by subtraction as

$$SS_E = SS_T - SS_{\text{Subtotals}}.$$

Although this experiment had one categorical factor (material type) and one continuous factor (temperature), both variables are treated as categorical in the ANOVA. Table 4.6 presents the basic ANOVA that would result from this assumption. Both main effects of material type and temperature are significant, and there is a strong two-factor interaction between these two factors.

It is useful to examine the model between life, material type, and temperature taking account of the continuous nature of the temperature variable. The display in Figure 4.5 shows the output from the JMP Fit Model analysis. The first graph in the output shows the three regression models that describe the relationship between life and temperature for each material type. These curves are very different, showing clearly the interaction between material type and temperature. There is a strong indication that Material type 3 is less sensitive to temperature than the other two. We also see that the quadratic effect of temperature is not significant by itself, but it is involved in one of the interaction components (see the Material type*Temperature*Temperature effect in the Effect Summary portion of the output), so we have elected to leave the quadratic effect on temperature along with the Material type*Temperature effect in the model to promote hierarchy; that is, if a higher-order term is in the model, then the corresponding lower-order terms should also be included. Finally we have targeted the prediction profiler at material type 3 and the nominal expected temperature of 70°, showing an expected life of 145.54 hours. Notice how flat the predicted life curve is over the range of temperature, except at the far upper end of the temperature range where the expected life drops more quickly (Figure 4.6).

In this analysis we assumed that the response battery lifetime had a normal distribution whose mean was a function of the material type and the testing temperature and whose variance is the same for all values of the predictor variables. As we've discussed in previous chapters, the normal distribution is not always an appropriate model for lifetimes. In this case, however, the data seem to be close to normally distributed and there is no censoring, so the assumptions seem reasonable.

Table 4.6 ANOVA table for battery lifetime experiment.

Source of variation	Sum of squares	Degree of freedom	Mean square	F statistic	P-value
Material type	10,683.72	2	5,341.86	7.91	0.0020
Temperature	39,118.72	2	19,559.36	28.97	<0.0001
Interaction	9,613.78	4	2,403.44	3.56	0.0186
Error	18,230.72	27	675.21		
Total	77,646.97	36			

Response Life
Regression Plot

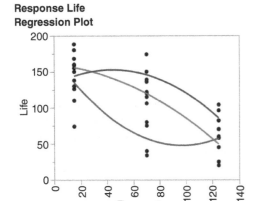

— Line of Fit for Material type [M1]
— Line of Fit for Material type [M2]
— Line of Fit for Material type [M3]

Actual by Predicted Plot

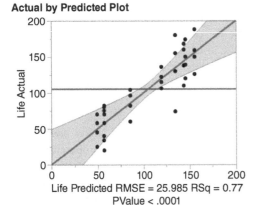

Life Predicted RMSE = 25.985 RSq = 0.77
PValue < .0001

Effect Summary

Source	LogWorth		PValue
Temperature	7.453		0.00000
Material type	3.788		0.00016
Material type*Temperature*Temperature	1.974		0.01061
Material type*Temperature	0.701		0.19911 ^
Temperature*Temperature	0.131		0.73975 ^

Figure 4.5 JMP fit model analysis for the battery life experiment.

We could also analyze the battery life data with R. The code to load the data file construct the basic plots can be done with the following code.

```
BatteryLifeExperiment = read.csv("BatteryLifeExperiment.csv")
LifeTime = BatteryLifeExperiment$Outcome
MaterialType = BatteryLifeExperiment$MaterialType
Temp = BatteryLifeExperiment$Temp

windows( 12, 7 )
par( mfrow=c(1,2) )
```

Effect Tests

Source	Nparm	DF	Sum of Squares	F Ratio	Prob > F
Material type	2	2	16552.667	12.2574	0.0002*
Temperature	1	1	39042.667	57.8227	<.0001*
Material type*Temperature	2	2	2315.083	1.7143	0.1991
Temperature*Temperature	1	1	76.056	0.1126	0.7398
Material type*Temperature*Temperature	2	2	7298.694	5.4047	0.0106*

Prediction Profiler

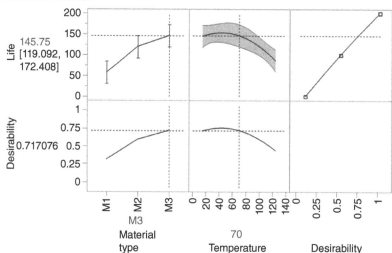

Figure 4.6 JMP fit model analysis for the battery life experiment (continued).

```
plot( MaterialType , LifeTime , pch=19 , ylim=c(0,200) )
pch = c(1,17,19)
colr = c("red","green","blue")
pch.plot = c()
pch.plot[ MaterialType == 1 ] = pch[1]
pch.plot[ MaterialType == 2 ] = pch[2]
pch.plot[ MaterialType == 3 ] = pch[3]
col.plot = c()
col.plot[ MaterialType == 1 ] = colr[1]
col.plot[ MaterialType == 2 ] = colr[2]
col.plot[ MaterialType == 3 ] = colr[3]

plot( Temp , LifeTime , pch=pch.plot , col=col.plot ,
      ylim=c(0,200 ) )
legend( "topright" , legend=c("Material 1","Material 2", "Material 3"),
col=colr, pch=pch)
```

The two scatter plots of the predictors `MaterialType` and `Temp` against the lifetime are shown in Figure 4.7. The R code to construct the ANOVA table for a two-factor model with interaction is as follows. Note that we must force `MaterialType` and `Temp` to be factor (categorical) variables.

```
MaterialType.f = as.factor( BatteryLifeExperiment$MaterialType )
Temp.f = as.factor( BatteryLifeExperiment$Temp )

battery.lm = lm( LifeTime ~ MaterialType.f + Temp.f +
                            MaterialType.f:Temp.f)
summary( battery.lm )
battery.anova = anova( battery.lm )

print( battery.anova )
print( sum( battery.anova$'Sum Sq' ) )
```

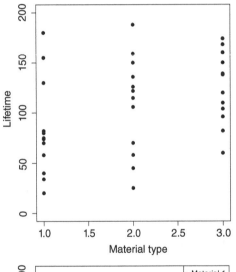

Figure 4.7 Scatter plots created in R for the battery life data. Plot on left shows the lifetime versus the material type. The one on the right shows the lifetime versus the temperature, with different characters for the material type.

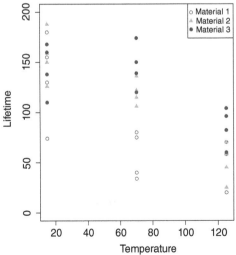

The results of the `summary` are shown in the following text.

```
Call:
lm(formula = LifeTime ~ MaterialType.f + Temp.f +
                MaterialType.f:Temp.f)

Residuals:

Min      1Q  Median     3Q     Max
-60.750 -14.625   1.375  17.938  45.250
Coefficients:
                         Estimate Std. Error t value Pr(>|t|)
(Intercept)               134.75      12.99  10.371 6.46e-11 ***
MaterialType.f2            21.00      18.37   1.143 0.263107
MaterialType.f3             9.25      18.37   0.503 0.618747
Temp.f70                  -77.50      18.37  -4.218 0.000248 ***
Temp.f125                 -77.25      18.37  -4.204 0.000257 ***
MaterialType.f2:Temp.f70   41.50      25.98   1.597 0.121886
MaterialType.f3:Temp.f70   79.25      25.98   3.050 0.005083 **
MaterialType.f2:Temp.f125 -29.00      25.98  -1.116 0.274242
MaterialType.f3:Temp.f125  18.75      25.98   0.722 0.476759
---
Signif. codes:  0 '***'0.001 '**'0.01 '*'0.05 '.'0.1 ' '1

Residual standard error: 25.98 on 27 degrees of freedom
Multiple R-squared:  0.7652,        Adjusted R-squared:  0.6956
F-statistic:     11 on 8 and 27 DF,  p-value: 9.426e-07
```

The symbol for interaction in the `lm` is the colon. The results of `anova` are

```
Analysis of Variance Table

Response: LifeTime
                      Df Sum Sq Mean Sq F value    Pr(>F)
MaterialType.f         2  10684  5341.9  7.9114  0.001976 **
Temp.f                 2  39119 19559.4 28.9677 1.909e-07 ***
MaterialType.f:Temp.f  4   9614  2403.4  3.5595  0.018611 *
Residuals             27  18231   675.2

---
Signif. codes:  0 '***'0.001 '**'0.01 '*'0.05 '.'0.1 ' '1

[1] 77646.97
```

Note that R does not print out the "total" line of the ANOVA. We can compute the total sum of squares by typing `print( sum( battery.anova$'Sum Sq' ) )`, which yields 77,646.97. The ANOVA table here matches (to the number of digits reported) the ANOVA table from JMP. We see that there are significant effects due to material type and temperature, and there is a significant interaction.

The previous analysis has assumed that both `MaterialType` and `Temp` are categorical variables. If we assume that `Temp` is continuous while `MaterialType` is categorical, we have a mix of continuous and categorical. We can still handle this with R's linear model function `lm`.

Figure 4.6 shows separate plots of `LifeTime` against `Temp` for each of the three material types. We see that the relationship is not linear. A second-order model in `Temp` may be reasonable, but it looks like different relationships exist between `LifeTime` and `Temp` for the three levels of `MaterialType`. The relationship seems to be concave up for material types 1 and 2, but concave down for material type 3.

The R code for a second-order model in `Temp` along with the categorical variable `MaterialType` and an interaction between them is as follows.

```
TempSq = Temp^2
battery.lm1 = lm(
      LifeTime ~ MaterialType.f + Temp + TempSq +
                 MaterialType.f:Temp +
                 MaterialType.f:TempSq )
summary( battery.lm1 )
battery.lm1.anova = anova( battery.lm1 )
print( battery.lm1.anova )
print( sum( battery.lm1.anova$'Sum Sq' ) )
```

The output from the `battery.lm1.anova` is

```
Analysis of Variance Table

Response: LifeTime
                    Df Sum Sq Mean Sq F value    Pr(>F)
MaterialType.f       2  10684    5342  7.9114  0.001976 **
Temp                 1  39043   39043 57.8227 3.525e-08 ***
TempSq               1     76      76  0.1126  0.739753
MaterialType.f:Temp  2   2315    1158  1.7143  0.199109
MaterialType.f:TempSq 2  7299    3649  5.4047  0.010612 *
Residuals           27  18231     675
---
Signif. codes:  0 '***'0.001 '**'0.01 '*'0.05 '.'0.1 ' '1
```

We can obtain the total sum of squares using the command

```
print( sum( battery.anova$'Sum Sq' ) )
```

which yields

```
[1] 77646.97.
```

This agrees with the total sum of squares we obtained previously.

4.5 The 2^k Factorial Design

Factorial designs are used extensively in experiments involving several factors where it is necessary to study the joint effect of the factors on a response. Section 4.3 presented methods for the construction and analysis of factorial designs focusing on the case of two factors. However, there are several special cases of the general factorial design that are important because they are widely used in industrial research and development work and because they form the basis of other designs of considerable practical value.

The most important of these special cases is when there are k factors, and each factor has only two levels. These levels may be quantitative, such as two values of temperature or pressure; or they may be *qualitative*, such as two machines or two operators, the "high" and "low" levels of a factor, or perhaps the presence and absence of a factor. A complete replicate of such a design requires $2 \times 2 \times \cdots \times 2 = 2^k$ observations and is called a 2^k *factorial design*. This section focuses on these designs. Throughout this presentation, we assume that (i) the factors are fixed, (ii) the designs are completely randomized, and (iii) the usual normality assumptions required for the analysis are satisfied.

The 2^k design is particularly useful in the early stages of experimental work when experimenters often need to study the effects of many factors and they suspect that not all factors are going to be important. In these situations, the experimenter is usually interested in discovering the set of active factors from the original group. The 2^k design provides the smallest number of runs with which k factors can be studied in a complete factorial design. Consequently, these designs are widely used in *factor screening experiments*. It is also easy to develop effective blocking schemes for these designs and to construct them in fractional versions.

4.5.1 The 2^2 Factorial Design

The first design in the 2^k family has two factors, say, A and B, each run at two levels. This design is called a 2^2 factorial design. The levels of the factors are arbitrarily called "low" and "high." As an example, suppose that a router is used to cut locating or registration notches on a printed circuit board. The vibration level at the surface of the board as it is cut is considered a major source of dimensional variation in the notches. This dimensional variation contributes to components that are not inserted or placed correctly on the unit and this can lead to early failure of the device in which these boards are used. Two factors potentially influence vibration: bit size (A) and cutting speed (B). Two different bit sizes (1/16 and 1/8 in.) and two speeds (40 and 90 rpm) are selected, and three boards are cut at each set of conditions as shown in Table 4.7. The response variable is vibration measured as the resultant vector of three accelerometers (x, y, and z) on each of the circuit board used in the experiment.

There is a convention for labeling the runs in a 2^k design shown in Table 4.7. The four treatment combinations in this design are shown in the first two columns of Table 4.7. We denote the effect of a factor by a capital Latin letter. Thus, "A" refers to the effect of factor A, "B" refers to the effect of factor B, and

Table 4.7 Data from three replicates in the vibration experiment.

A Bit size	B Speed	Run Labels	Replicate		
			Rep. 1	Rep. 2	Rep. 3
−	−	(1) (A low, B low)	18.2	18.9	12.9
+	−	a (A high, B low)	27.2	24.0	22.4
−	+	b (A low, B high)	15.9	14.5	15.1
+	+	ab (A high, B high)	41.0	43.9	36.3

"*AB*" refers to the *AB* interaction. In the 2^2 design, the low and high levels of *A* and *B* are denoted by "−" and "+" respectively. Thus, − for the factor *A* axis represents the low level of bit size, and + represents the high level; similarly − for factor *B* axis represents the low level of cutting speed, and + denotes the high level. This is sometimes called the "orthogonal coding" for a 2^k design. The four treatment combinations in the design are also represented by lowercase letters, as shown in Table 4.7. The high level of any factor in the treatment combination is denoted by the corresponding lowercase letter and that the low level of a factor in the treatment combination is denoted by the absence of the corresponding letter. Thus, *a* represents the treatment combination of *A* at the high level and *B* at the low level, *b* represents *A* at the low level and *B* at the high level, and *ab* represents both factors at the high level. By convention, (1) is used to denote both factors at the low level. This notation is used throughout the 2^k series. Often these lowercase letters represent the total of all observations that are taken at the corresponding treatment combinations.

The analysis of a replicated 2^k design is similar to the analysis of general factorials discussed previously. Figures 4.8 and 4.9 give the output from the Fit Model platform in JMP for the 2^2 factorial for the router experiment in Table 4.7. The ANOVA portion of the display shows that the overall model is significant, implying that at least one of the two main effects and their interaction are active. The following sections showing the individual effect tests and the parameter estimates indicate that both of the main effects and the bit size-speed interaction are significant. The residuals versus predicted value plot does not indicate any obvious problems with model adequacy and the model R^2 is 95%. The two-factor interaction plot shows that when the small-bit is used cutting speed has very little effect on vibration, but when the larger bit is used, increasing cutting speed leads to large increases in vibration. This is confirmed by the prediction profiler, which is set at the smallest bit size. This display shows that when using the smaller bit any cutting speed can be used without increasing the vibration level.

How Much Replication is Necessary? A standard question that arises in almost every experiment is how much replication is necessary? We have discussed this in previous chapters, but there are some aspects of this topic that are particularly useful in 2^k designs, which are used extensively for factor screening. That is, studying a group of *k* factors to determine which ones are active. Recall from our previous discussions that the choice of an appropriate sample size in a designed experiment depends on how large the effect of interest is, the power of the statistical test, and the choice of type I error. While the size of an important effect is obviously problem-dependent, in many practical situations experimenters are interested in detecting effects that are at least as large as twice the error standard deviation (2σ). Smaller effects are usually of less interest because changing the factor associated with such a small effect often results in a change in response that is very small relative to the background noise in the system. Adequate power is also problem-dependent, but in many practical situations achieving power of at least 0.80 or 80% should be the goal.

We will illustrate how an appropriate choice of sample size can be determined using the 2^2 router experiment. Suppose that we are interested in detecting effects of size 2σ. If the basic 2^2 design is replicated twice for a total of eight runs, there will be four degrees of freedom for estimating a model-independent estimate of error (pure error). If the experimenter uses a significance level or Type I error rate of $\alpha = 0.05$,

Response Vibration
Actual by Predicted Plot

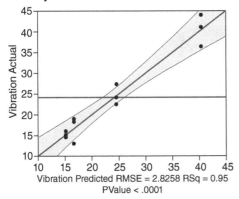

Vibration Predicted RMSE = 2.8258 RSq = 0.95
PValue < .0001

Effect Summary

Source	LogWorth		PValue
Bit Size(0.0625,0.125)	5.118		0.00001
Bit Size*Speed	3.149		0.00071
Speed(40,90)	2.643		0.00228 ^

Residual by Predicted Plot

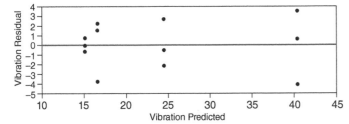

Studentized Residuals

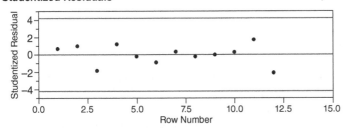

Externally Studentized Residuals with 95% Simultaneous Limits (Bonferroni)

Figure 4.8   JMP fit model analysis for the vibration experiment.

Prediction Profiler

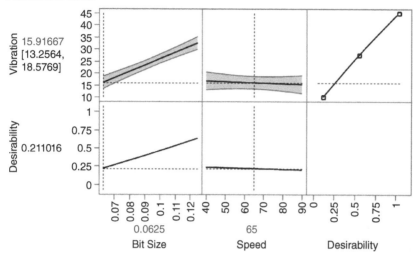

Interaction Profiles

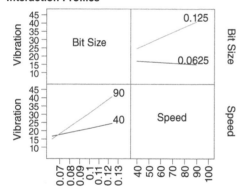

Figure 4.9 JMP fit model output for the router experiment in Table 4.7.

this design results in a power of 0.572 or 57.2%. This is too low, and the experimenter should consider more replication. There is another alternative that could be useful in screening experiments, using a higher Type I error rate. In screening experiments Type I errors (thinking a factor is active when it really isn't) usually does not have the same impact than a Type II error (failing to identify an active factor). If a factor is mistakenly thought to be active, that error will be discovered in further work and so the consequences of this Type I error is usually small. However, failing to identify an active factor is usually very problematic because that factor is set aside and typically never considered again. So in screening experiments experimenters are often willing to consider higher Type I error rates, say, 0.10 or 0.20. Suppose that we use $\alpha = 0.10$ in our router experiment. This would result in power of 75%. Using $\alpha = 0.20$

Evaluate Design

Model

Intercept

X1

X2

X1*X2

Power Analysis

Significance Level	0.05
Anticipated RMSE	1

Term	Anticipated Coefficient	Power
Intercept	1	0.857
X1	1	0.857
X2	1	0.857
X1*X2	1	0.857

Figure 4.10 JMP power calculations.

increases the power to 89%, a very reasonable value. The other alternative is to increase the sample size by using additional replicates. If we use three replicates there will be eight degrees of freedom for pure error and if we want to detect effects of size σ with $\alpha = 0.05$, this design will result in power of 85.7%. This is a very good value for power, so the experimenters decided to use three replicates of the 2^2 design (Figure 4.5.1).

Software packages can be used to produce the power calculations given earlier. Figure 4.10 shows the power calculations from JMP. The model has both main effects and the two-factor interaction and the effects of size 2σ is chosen by setting the square root of mean square error (Anticipated RMSE) to 1 and setting the size of each anticipated coefficient in the prediction (regression) model coefficient to 1.

4.5.2 The 2^3 Factorial Design

Suppose that three factors, A, B, and C, each at two levels, are of interest. This design is called a 2^3 factorial design, and the eight treatment combinations can be displayed geometrically as a cube, as shown in Figure 4.11a. Using the "+" and "−" coding to represent the low and high levels of the factors, we show the eight runs in the 2^3 design in Figure 4.11b. This table is called the design matrix. Extending the label notation discussed from Section 4.5.1, we write the treatment combinations in standard order as (1), a, b, ab, c, ac, bc, and abc. Remember that these symbols also often represent the total of all n observations

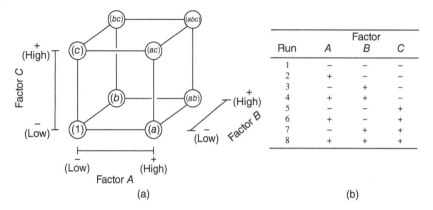

Figure 4.11 Illustration of the 2^3 factorial design. (a) Geometric view and (b) design matrix.

taken at those specific treatment combinations. Figure 4.12 is a geometric presentation of the contrasts defining the main effects and interactions in the 2^3 factorial design.

Carbonation percent	Pressure psi	Speed bottles/min	Treatment label	Replicate	
				Rep. 1	Rep. 2
−	−	−	(1)	−3	−1
+	−	−	a	0	1
−	+	−	b	−1	0
+	+	−	ab	2	3
−	−	+	c	−1	0
+	−	+	ac	2	1
−	+	+	bc	1	1
+	+	+	abc	6	5

As an example of an experiment involving a 2^3 design, suppose that a beverage bottler is investigating the performance of a bottle-filling machine. These fillers are usually designed to fill to a specific target height, and bottles that deviate significantly from target are either underfilled or overfilled. The filler has three controllable factors; carbonation level of the product, filling pressure, and bottling speed. These three factors were investigated using the 2^3 design shown in Table 4.6. The low and high levels of the three factors are as follows: carbonation level: 10%, 12%, Pressure: 25, 30 psi, speed: 200, 250 bottles/min. The response variable is fill deviation as measured on a coded scale with 0 indicating a fill height that is exactly on the desired target. Positive values of the response indicate overfilled bottles and negative values indicate underfilled bottles.

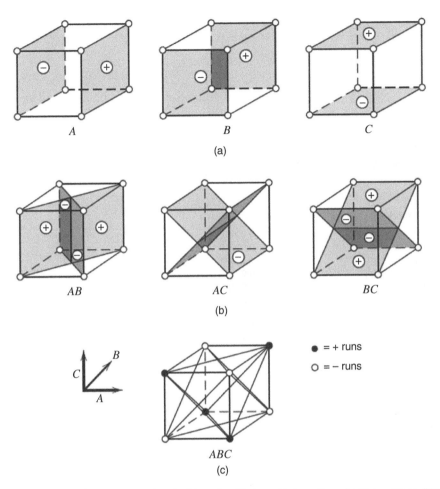

Figure 4.12 Contrasts associated with main effects and interactions. (a) Main effects, (b) two-factor interaction, and (c) three-factor interaction.

We will analyze this experiment by fitting the full three-factor factorial model. This model contains all three main effects, the three two-factor interactions, and the three-factor interaction. Figures 4.13 and 4.14 show the output from the JMP fit model platform.

The ANOVA indicates that the overall model is significant, and the effect tests show that all three main effects are highly significant and that one of the two-factor interactions, Carbonation–Pressure, is significant at approximately the 10% level. The size of this interaction effect is less than half of the smallest main effect.

Figure 4.14 is the result of fitting a reduced model containing only the three main effects and the Carbonation–Pressure interaction. One again, we see that all three main effects are highly significant

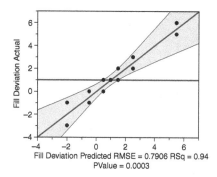

Effect Summary

Source	LogWorth		PValue
Carbonation(10,12)	4.196		0.00006
Pressure(25,30)	3.339		0.00046
Speed(200,250)	2.657		0.00221
Carbonation*Pressure	1.025		0.09435
Pressure*Speed	0.617		0.24150
Carbonation*Pressure*Speed	0.617		0.24150
Carbonation*Speed	0.264		0.54474 ^

Residual by Predicted Plot

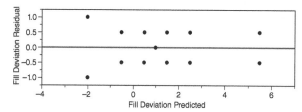

Studentized Residuals

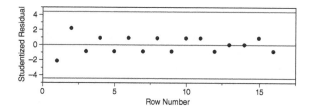

Externally Studentized Residuals with 95% Simultaneous Limits (Bonferroni)

Figure 4.13   Additional JMP fit model output.

Analysis of Variance

Source	DF	Sum of Squares	Mean Square	F Ratio
Model	7	73.000000	10.4286	16.6857
Error	8	5.000000	0.6250	**Prob > F**
C. Total	15	78.000000		0.0003*

Parameter Estimates

Term	Estimate	Std Error	t Ratio	Prob > |t|
Intercept	1	0.197642	5.06	0.0010*
Carbonation(10,12)	1.5	0.197642	7.59	<.0001*
Pressure(25,30)	1.125	0.197642	5.69	0.0005*
Speed(200,250)	0.875	0.197642	4.43	0.0022*
Carbonation*Pressure	0.375	0.197642	1.90	0.0943
Carbonation*Speed	0.125	0.197642	0.63	0.5447
Pressure*Speed	0.25	0.197642	1.26	0.2415
Carbonation*Pressure*Speed	0.25	0.197642	1.26	0.2415

Effect Tests

Source	Nparm	DF	Sum of Squares	F Ratio	Prob > F
Carbonation(10,12)	1	1	36.000000	57.6000	<.0001*
Pressure(25,30)	1	1	20.250000	32.4000	0.0005*
Speed(200,250)	1	1	12.250000	19.6000	0.0022*
Carbonation*Pressure	1	1	2.250000	3.6000	0.0943
Carbonation*Speed	1	1	0.250000	0.4000	0.5447
Pressure*Speed	1	1	1.000000	1.6000	0.2415
Carbonation*Pressure*Speed	1	1	1.000000	1.6000	0.2415

Figure 4.14 JMP fit model output for the bottle filling experiment.

and that the Carbonation–Pressure interaction is only borderline significant. Since the size of the interaction effect is less than half the size of the smallest main effect it is unlikely that it will play an important role in determining the settings for the bottle filling machine (Figure 4.15).

It is desirable to set up the machine so that the average fill deviation is close to zero. The prediction profiler in Figure 4.10 displays one possible configuration. In this solution speed is set at the high level (this is a throughput consideration so that as much product as possible can be filled). Filling pressure is also at the low level (this is known to reduce foaming, a major source of under- and over-filled bottles) and carbonation is at the midpoint of the range since this is specification for the characteristic (Figure 4.16).

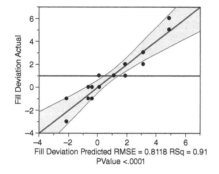

Fill Deviation Predicted RMSE = 0.8118 RSq = 0.91
PValue <.0001

Effect Summary

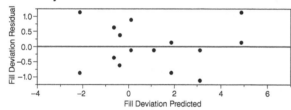

Source	LogWorth		PValue
Carbonation(10,12)	4.861		0.00001
Pressure(25,30)	3.758		0.00017
Speed(200,250)	2.909		0.00123
Carbonation*Pressure	1.038		0.09170

Residual by Predicted Plot

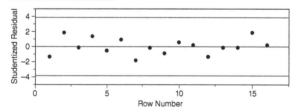

Studentized Residuals

Externally Studentized Residuals with 95% Simultaneous Limits (Bonferroni)

Figure 4.15 JMP fit model output for the bottle filling experiment continued.

4.5.3 A Singe Replicate of the 2^k Design

For even a moderate number of factors, the total number of treatment combinations in a 2^k factorial design is large. For example, a 2^5 design has 32 treatment combinations, a 2^6 design has 64 treatment combinations, a 2^{10} has 1024 treatment combinations, and so on. Because resources are usually limited, the number of replicates that the experimenter can employ may be restricted. Frequently, available resources only allow a single replicate of the design to be run, unless the experimenter is willing to omit some of the original factors. This is usually an undesirable solution, since it runs the risk of missing an important factor.

Analysis of Variance

Source	DF	Sum of Squares	Mean Square	F Ratio
Model	4	70.750000	17.6875	26.8362
Error	11	7.250000	0.6591	**Prob > F**
C. Total	15	78.000000		<.0001*

Parameter Estimates

Term	Estimate	Std Error	t Ratio	Prob>|t|
Intercept	1	0.202961	4.93	1.0005*
Carbonation(10,12)	1.5	0.202961	7.39	<.0001*
Pressure(25,30)	1.125	0.202961	5.54	0.0002*
Speed(200,250)	0.875	0.202961	4.31	0.0012*
Carbonation*Pressure	0.375	0.202961	1.85	0.0917

Effect Tests

Source	Nparm	DF	Sum of Squares	F Ratio	Prob > F
Carbonation(10,12)	1	1	36.000000	54.6207	<.0001*
Pressure(25,30)	1	1	20.250000	30.7241	0.0002*
Speed(200,250)	1	1	12.250000	18.5862	0.0012*
Carbonation*Pressure	1	1	2.250000	3.4138	0.0917

Prediction Profiler

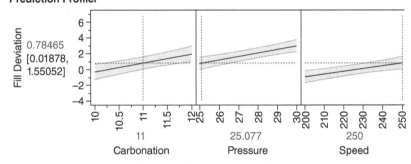

Figure 4.16 The reduced model for the bottle filling experiment.

An obvious risk when conducting an experiment that has only one run at each test combination is that we may be fitting a model to noise. That is, if the response variable y is highly variable, misleading conclusions may result from the experiment. Of course, if there is less variability in the response, the likelihood of an erroneous conclusion is smaller. One way to ensure that reliable effect estimates are obtained is to increase the distance between the low (−) and high (+) levels of the factor.

The single-replicate strategy is often used in screening experiments when there are relatively many factors under consideration. Because we can never be entirely certain in such cases that the experimental error is small, a good practice in these types of experiments is to spread out the factor levels aggressively.

A single replicate of a 2^k design is sometimes called an unreplicated factorial. With only one replicate, there is no internal estimate of error (or "pure error"). One approach to the analysis of an unreplicated factorial is to assume that certain high-order interactions are negligible and combine their mean squares to estimate the error. This is an appeal to the *sparsity of effects* principle; that is, most systems are dominated by some of the main effects and low-order interactions, and most high-order interactions are negligible.

While the effect sparsity principle has been observed by experimenters for many decades, only recently has it been studied more objectively. Li et al. (2006) studied 113 response variables obtained from 43 published experiments from a wide range of science and engineering disciplines. All of the experiments were full factorials with between three and seven factors, so no assumptions had to be made about interactions. Most of the experiments had either three or four factors. The authors found that about 40% of the main effects in the experiments they studied were significant, while only about 11% of the two-factor interactions were significant. Three-factor interactions were very rare, occurring only about 5% of the time. The authors also investigated the absolute values of factor effects for main effects, two-factor interactions, and three-factor interactions. The median size of main effects was found to be about four times larger than the median size of two-factor interactions. The median size of two-factor interactions was more than two times larger than the median size of three-factor interactions. However, there were many two- and three-factor interactions that were larger than the median main effect. Another paper by Bergquist et al. (2011) also studied the effect of the sparsity question using 22 different experiments with 35 responses. They considered both full factorial and fractional factorial designs with factors at two levels. Their results largely agree with those of Li et al. (2006), with the exception that three-factor interactions were less frequent, occurring only about 2% of the time. This difference may be partially explained by the inclusion of experiments with indications of curvature and the need for transformations in the Li et al. (2006) study. Bergquist et al. (2011) excluded such experiments. Overall, both of these studies confirm the validity of the sparsity of effects principle.

When analyzing data from unreplicated factorial designs, occasionally real high-order interactions occur. The use of an error mean square obtained by pooling high-order interactions is inappropriate in these cases as it will make the error term too large and can lead to failure to identify active effects. A method of analysis proposed by Daniel (1959) provides a simple way to overcome this problem. Daniel suggests examining a normal probability plot of the estimates of the effects. The effects that are negligible are normally distributed, with mean zero and variance σ^2 and will tend to fall along a straight line on this plot, whereas significant effects will have nonzero means and will not lie along the straight line. Alternatively, a half-normal plot can also be used. The apparently nonzero effects will be used to form the preliminary model and the remaining negligible effects are then combined as an estimate of error.

As an example of an unreplicated factorial, consider an experiment involving a nickel–titanium alloy that is used to make components for jet turbine aircraft engines. Cracking is a potentially serious defect in the final part because it can lead to nonrecoverable failure. The parts producer performed an experiment to determine the effects of four factors on crack length. The four factors are pouring temperature (A), titanium content (B), heat treatment method (C), and amount of grain refiner used (D). Two replicates of

Table 4.8 The alloy experiment is a 2^4 design.

| | Factor | | | Treatment | Crack |
A	B	C	D	combination	length
−	−	−	−	(1)	8.4
+	−	−	−	*a*	13.3
−	+	−	−	*b*	13.3
+	+	−	−	*ab*	19.1
−	−	+	−	*c*	8.6
+	−	+	−	*ac*	5.8
−	+	+	−	*bc*	10.9
+	+	+	−	*abc*	11.6
−	−	−	+	*d*	10.3
+	−	−	+	*ad*	15.3
−	+	−	+	*bd*	12.2
+	+	−	+	*abd*	20.3
−	−	+	+	*cd*	10.5
+	−	+	+	*acd*	7.6
−	+	+	+	*bcd*	12.7
+	+	+	+	*acbd*	13.8

a 2^4 factorial design are run, and the length of crack (in mm $\times 10^{-2}$) induced in a sample coupon subjected to a standard test is measured. The design and the response data are shown in Table 4.8.

Figure 4.17 is the output from the JMP screening analysis for this experiment. The bottom portion of the display is a half-normal plot of the factor effects. This plot indicates that all four main effects are active, and that two of the two-factor interactions, Titanium Content–Temperature and Heat Treatment Method–Temperature, are also active (Figure 4.18).

The normal (or half-normal) probability plotting method has been subjected to some criticism because it's subjective, and while it is usually effective in identifying large active effects, it may not detect smaller ones as effectively. A very popular analytic approach has been developed by Lenth (1989) that has good power to detect active factors. Lenth's method uses the smallest half of the effect estimates to form a robust "pseudo standard error" that can be used to form a t-like ratio for testing the significance of individual effects. The JMP software package implements Lenth's method as part of the screening platform analysis procedure for two-level designs. In their implementation, P-values for each factor and interaction are computed from a "real-time" simulation. This simulation assumes that none of the factors in the experiment are significant and calculates the observed value of the Lenth statistic 10,000 times for this

Screening for Crack Length
Contrasts

Term	Contrast		Lenth t-Ratio	Individual p-Value	Simultaneous p-Value
Titanium content	1.94375		29.62	<.0001*	<.0001*
Heat treatment method	-1.73125		-26.38	<.0001*	<.0001*
Temperature	1.43125		21.81	<.0001*	<.0001*
Amt of grain refiner	0.91875		14.00	0.0002*	0.0005*
Titanium content*Heat treatment method	0.11875		1.81	0.0853	0.6011
Titanium content*Temperature	0.90625		13.81	0.0002*	0.0005*
Heat treatment method*Temperature	-1.91875		-29.24	<.0001*	<.0001*
Titanium content*Amt of grain refiner	-0.03125		-0.48	0.6627	1.0000
Heat treatment method*Amt of grain refiner	0.04375		0.67	0.5002	1.0000
Temperature*Amt of grain refiner	-0.01875		-0.29	0.7870	1.0000
Titanium content*Heat treatment method*Temperature	0.03125		0.48	0.6627	1.0000
Titanium content*Heat treatment method*Amt of grain refiner	0.06875		1.05	0.2729	0.9904
Titanium content*Temperature*Amt of grain refiner	-0.01875		-0.29	0.7870	1.0000
Heat treatment method*Temperature*Amt of grain refiner	0.05625		0.86	0.3644	0.9999
Titanium content*Heat treatment method*Temperature*Amt of grain refiner	0.08125		1.24	0.2022	0.9405

Figure 4.17 Crack length contrasts for screening design.

Half Normal Plot

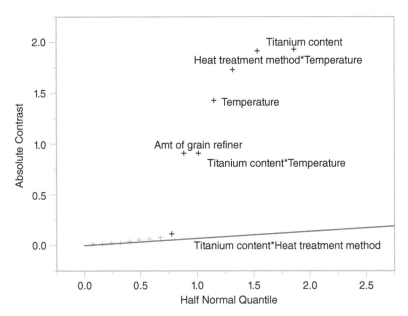

Length PSE = 0.06563
P-Values derived from a simulation of 10000 Length t ratios.

Figure 4.18   Half normal plot for crack length experiment.

null model. Then *P*-values are obtained by determining where the observed Lenth statistics fall relative to the tails of these simulation-based reference distributions. These *P*-values can be used as guidance in selecting factors for the model (Figure 4.19). The top portion of Figure 4.11 shows the JMP output from the screening analysis platform for the alloy experiment. When the factors are entered into the model, the Lenth procedure would recommend including the same factors in the model that we identified previously from the half normal plot. Figure 4.20 presents the analysis of the final model. Temperature is a very important variable, both as a main effect and because it is involved in two of the active two-factor interactions. The prediction profiler at the bottom of Figure 4.20 indicates a set of process operating conditions that results in small crack lengths. The recommendations are to set temperature at the high level, titanium content at the low level, and use the second of the two heat treating methods. The amount of grain refiner at the aforementioned set of conditions on the first three factors seems to have a very small effect, but from a crack length viewpoint and perhaps an operating cost perspective, the lower level is best.

4.5.4 2^k Designs are Optimal Designs

Two-level factorial designs have many useful properties. For example, they are orthogonal designs, which make interpretation of the results of the experiment simpler. They also have several optimality properties,

Parameter Estimates

Term	Estimate	Std Error	t Ratio	Prob>|t|
Intercept	11.91875	0.060703	196.34	<.0001*
Titanium content	1.94375	0.060703	32.02	<.0001*
Heat treatment method	−1.73125	0.060703	−28.52	<.0001*
Temperature	1.43125	0.060703	23.58	<.0001*
Amt of grain refiner	0.91875	0.060703	15.14	<.0001*
Titanium content*Temperature	0.90625	0.060703	14.93	<.0001*
Heat treatment method*Temperature	−1.91875	0.060703	−31.61	<.0001*

Prediction Profiler

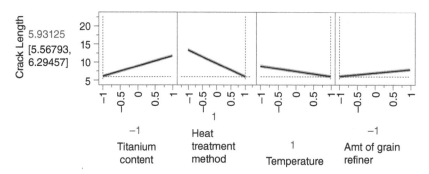

Figure 4.19 JMP output for crack length experiment.

which is the subject of this section. We will discuss and illustrate these properties using the 2^2 factorial design, but the results hold for all members of the 2^k full factorial family.

The full factorial model for the 2^2 design is

$$y = \beta_0 + \beta_1 x_1 + \beta_2 x_2 + \beta_{12} x_1 x_2 + \epsilon,$$

where x_1 and x_2 are the main effects of the two factors on the ± 1 scale and $x_1 x_2$ is the two-factor interaction. The error term ϵ is assumed to have a mean of 0 and a variance of σ^2 that is constant for all runs regardless of the levels of x_1 and x_2. In matrix form this model can be written as

$$y = X\beta + \epsilon,$$

where the model matrix is, assuming a single replicate (only four runs),

$$X = \begin{bmatrix} 1 & -1 & -1 & 1 \\ 1 & 1 & -1 & -1 \\ 1 & -1 & 1 & -1 \\ 1 & 1 & 1 & 1 \end{bmatrix}.$$

Response Crack Length

Actual by Predicted Plot

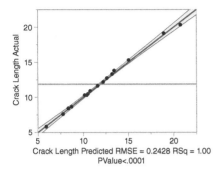

Crack Length Predicted RMSE = 0.2428 RSq = 1.00
PValue<.0001

Effect Summary

Source	LogWorth		PValue
Titanium content	9.858		0.00000
Heat treatment method*Temperature	9.807		0.00000
Heat treatment method	9.409		0.00000 ^
Temperature	8.674		0.00000 ^
Amt of grain refiner	6.981		0.00000
Titanium content*Temperature	6.930		0.00000

Residual by Predicted Plot

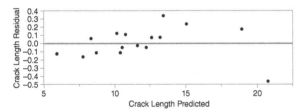

Studentized Residuals

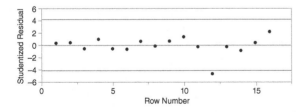

Externally Studentized Residuals with 95% Simultaneous Limits (Bonferroni)

Figure 4.20 Additional JMP output for crack length experiment.

The least squares estimates of the model parameters are

$$\hat{\beta} = (X'X)^{-1}X'y,$$

where

$$X'X = \begin{bmatrix} 4 & 0 & 0 & 0 \\ 0 & 4 & 0 & 0 \\ 0 & 0 & 4 & 0 \\ 0 & 0 & 0 & 4 \end{bmatrix}$$

and

$$(X'X)^{-1} = \begin{bmatrix} 1/4 & 0 & 0 & 0 \\ 0 & 1/4 & 0 & 0 \\ 0 & 0 & 1/4 & 0 \\ 0 & 0 & 0 & 1/4 \end{bmatrix}.$$

Both the $(X'X)$ matrix and its inverse are diagonal because the design is *orthogonal*. If the design were replicated n times the diagonal elements of $(X'X)^{-1}$ are $n2^k$. Now, the determinant of $(X'X)$ and variance of any estimated model regression coefficient are $4^4 = 256$ and $\sigma^2/4$, respectively. This is the maximum possible value of the determinant for a four-run design on the design space bounded by ± 1 and the smallest possible variance of any regression coefficient. Consequently, the 2^2 design (and all 2^k designs) is D-optimal. Furthermore, the sum of the main diagonal elements of $(X'X)^{-1}$ is 1, which is the minimal possible value, so the 2^2 design is also A-optimal.

Now consider the variance of the predicted response in the 2^2 design

$$V(\hat{y}) = V(\hat{\beta}_0 + \hat{\beta}_1 x_1 + \hat{\beta}_2 x_2 + \hat{\beta}_{12} x_1 x_2).$$

The variance of the predicted response is a function of the point in the design space where the prediction is made (x_1 and x_2) and the variance of the model regression coefficients. The estimates of the regression coefficients are independent because the 2^2 design is orthogonal and they all have variance $\sigma^2/4$, so

$$V(\hat{y}) = V(\hat{\beta}_0 + \hat{\beta}_1 x_1 + \hat{\beta}_2 x_2 + \hat{\beta}_{12} x_1 x_2) = \frac{\sigma^2}{4}\left(1 + x_1^2 + x_2^2 + x_1^2 x_2^2\right).$$

The maximum prediction variance occurs when $x_1 = x_2 = \pm 1$ and is equal to σ^2. To determine how good this is, we need to know the best possible value of prediction variance that we can attain. It turns out that the smallest possible value of the maximum prediction variance over the design space is $p\sigma^2 N$, where p is the number of model parameters and N is the number of runs in the design. The 2^2 design has $N = 4$ runs and the model has $p = 4$ parameters, so the model that we fit to the data from this experiment minimizes the maximum prediction variance over the design region. A design that has this property is called a G-optimal design. In general, 2^k designs are G-optimal designs for fitting the first-order model or the first-order model with interaction.

We can also evaluate the prediction variance at any point of interest in the design space. For example, when we are at the center of the design where $x_1 = x_2 = 0$, the prediction variance is $\sigma^2/4$ and when we are at the point where $x_1 = 1$ and $x_2 = 0$ the prediction variance is $\sigma^2/2$. Instead of evaluating the prediction variance at many points in the design space, it is better to consider the average prediction variance over the design space. One way to calculate this average prediction variance is

$$I = \frac{1}{A} \int_{-1}^{1} \int_{-1}^{1} V(\hat{\beta}_0 + \hat{\beta}_1 x_1 + \hat{\beta}_2 x_2 + \hat{\beta}_{12} x_1 x_2) \, dx_1 \, dx_2,$$

where A is the area (or the volume or "hypervolume" if there are three or more factors) of the design space. To compute the average, we are integrating the variance function over the design space and dividing by the area (volume) of the region.

Sometimes I is called the *integrated variance* or *average variance* criterion. Now for a 2^2 design, the area of the design region is $A = 4$, and $I = 4\sigma^2/9$. It turns out that this is the smallest possible value of the average prediction variance that can be obtained from a four-run design used to fit a first-order model with interaction on this design space. A design with this property is called an *I*-optimal design. In general, 2^k designs are *I*-optimal designs for fitting the first-order model or the first-order model with interaction.

The JMP software will construct *D*-, *A*-, and *I*-optimal designs. This can be very useful in constructing designs when a standard design isn't readily available.

For example, suppose that we have six continuous two-level factors and we are unsure about the two-factor interactions as well as the main effects. A design that allows us to estimate all of 21 of these effects is a one-half fraction of the 2^6, or a 2^{6-1} fractional factorial with 32 runs. However, the experimenter cannot afford to do 32 runs. The maximum possible number of runs is 28. The smallest possible design that could estimate all of these effects would have 22 runs. However, the experimenters decided to use a *D*-optimal design with 28 runs. The design found from the JMP Custom Designer is shown in Table 4.9. The *D*-efficiency of this design is 93.4%. It isn't 100% because the design isn't orthogonal. The next segment of the display shows some design diagnostic information. Notice that all of the effect estimates do not have the same relative standard error. If the design were orthogonal all of the standard errors would be equal to $0.1890 = 1/\sqrt{28}$. The last segment of the output is a color map of the correlations between all pairs of estimated effects. If the design were orthogonal, all off-diagonal elements of this matrix would be zero. However, notice that almost all of the correlations are quite small, so it should be possible to obtain reliable estimates of the 6 main effects and 15 two-factor interactions from this experimental design (see Tables 4.10 and 4.11).

4.5.5 Adding Center Runs to a 2^k Design

Another strategy for adding replicate runs to a 2^k design instead of replicating all of the runs is to add between three and five runs at the center of the design region (all factors at the 0 level in coded units). This strategy usually assumes that all of the design factors are continuous so that a "middle" level exists, but it can be employed if some variables are categorical. In that case, the experimenter adds the center runs in the subspaces of the continuous factors at each level of the categorical variable. Experimenters often use this approach to obtain an estimate of pure error in single-replicate designs with four or more factors. This also allows the experimenter to test for curvature or nonlinearity in the response function over the design region. Comparing the average of the center runs to the average of all of the factorial runs with a two-sample *t*-test is a simple way to test for the presence of curvature. If the test of equal means indicates

Table 4.9 Custom 28-run design.

Run	X_1	X_2	X_3	X_4	X_5	X_6
1	−1	−1	−1	1	1	1
2	−1	1	−1	−1	−1	1
3	1	1	1	1	−1	1
4	−1	−1	−1	−1	1	1
5	−1	1	1	1	−1	1
6	1	1	−1	1	1	1
7	−1	−1	1	−1	−1	1
8	−1	1	−1	1	1	−1
9	−1	1	−1	−1	1	−1
10	−1	1	1	−1	1	1
11	1	−1	−1	1	−1	1
12	1	1	−1	−1	−1	−1
13	−1	−1	1	1	1	−1
14	1	−1	−1	1	1	−1
15	1	1	−1	−1	1	1
16	1	1	1	−1	1	−1
17	−1	−1	−1	−1	−1	−1
18	1	−1	1	−1	1	1
19	−1	1	1	−1	−1	−1
20	1	1	1	−1	−1	1
21	1	1	−1	1	−1	−1
22	1	−1	1	−1	−1	−1
23	−1	−1	−1	1	−1	−1
24	1	−1	−1	−1	−1	1
25	−1	−1	1	−1	1	−1
26	1	−1	1	1	−1	−1
27	1	1	1	1	1	−1
28	1	−1	1	1	1	1

that there is curvature, this implies that the underlying model with only main effects and interactions is probably inadequate and a model with higher-order terms, such as quadratic, should be considered. Fitting such a model would require that the design be augmented with a sufficient number of runs to allow estimation of all of the parameters in a quadratic model.

Table 4.10 Design evaluation and estimation efficiency.

Term	Fractional increase in CI length	Relative Std error of estimate
Intercept	0.057	0.200
X_1	0.077	0.203
X_2	0.099	0.208
X_3	0.099	0.208
X_4	0.077	0.203
X_5	0.099	0.208
X_6	0.099	0.208
X_1*X_2	0.077	0.203
X_1*X_3	0.077	0.203
X_1*X_4	0.057	0.200
X_1*X_5	0.077	0.203
X_1*X_6	0.077	0.203
X_2*X_3	0.099	0.208
X_2*X_4	0.077	0.203
X_2*X_5	0.099	0.208
X_2*X_6	0.099	0.208
X_3*X_4	0.077	0.203
X_3*X_5	0.099	0.208
X_3*X_6	0.099	0.208
X_4*X_5	0.077	0.203
X_4*X_6	0.077	0.203
X_5*X_6	0.099	0.208

4.6 Fractional Factorial Designs

We observed previously that as the number of factors in a 2^k factorial design increases, the number of runs required for a complete replicate of the design rapidly outgrows the resources of most experimenters. For example, a complete replicate of the 2^6 design requires 64 runs. In this design, only 6 of the 63 degrees of freedom correspond to main effects, and only 15 degrees of freedom correspond to two-factor interactions. There are only 21 degrees of freedom associated with effects that are likely to be of major interest. The remaining 42 degrees of freedom are associated with three-factor and higher-order interactions.

Fractional factorial designs are usually effective in factor screening because of the *sparsity of effects principle* discussed previously. The principle stares that only a fraction (often less than a half) of the

Table 4.11 Correlation matrix for estimates of main effects and two-factor interactions for data in Table 4.9.

Effect	A	B	C	D	E	F	AB	AC	AD	AE	AF	BC	BD	BE	BF	CD	CE	CF	DE	DF	EF
A	1	0.07	0.07	0.07	0.15	0.07	0.01	0.01	0.01	0.08	0.01	0.07	0.07	0.01	0.07	0.07	0.01	0.07	0.01	0.07	0.01
B	0.07	1	0	0	0.07	0	0.07	0.07	0.07	0	0.07	0	0	0.07	0	0.14	0.07	0.14	0.07	0.14	0.07
C	0.07	0	1	0	0.07	0	0.07	0.07	0.07	0	0.07	0	0.14	0.07	0.14	0	0.07	0	0.07	0.14	0.07
D	0.07	0	0	1	0.07	0	0.07	0.07	0.07	0	0.07	0.14	0	0.07	0.14	0	0.07	0.14	0.07	0	0.07
E	0.15	0.07	0.07	0.07	1	0.07	0.01	0.01	0.01	0.08	0.01	0.07	0.07	0.01	0.07	0.07	0.01	0.07	0.01	0.07	0.01
F	0.07	0	0	0	0.07	1	0.07	0.07	0.07	0	0.07	0.07	0.14	0.07	0	0.14	0.07	0	0.07	0	0.07
AB	0.01	0.07	0.07	0.07	0.01	0.07	1	0.01	0.01	0.08	0.01	0.07	0.07	0.15	0.07	0.07	0.01	0.07	0.01	0.07	0.01
AC	0.01	0.07	0.07	0.07	0.01	0.07	0.01	1	0.01	0.08	0.01	0.07	0.07	0.01	0.07	0.07	0.15	0.07	0.01	0.07	0.01
AD	0.01	0.07	0.07	0.07	0.01	0.07	0.01	0.01	1	0.08	0.01	0.07	0.07	0.01	0.07	0.07	0.01	0.07	0.15	0.07	0.01
AE	0.08	0	0	0	0.08	0	0.08	0.08	0.08	1	0.08	0	0	0.08	0	0	0.08	0	0.08	0	0.08
AF	0.01	0.07	0.07	0.07	0.01	0.07	0.01	0.01	0.01	0.08	1	0.07	0.07	0.01	0.07	0.07	0.01	0.07	0.01	0.07	0.15
BC	0.07	0	0	0.14	0.07	0	0.07	0.07	0.07	0	0.07	1	0	0.07	0	0	0.07	0	0.07	0.14	0.07
BD	0.07	0	0.14	0	0.07	0.14	0.07	0.07	0.07	0	0.07	0	1	0.07	0	0	0.07	0.14	0.07	0	0.07
BE	0.01	0.07	0.07	0.07	0.01	0.07	0.15	0.01	0.01	0.08	0.01	0.07	0.07	1	0.07	0.07	0.01	0.07	0.01	0.07	0.01
BF	0.07	0	0.14	0.14	0.07	0	0.07	0.07	0.07	0	0.07	0	0	0.07	1	0.14	0.07	0	0.07	0	0.07
CD	0.07	0.14	0	0	0.07	0.14	0.07	0.07	0.07	0	0.07	0	0	0.07	0.14	1	0.07	0	0.07	0	0.07
CE	0.01	0.07	0.07	0.07	0.01	0.07	0.01	0.15	0.01	0.08	0.01	0.07	0.07	0.01	0.07	0.07	1	0.07	0.01	0.07	0.01
CF	0.07	0.14	0	0.14	0.07	0	0.07	0.07	0.07	0	0.07	0	0.14	0.07	0	0	0.07	1	0.07	0	0.07
DE	0.01	0.07	0.07	0.07	0.01	0.07	0.01	0.01	0.15	0.08	0.01	0.07	0.07	0.01	0.07	0.07	0.01	0.07	1	0.07	0.01
DF	0.07	0.14	0.14	0	0.07	0	0.07	0.07	0.07	0	0.07	0.14	0	0.07	0	0	0.07	0	0.07	1	0.07
EF	0.01	0.07	0.07	0.07	0.01	0.07	0.01	0.01	0.01	0.08	0.15	0.07	0.07	0.01	0.07	0.07	0.01	0.07	0.01	0.07	1

potential factors of interest in any system are actually important. There are a few other concepts that have proven useful in factor screening. First is the assumption of *effect hierarchy*. The hierarchy of effects assumption states that main effects tend to be more common and larger than second-order effects like two-factor interactions and quadratic effects, which, in turn, tend to be larger and more common than three-factor interactions. Second is the assumption of *effect heredity*. The effects heredity assumption is that a two-factor interaction is most probable if both of its associated factors have significant main effects. Such models exhibit *strong heredity*. Much less probable is a two-factor interaction where only one of the associated main effects is significant. These models exhibit *weak heredity*. Least probable are two-factor interactions that present themselves when neither associated main effect is significant. Every experienced experimenter can describe at least one incident where this does not happen and a high-order interaction such as a three-factor interaction is active. However, it is important to remember that this happenstance is an exception that doesn't occur frequently.

If the experimenter can reasonably assume that certain high-order interactions are negligible (that is, sparsity of effects), information on the main effects and low-order interactions may be obtained by running only a fraction of the complete factorial experiment. In this section we present the class of 2^{k-p} *fractional factorial designs*. These are fractional factorials for k factors that are $1/2^p$ fractions of the original full factorial. These are usually called one-half fractions ($p = 1$), one-quarter fractions ($p = 2$), one-eighth fractions ($p = 3$), and so on, and they are among the most widely used types of designs for product and process design, process improvement, and industrial/business experimentation.

An important application of fractional factorials is in screening experiments–experiments in which many factors are considered and the objective is to identify those factors (if any) that have large effects. Screening experiments are usually performed in the early stages of a project when many of the factors initially considered likely have little or no effect on the response. The factors identified as important are then investigated more thoroughly in subsequent experiments.

As an example of a fractional factorial design, Table 4.12 shows the 2^{5-1}, a one-half fraction of the 2^5. This design was constructed by first writing down the full factorial that has the desired number of runs (in this case, the 16-run full 2^4 factorial) and then equating the levels of the fifth factor to the product of the signs in the first four columns (that is, the four-factor interaction of those main effects). The equation $E = ABCD$ is called the generator of the fraction, and this approach to design construction is called the column generator approach. Imagine writing down the 16 runs, i.e. rows of ±1, for all of the runs in a 2^4 design; then for column E simply multiply the -1s or $+1$s, which will give $+1$ if the number of -1s is even, and -1 if the number of -1s is odd. This design is D-optimal for either the main effects model or the model containing the main effects and any number of two-factor interactions, so it can also be constructed algorithmically.

Multiplying both sides of the column generator by E produces

$$E^2 = ABCDE,$$

but since all the entries in the column E are -1 or $+1$, E^2 will always be a column of ones, which is called the *identity column* and is denoted I. Thus, we can write

$$I = ABCDE.$$

since the square of column E contains only plus signs and by convention a column of all plus signs is called the identity column I. The relationship $I = ABCDE$ is called the **defining relation** of the fraction.

Table 4.12 The one-half fraction of the 2^5 design (denoted 2^{5-1}) defined by $E = ABCD$.

Run	Basic design A	B	C	D	$E = ABCD$	Treatment combination
1	−	−	−	−	+	e
2	+	−	−	−	−	a
3	−	+	−	−	−	b
4	+	+	−	−	+	abe
5	−	−	+	−	−	c
6	+	−	+	−	+	ace
7	−	+	+	−	+	bce
8	+	+	+	−	−	abc
9	−	−	−	+	−	d
10	+	−	−	+	+	ade
11	−	+	−	+	+	bde
12	+	+	−	+	−	abd
13	−	−	+	+	+	cde
14	+	−	+	+	−	acd
15	−	+	+	+	−	bcd
16	+	+	+	+	+	$abcde$

This 16-run design has 15 degrees of freedom that can be used to estimate factor effects. However, with 5 factors, there are 5 main effects, 10 two-factor interactions, 10 three-factor interactions, 5 four-factor interactions, and 1 five-factor interaction. Clearly, not all of these effects can be uniquely estimated. Suppose that we estimate the five main effects. We can show that when we estimate the main effects we also estimate one of the four-factor interactions along with each main effect:

$$A = A + BCDE,$$
$$B = B + ACDE,$$
$$C = C + ABDE,$$
$$D = D + ABCE,$$
$$E = E + ABCD.$$

Thus, each main effect estimate is **aliased** by a four-factor interaction. Notice that each alias is the four-factor interaction that does not contain that main effect. The alias relationships can be found from the defining relation for the design. Simply multiply each main effect by both sides of the defining relation by the main effect and simplify, as in:

$$A(I) = A(ABCDE),$$

$$A = A^2BCDE,$$

$$A = BCDE.$$

We say that the A main effect is *confounded* with the $BCDE$ four-factor interaction.

Now suppose that we estimate the two-factor interactions. The aliases can be found by the same technique. For example, the alias of the AB interaction is $AB(I) = AB(ABCDE)$, or $AB = A^2B^2CDE = CDE$. It is easy to see that the alias of any two-factor interaction is the three-factor interaction that does not involve the two letters that define the original two-factor interaction.

The 2^{5-1} fractional factorial design is an example of a **Resolution V** design; that is, a design in which the main effects are clear of **aliasing** or **confounding** with two-factor interactions, and two-factor interactions are aliased or confounded with interactions of order three or higher. If cost is not an object, a Resolution V design is ideal for factor screening, as in many of these studies interest is focused on main effects and two-factor interactions and these are estimated clear of any other effects assuming that interactions of order three and higher and not active. The design in Table 4.12 is a **principal fraction**. An **alternate fraction** could be obtained by switching the signs in column E. This alternate fraction would have a column generator $E = -ABCD$, and a defining relation $I = -ABCDE$. Using the alternate fraction simply changes the sign in the alias relationship; for example, in the alternate fraction the alias of A is $A = -BCDE$. The two fractions of the 2^5 with $I = \pm ABCDE$ are **regular** fractions, because the constants in the alias relationships are either ± 1. This means that aliased terms are **completely confounded**. If two effects are confounded, there is no data driven way to determine which effect is causal.

There is no six-factor regular fractional factorial design that is Resolution V. However, the six-factor 2^{6-1} design, which is a one-half fraction with 32 runs, is Resolution VI. This design is constructed by starting with the 2^5 full factorial as the basic design and using the column generator $F = \pm ABCDE$. The defining relation is $I = \pm ABCDEF$, and each main effect is aliased with a single five-factor interaction. The two-factor interactions are aliased with one of the four-factor interactions. Similarly, there is no seven-factor Resolution V design, but the 64 run seven-factor 2^{7-1} one-half fraction is Resolution VII. This design starts with the full 2^6 as the basic design and using the column generator $G = \pm ABCDEF$, resulting in the defining relation $I = \pm ABCDEFG$.

The column generators that we have shown are selected to ensure that the designs created have the highest possible resolution for the specified number of runs. These designs are widely available in software. Most software packages default to the principal fraction of highest possible resolution when a regular fractional factorial design is selected.

As another example of a fractional factorial design, consider the eight-run design for five two-level factors in Table 4.9. This design was constructed using the column generator method starting with the full 2^3 factorial design and using the generators $D = +ABC$ and $E = +BC$, or $I = +ABCD$ and $I = +BCE$. The complete defining relation consists of $I = +ABCD$ and $I = +BCE$ and their product, or

$$I = +ABCD = +BCE = +ADE.$$

The complete defining relation has three "words" so each effect will have three aliases. For example, the alias of A is $A = +BCD = +ABCE = +DE$, so the main effect of A is aliased with a two-factor interaction, DE. It is easy to see that every main effect is aliased with at least one two-factor interaction. This design is shown in Table 4.13.

Table 4.13 The 2^{5-2} design defined by $D = +ABC$ and $E = +BC$.

A	B	C	D = +ABC	E = +BC
1	1	−1	−1	−1
1	−1	1	−1	−1
−1	1	−1	1	−1
−1	−1	1	1	−1
−1	−1	−1	−1	1
−1	1	1	−1	1
1	−1	−1	1	1
1	1	1	1	1

This is an example of a Resolution III design. In general, a Resolution III design aliases at least one two-factor interaction with at least one main effect. The aliases of the design in Table 4.9 are

$$A = +DE,$$
$$B = +CE,$$
$$C = +BE,$$
$$D = +AE,$$
$$E = +AD = +BC,$$
$$AB = +CD,$$
$$AC = +BD,$$

where we ignore interactions of order three and higher. All the column generators used in this example were positive so this is a *principal fraction*. There are three other alternate fractions that are obtained by simply changing the signs on the column generators. This would produce corresponding sign changes on some of the aliases.

The resolution of a regular fractional factorial design can be found directly from the number of letters in the shortest word in the complete defining relation. The generators selected in this example guarantee that the design has the highest possible resolution. Note also that the three word lengths in the defining relation for this design have the **pattern** 4,3,3. This design minimizes the number of words in the defining relation that are of minimum length. Such a design is called a **minimum aberration design**. Minimizing aberration in a design of resolution R ensures that the design has the minimum number of main effects aliased with interactions of order $R - 1$, the minimum number of two-factor interactions aliased with interactions of order $R - 2$, and so on. It is a very nice property. When computer software is used to select a regular fractional factorial design, the generators are chosen to maximize resolution and minimize aberration.

Regular Resolution III designs are available for 5, 6, and 7 factors in 8 runs and for 9 to 15 factors in 16 runs. The column generators for these designs are shown in Table 4.14. These generators ensure that the resulting deigns maximize resolution and minimize aberration. In some cases, the generators are

Table 4.14 Regular Resolution III or higher designs in 8 and 16 runs.

Number of factors	Fraction	Number of runs	Column generators
5	2^{5-2}	8	$D = \pm ABC,\ E = \pm BC$
6	2^{6-3}	8	$D = \pm AB,\ E = \pm AC,\ F = \pm BC$
7	2^{7-4}	8	$D = \pm AB,\ E = \pm AC,$ $F = \pm BC,\ G = \pm ABCD$
8	2^{8-4}	16	$E = \pm BCD, F = \pm ACD$ $G = \pm ABC,\ H = \pm ABCD$
9	2^{9-5}	16	$E = \pm ABC,\ F = \pm BCD,$ $G = \pm ACD,\ H = \pm ABD,$ $J = \pm ABCD$
10	2^{10-6}	16	$E = \pm ABC,\ F = \pm BCD,$ $G = \pm ACD,\ H = \pm ABD,$ $J = \pm ABCD,\ K = \pm AB$
11	2^{11-7}	16	$F = \pm BCD,\ G = \pm ACD,$ $H = \pm ABD,\ J = \pm ABCD,$ $K = \pm AB,\ L = \pm AC$
12	2^{12-8}	16	$E = \pm ABC,\ F = \pm ABD,$ $G = \pm ACD,\ H = \pm BCD,$ $J = \pm ABCD,\ K = \pm AB,$ $L = \pm AC,\ M = \pm AD$
13	2^{13-9}	16	$E = \pm ABC,\ F = \pm ABD,\ G = \pm ACD,$ $H = \pm BCD,\ J = \pm ABCD,\ K = \pm AB,$ $L = \pm AC,\ M = \pm AD,\ N = \pm BC$
14	2^{14-10}	16	$E = \pm ABC, F = \pm ABD,$ $G = \pm ACD,\ H = \pm BCD,$ $J = \pm ABCD,\ K = \pm AB,\ L = \pm AC,$ $M = \pm AD,\ N = \pm BC,\ O = \pm BD$
15	2^{15-11}	16	$E = \pm ABC,\ F = \pm ABD,\ G = \pm ACD,$ $H = \pm BCD,\ J = \pm ABCD$ $K = \pm AB,\ L = \pm AC,\ M = \pm AD,$ $N = \pm BC,\ O = \pm BD,\ P = \pm CD$

Table 4.15 The 2^{4-1} fractional factorial design.

Run	Basic design			$D = ABC$
	A	*B*	*C*	
1	−1	−1	−1	−1
2	1	−1	−1	1
3	−1	1	−1	1
4	1	1	−1	−1
5	−1	−1	1	1
6	1	−1	1	−1
7	−1	1	1	−1
8	1	1	1	1

not unique; that is, a different set of generators will produce a design having the same resolution and aberration.

As another example of a fractional factorial design, consider the one-half fraction of the 2^4 (a 2^{4-1} design) in Table 4.15. This design was constructed using the column generator $D = ABC$, so the defining contrast is $I = ABCD$. Therefore, the design is Resolution IV. In Resolution IV designs, main effects are aliased with three-factor and higher-order interactions, and two-factor interactions are aliased with each other. For example, in the design in Table 4.15 the alias of the main effect A is $A = BCD$ and the alias of the AB interaction is $AB = CD$.

The 2^{4-1} is the only regular fractional factorial with eight runs. Resolution IV fractions with 16 runs are available for 6, 7, and 8 factors. The 2^{9-3} design, the 2^{10-4} and the 2^{11-5} design with 64 runs and the 2^{7-3} design, the 2^{8-3} design, the 2^{9-4} design, the 2^{10-5} design, and the 2^{11-6} design with 32 runs are Resolution IV designs. The generators for these designs are shown in Table 4.16. These generators assure maximum resolution and minimum aberration, although other choices of generators will produce equivalent designs.

4.6.1 A General Method for Finding the Alias Relationships in Fractional Factorial Designs

We have shown how to find the alias relationships in a 2^{k-p} fractional factorial design by use of the complete defining relation. This method works well in regular fractional factorial designs, but it doesn't work in other cases, such as fractional factorials that do not have defining relations or designs that are not fractions but these designs are used in situations where the experimenter may be concerned about effects of higher-order than are supported by the design that was used (Table 4.16).

Fortunately, there is a general method available that works satisfactorily in many situations. The method uses the polynomial or regression model representation of the model, say,

$$y = X_1\beta_1 + \epsilon,$$

Table 4.16 Some regular Resolution IV or higher fractional factorial designs.

Number of factors	Fraction	Number of runs	Column generators
4	2^{4-1}	8	$D = \pm ABC$
6	2^{6-2}	16	$E = \pm ABC,\ F = \pm BCD$
7	2^{7-2}	32	$F = \pm ABCD,\ G = \pm ABDE$
7	2^{7-3}	16	$E = \pm ABC,\ F = \pm BCD,\ G = \pm ACD$
8	2^{8-3}	32	$F = \pm ABC,\ G = \pm ABD,\ H = \pm BCDE$
8	2^{8-4}	16	$E = \pm BCD,\ F = \pm ACD,$ $G = \pm ABC,\ H = \pm ABD,$
9	2^{9-3}	64	$G = \pm ABCD,\ H = \pm ACEF,$ $J = \pm CDEF$
9	2^{9-4}	32	$F = \pm BCDE,\ G = \pm ACDE,$ $H = \pm ABDE,\ J = \pm ABCE$
10	2^{10-4}	64	$G = \pm BCDF,\ H = \pm ACDF,$ $J = \pm ABDE,\ K = \pm ABCE$
10	2^{10-5}	32	$F = \pm ABCD,\ G = \pm ABCE,$ $H = \pm BCD,\ J = \pm ABCD$ $K = \pm BCDE$
11	2^{11-5}	64	$G = \pm CDE,\ H = \pm ABCD,\ J = \pm ABF$ $K = \pm BDEF,\ L = \pm ADEF$
11	2^{11-6}	32	$F = \pm ABC,\ G = \pm BCD,\ H = \pm CDE,$ $J = \pm ACD,\ K = \pm ADE,\ L = \pm BDE$

where $\boldsymbol{y}$ is an $n \times 1$ vector of the responses, $\boldsymbol{X}$ is an $n \times p_1$ matrix containing the design matrix expanded to the form of the model that the experimenter is fitting, $\boldsymbol{\beta}_1$ is a $p_1 \times 1$ vector of the model parameters, and ϵ is an $n \times 1$ vector of errors. The least squares estimate of $\boldsymbol{\beta}$ is

$$\hat{\boldsymbol{\beta}} = (\boldsymbol{X}_1' \boldsymbol{X}_1)^{-1} \boldsymbol{X}_1' \boldsymbol{y}.$$

Suppose that the true model is

$$\boldsymbol{y} = \boldsymbol{X}_1 \boldsymbol{\beta}_1 + \boldsymbol{X}_2 \boldsymbol{\beta}_2 + \epsilon,$$

where X_2 is an $n \times p_2$ matrix containing additional variables that are not in the model and $\boldsymbol{\beta_2}$ is a $p_2 \times 1$ vector of the parameters associated with these variables. It can be shown that

$$E(\hat{\boldsymbol{\beta}}_1) = \boldsymbol{\beta}_1 + (\boldsymbol{X}_1'\boldsymbol{X}_1)^{-1}\boldsymbol{X}_1'\boldsymbol{X}_2\boldsymbol{\beta}_2$$
$$= \boldsymbol{\beta}_1 + \boldsymbol{A}\boldsymbol{\beta}_2\boldsymbol{\beta}_2.$$

The matrix $\boldsymbol{A}$ is called the **alias matrix**. The elements of this matrix operating on $\boldsymbol{\beta}_2$ identify the alias relationships for the parameters in the vector $\boldsymbol{\beta}_1$.

We illustrate the application of this procedure with a simple example. Suppose that we have conducted a 2^{3-1} design with defining relation $I = ABC$. The model that the experimenter plans to fit is the main-effects-only model

$$y = \beta_0 + \beta_1 x_1 + \beta_2 x_2 + \beta_3 x_3 + \epsilon.$$

In the notation defined earlier

$$\boldsymbol{\beta_1} = \begin{bmatrix} \beta_0 \\ \beta_1 \\ \beta_2 \\ \beta_3 \end{bmatrix} \quad \text{and} \quad \boldsymbol{X_1} = \begin{bmatrix} 1 & -1 & -1 & 1 \\ 1 & 1 & -1 & -1 \\ 1 & -1 & 1 & -1 \\ 1 & 1 & 1 & 1 \end{bmatrix}.$$

Suppose that the true model contains all the two-factor interactions, so that

$$y = \beta_0 + \beta_1 x_1 + \beta_2 x_2 + \beta_3 x_3 + \beta_{12} x_1 x_2 + \beta_{13} x_1 x_3 + \beta_{23} x_2 x_3 + \epsilon,$$

and

$$\boldsymbol{\beta_2} = \begin{bmatrix} \beta_{12} \\ \beta_{13} \\ \beta_{23} \end{bmatrix} \quad \text{and} \quad \boldsymbol{X_2} = \begin{bmatrix} 1 & -1 & -1 \\ -1 & -1 & 1 \\ -1 & 1 & -1 \\ 1 & 1 & 1 \end{bmatrix}.$$

Now

$$(\boldsymbol{X}_1'\boldsymbol{X}_1) = 4\boldsymbol{I}_4 \quad \text{and} \quad (\boldsymbol{X}_1'\boldsymbol{X}_2) = \begin{bmatrix} 0 & 0 & 0 \\ 0 & 0 & 4 \\ 0 & 4 & 0 \\ 4 & 0 & 0 \end{bmatrix}.$$

Therefore,

$$(\boldsymbol{X}_1'\boldsymbol{X}_1)^{-1} = \frac{1}{4}\boldsymbol{I}_4$$

and

$$E(\boldsymbol{\beta}_1) = \boldsymbol{\beta}_1 + \boldsymbol{A}\boldsymbol{\beta}_2$$

$$= \begin{bmatrix} \beta_0 \\ \beta_1 \\ \beta_2 \\ \beta_3 \end{bmatrix} + \frac{1}{4}\boldsymbol{I}_4 \begin{bmatrix} 0 & 0 & 0 \\ 0 & 0 & 4 \\ 0 & 4 & 0 \\ 4 & 0 & 0 \end{bmatrix} \begin{bmatrix} \beta_{12} \\ \beta_{13} \\ \beta_{23} \end{bmatrix}$$

$$
= \begin{bmatrix} \beta_0 \\ \beta_1 \\ \beta_2 \\ \beta_3 \end{bmatrix} + \begin{bmatrix} 0 & 0 & 0 \\ 0 & 0 & 1 \\ 0 & 1 & 0 \\ 1 & 0 & 0 \end{bmatrix} \begin{bmatrix} \beta_{12} \\ \beta_{13} \\ \beta_{23} \end{bmatrix}
$$

$$
= \begin{bmatrix} \beta_0 \\ \beta_1 \\ \beta_2 \\ \beta_3 \end{bmatrix} + \begin{bmatrix} 0 \\ \beta_{12} \\ \beta_{13} \\ \beta_{23} \end{bmatrix}
$$

$$
= \begin{bmatrix} \beta_0 \\ \beta_1 + \beta_{12} \\ \beta_2 + \beta_{13} \\ \beta_3 + \beta_{23} \end{bmatrix}.
$$

The interpretation of this, of course, is that each of the main effects is aliased with one of the two-factor interactions, which we know to be the case for this design. Notice that every row of the alias matrix represents one of the factors in β_1 and every column represents one of the factors in β_2. While this is a very simple example, the method is very general and can be applied to much more complex designs.

Software can display the alias matrix for fractional factorial designs. Table 4.17 shows the alias matrix from JMP for the 2^{3-1} fractional factorial design with $I = ABC$. Notice that as we saw earlier that every main effect is aliased with the two-factor interaction with which it is not involved. Additionally, we see that the intercept estimate is biased by the three-factor interaction ABC.

4.6.2 De-aliasing Effects

When analyzing a regular fractional factorial design, it is not unusual to find that one or more factor effects of potential interest are aliased. In regular designs aliased effects are completely confounded, and their individual effects cannot be estimated from the data supplied by the design. For example, in the 2^{5-2} fractional factorial design in Table 4.9, suppose that the largest effect estimates are for A, B, C, and the AB interaction. A relatively simple interpretation of these results is that these are the active factors, but AB is aliased with CD, so it is certainly possible that the large AB effect is attributable to either AB or CD, or to both AB and CD. It is possible to de-alias effects in situations like this by augmenting the original fraction with additional runs.

Table 4.17 Alias matrix for the 2^{3-1} design with $I = ABC$.

Effect	*AB*	*AC*	*BC*	*ABC*
Intercept	0	0	0	1
A	0	0	1	0
B	0	1	0	0
C	1	0	0	0

A simple type of augmentation scheme that can be very effective for some Resolution III designs is fold-over. A full fold-over of a Resolution III design consists of performing an additional group of runs in which the sign of every run in the original fraction is reversed. This produces a combined design with $2(2^{k-p}) = 2^{k-p+1}$ runs. In this combined set of runs all main effects are completely separated from their two-factor interaction aliases. This procedure results in a combined design that is now Resolution IV. Unfortunately, this method will not be useful here, because we have a pair of aliased two-factor interactions, so after a full fold-over they will still be aliased. Also, the full fold-over technique will only work with Resolution III designs. There are variations of the fold-over technique that can be useful. The single-factor fold-over reverses signs in a single column only. This can be useful in Resolution IV designs to de-alias all of the two-factor interactions associated with that column. A significant disadvantage of fold-overs is that they require a follow-on experiment that is as large as the original one. Partial fold-overs can alleviate that problem, as they use only half of the runs associated with a full fold-over. For a complete discussion of the fold-over technique and some examples, see Montgomery (2020).

The runs to add to a fractional factorial design to de-alias terms of interest can be determined by using a D-optimal augmentation procedure. This uses the D-criterion to determine the set of runs to add to the original fraction to fit the desired model, which would not include the terms originally aliased. The output from JMP illustrating the original and the augmented design, where we have elected to add four runs to the original eight-run fraction and to place the additional four runs in a separate block, is shown in Table 4.18. The alias matrix for this procedure, shown in Table 4.19, indicates that AB and CE are no longer completely confounded, and that all of the main effects are partially confounded with two-factor interactions. The power to detect effect sizes of two standard deviations is shown in Table 4.20 and is approximately 0.6. This is relatively low power, but about all one can expect from an eight-run design.

Table 4.18 Augmented design, including original eight runs and the four added runs.

Run	A	B	C	D	E	Block
1	1	1	−1	−1	−1	1
2	1	−1	1	−1	−1	1
3	−1	1	−1	1	−1	1
4	−1	−1	1	1	−1	1
5	−1	−1	−1	−1	1	1
6	−1	1	1	−1	1	1
7	1	−1	−1	1	1	1
8	1	1	1	1	1	1
9	−1	−1	1	−1	1	2
10	−1	1	1	1	−1	2
11	1	−1	−1	1	−1	2
12	1	1	−1	−1	1	2

Table 4.19 Alias matrix for augmented design.

Effect	$A \times C$	$A \times D$	$A \times E$	$B \times C$	$B \times D$	$B \times E$	$C \times D$	$D \times E$	$A \times B$	$C \times E$
Intercept	−0.5	0	0	0	0	0	0	−0.5	0	0
A	0	0	0	0	−0.25	0.5	0	0.75	0	0
B	0	−0.5	0.5	0	0	0	0.5	0	0	0
C	0	0	0	0	0	0.25	0.5	0.25	0	0
D	0	0.2	0.8	0.4	0	0	0.2	0	0	0
E	0	0.8	0.2	0.6	0	0	−0.2	0	0	0
$A \times B$	0	−0.2	0.2	−0.4	0	0	0.8	0	1	0
$C \times E$	0	0.5	−0.5	0	0	0	−0.5	0	0	1
Block	0.5	0	0	0	0	0	0	0.5	0	0

Table 4.20 Power analysis assuming $\alpha = 0.05$ and an RMSE equal to 1.

Term	Anticipated coefficient	Power
Intercept	1	0.597
A	1	0.597
B	1	0.597
C	1	0.597
D	1	0.573
E	1	0.573
$A \times B$	1	0.573
$C \times E$	1	0.597
Block	1	0.597

Computed using JMP.

Problems

4.1 An engineer is interested in the effects of cutting speed (A), tool geometry (B), and cutting angle (C) on the life (in hours) of a machine tool. Two levels of each factor are chosen, and three replicates of a 2^3 factorial design are run. The results are as follows:

A	B	C	Treatment combination	Replicate		
				Rep. 1	Rep. 2	Rep. 3
−	−	−	(1)	22	31	25
+	−	−	a	32	43	29
−	+	−	b	35	34	50
+	+	−	ab	55	47	46
−	−	+	c	44	45	38
+	−	+	ac	40	37	36
−	+	+	bc	60	50	54
+	+	+	abc	39	41	47

(a) Estimate all of the factor main effects. Which effects appear to be large?

(b) Use the analysis of variance to confirm your conclusions for part (a).

(c) Write down a regression model for predicting tool life in hours based on the results of this experiment.

(d) Analyze the residuals. Are there any obvious problems?

(e) On the basis of analysis of main effect and interaction plots, what coded factor levels of A, B, and C would you recommend using?

4.2 Reconsider part (c) of Problem 4.1. Use the regression model to generate response surface and contour plots of the tool life response. Interpret these plots. Do they provide insight regarding the desirable operating conditions for this process?

4.3 An experiment was performed to improve the yield of a chemical process. Four factors were selected, and two replicates of a completely randomized experiment were run. The results are shown in the following table:

Treatment combination	Replicate		Treatment condition	Replicate	
	Rep. 1	Rep. 2		Rep. 1	Rep. 2
(1)	90	93	d	98	95
a	74	78	ad	72	76
b	81	85	bd	87	83
ab	83	80	abd	85	86
c	77	78	cd	99	90
ac	81	80	acd	79	75
bc	88	82	bcd	87	84
abc	73	70	abcd	80	80

(a) Estimate the factor effects. Prepare an analysis of variance table and determine which factors are important in explaining yield.

(b) Write down a regression model for predicting yield, assuming that all four factors were varied over the range from -1 to $+1$ (in coded units).

(c) Plot the residuals versus the predicted yield and on a normal probability scale. Does the residual analysis appear satisfactory?

(d) Two three-factor interactions, ABC and ABD, apparently have large effects. Draw a cube plot in the factors A, B, and C with the average yields shown at each corner. Repeat using the factors A, B, and D. Do these two plots aid in data interpretation? Where would you recommend running the process with respect to the four variables?

4.4 Suppose that you need to plan an experiment with six factors, each at two levels. You need to estimate all 6 main effects, and all 15 two-factor interactions. The smallest regular fractional factorial that can do this is the 2^{6-1} design, with 32 runs. However, the experimental budget is not sufficient for this many runs. The minimum number of runs that can accomplish the objectives is 22.

(a) Using software, find a D-optimal design for this problem with 22 runs.

(b) Suppose that the budget will allow two more runs. Using software, find a D-optimal design with 24 runs. Compare this design to the one you found in part a. Explain how you might justify the additional expense associated with the larger experiment.

4.5 Snee et al. (1985) describe an experiment in which a 2^{5-1} design with $I = ABCDE$ was used to investigate the effects of five factors on the color of a chemical product. The factors are $A = $ solvent/reactant, $B = $ catalyst/reactant, $C = $ temperature, $D = $ reactant purity, and $E = $ reactant pH. The responses obtained are as follows:

$e = -0.63$	$d = 6.79$
$a = 2.51$	$ade = 5.47$
$b = -2.68$	$bde = 3.45$
$abe = 1.66$	$abd = 5.68$
$c = 2.06$	$cde = 5.22$
$ace = 1.22$	$acd = 4.38$
$bce = -2.09$	$bcd = 4.30$
$abc = 1.93$	$abcde = 4.05$

(a) Which effects seem active?

(b) Calculate the residuals. Construct a normal probability plot of the residuals and plot the residuals versus the fitted values. Comment on the plots.

(c) If any factors are negligible, collapse the fractional design into a full factorial in the active factors. Comment on the resulting design, and interpret the results.

4.6 An article in *Industrial and Engineering Chemistry* (Hill and Demler 1970) uses a 2^{5-2} design to investigate the effect of $A = $ condensation temperature, $B = $ amount of material 1, $C = $ solvent volume, $D = $ condensation time, and $E = $ amount of material 2 on the response variable yield.

The results obtained are as follows:

$$e = 23.2, \quad ad = 16.9, \quad cd = 23.8, \quad bde = 16.8$$
$$ab = 15.5, \quad bc = 16.2, \quad ace = 23.4, \quad abcde = 18.1$$

(a) Verify that the design generators used were $I = ACE$ and $I = BDE$.
(b) Write down the complete defining relation and the aliases for this design.
(c) Estimate the main effects.
(d) Prepare an analysis of variance table. Verify that the AB and AD interactions are available to use as error.

4.7 Construct a 2^{7-2} design by choosing two four-factor interactions as the independent generators. Write down the complete alias structure for this design. Outline the analysis of variance table. What is the resolution of this design?

4.8 Construct a 2^{5-1} design. Show how the design may be run in two blocks of eight observations each. Are any main effects or two-factor interactions confounded with blocks?

4.9 Consider a 2^4 design. We must estimate the four main effects and the six two-factor interactions, but the full 16-run factorial cannot be run. The largest possible block size contains 12 runs. These 12 runs can be obtained from the four one-quarter replicates defined by $I = \pm AB = \pm ACD = \pm BCD$ by omitting the principal fraction. Show how the remaining three fractions can be combined to estimate the required effects, assuming three-factor and higher interactions are negligible. This design could be thought of as a three-quarter fraction.

4.10 Reconsider the design situation in Problem 4.9. Construct a D-optimal design for this situation. Compare your design with the one from Problem 4.9.

4.11 A 16-run experiment was performed in a semiconductor manufacturing plant to study the effects of six factors on the curvature or camber of the substrate devices produced. The six variables and their levels are shown in the following tables.
(a) What type of design did the experimenters use?
(b) What are the alias relationships in this design?
(c) Do any of the process variables affect average camber?
(d) Do any of the process variables affect the variability in camber measurements?
(e) If it is important to reduce camber as much as possible, what recommendations would you make?

Run	Lamination temp (°C)	Lamination time (s)	Lamination pressure (tn)	Firing temp (°C)	Firing cycle time (h)	Firing dew point (°C)
1	55	10	5	1580	17.5	20
2	75	10	5	1580	29	26
3	55	25	5	1580	29	20
4	75	25	5	1580	17.5	26
5	55	10	10	1580	29	26
6	75	10	10	1580	17.5	20
7	55	25	10	1580	17.5	26
8	75	25	10	1580	29	20
9	55	10	5	1620	17.5	26
10	75	10	5	1620	29	20
11	55	25	5	1620	29	26
12	75	25	5	1620	17.5	20
13	55	10	10	1620	29	20
14	75	10	10	1620	17.5	26
15	55	25	10	1620	17.5	20
16	75	25	10	1620	29	26

	Camber for replicate (in./in.)				Total (10^{-4} in./in)	Mean (10^{-4} in./in.)	Standard deviation
Run	1	2	3	4			
1	0.0167	0.0128	0.0147	0.0185	629	157.25	24.418
2	0.0062	0.0066	0.0044	0.0020	192	48.00	20.976
3	0.0041	0.0043	0.0042	0.0050	176	44.00	4.083
4	0.0073	0.0081	0.0039	0.0030	223	55.75	25.025
5	0.0047	0.0047	0.0040	0.0089	223	55.75	22.410
6	0.0219	0.0258	0.0147	0.0296	920	230.00	63.639
7	0.0121	0.0090	0.0092	0.0086	389	97.25	16.029
8	0.0255	0.0250	0.0226	0.0169	900	225.00	39.420
9	0.0032	0.0023	0.0077	0.0069	201	50.25	26.725
10	0.0078	0.0158	0.0060	0.0045	341	85.25	50.341

(Continued)

	Camber for replicate (in./in.)				Total	Mean	Standard
Run	1	2	3	4	$(10^{-4}$ in./in)	$(10^{-4}$ in./in.)	deviation
11	0.0043	0.0027	0.0028	0.0028	126	31.50	7.681
12	0.0186	0.0137	0.0158	0.0159	640	160.00	20.083
13	0.0110	0.0086	0.0101	0.0158	455	113.75	31.12
14	0.0065	0.0109	0.0126	0.0071	371	92.75	29.51
15	0.0155	0.0158	0.0145	0.0145	603	150.75	6.75
16	0.0093	0.0124	0.0110	0.0133	460	115.00	17.45

4.12 A spin coater is used to apply photoresist to a bare silicon wafer. This operation usually occurs early in the semiconductor manufacturing process, and the average coating thickness and the variability in the coating thickness have an important impact on downstream manufacturing steps. Six variables are used in the experiment. The variables and their high and low levels are as follows:

Factor	Low level	High level
Final spin speed	7350 rpm	6650 rpm
Acceleration rate	5	20
Volume of resist applied	3 cc	5 cc
Time of spin	14 s	6 s
Resist batch variation	Batch 1	Batch 2
Exhaust pressure	Cover off	Cover on

The experimenter decides to use a 2^{6-1} design and to make three readings on resist thickness on each test wafer. The design and the data are shown in the following table.

	A	B	C	D	E	F	Resist thickness				
Run	Vol.	Batch	Time (s)	Speed	Acc.	Cover	Left	Center	Right	Mean	Range
1	5	2	14	7350	5	Off	4531	4531	4515	4525.7	16
2	5	1	6	7350	5	Off	4446	4464	4428	4446.0	36
3	3	1	6	6650	5	Off	4452	4490	4452	4464.7	38
4	3	2	14	7350	20	Off	4316	4328	4308	4317.3	20
5	3	1	14	7350	5	Off	4307	4295	4289	4297.0	18
6	5	1	6	6650	20	Off	4470	4492	4495	4485.7	25

	A	B	C	D	E	F	Resist thickness				
Run	Vol.	Batch	Time (s)	Speed	Acc.	Cover	Left	Center	Right	Mean	Range
7	3	1	6	7350	5	On	4496	4502	4482	4493.3	20
8	5	2	14	6650	20	Off	4542	4547	4538	4542.3	9
9	5	1	14	6650	5	Off	4621	4643	4613	4625.7	30
10	3	1	14	6650	5	On	4653	4670	4645	4656.0	25
11	3	2	14	6650	20	On	4480	4486	4470	4478.7	16
12	3	1	6	7350	20	Off	4221	4233	4217	4223.7	16
13	5	1	6	6650	5	On	4620	4641	4619	4626.7	22
14	3	1	6	6650	20	On	4455	4480	4466	4467.0	25
15	5	2	14	7350	20	On	4255	4288	4243	4262.0	45
16	5	2	6	7350	5	On	4490	4534	4523	4515.7	44
17	3	2	14	7350	5	On	4514	4551	4540	4535.0	37
18	3	1	14	6650	20	Off	4494	4503	4496	4497.7	9
19	5	2	6	7350	20	Off	4293	4306	4302	4300.3	13
20	3	2	6	7350	5	Off	4534	4545	4512	4530.3	33
21	5	1	14	6650	20	On	4460	4457	4436	4451.0	24
22	3	2	6	6650	5	On	4650	4688	4656	4664.7	38
23	5	1	14	7350	20	On	4231	4244	4230	4235.0	14
24	3	2	6	7350	20	On	4225	4228	4208	4220.3	20
25	5	1	14	7350	5	On	4381	4391	4376	4382.7	15
26	3	2	6	6650	20	Off	4533	4521	4511	4521.7	22
27	3	1	14	7350	20	On	4194	4230	4172	4198.7	58
28	5	2	6	6650	5	Off	4666	4695	4672	4677.7	29
29	5	1	6	7350	20	On	4180	4213	4197	4196.7	33
30	5	2	6	6650	20	On	4465	4496	4463	4474.7	33
31	5	2	14	6650	5	On	4653	4685	4665	4667.7	32
32	3	2	14	6650	5	Off	4683	7412	4677	4690.7	35

(a) Verify that this is a 2^{6-1} design. Discuss the alias relationships in this design.

(b) What factors appear to affect average resist thickness?

(c) Because the volume of resist applied has little effect on average thickness, does this have any important practical implications for the process engineers?

4.13 An article in *Soldering and Surface Mount Technology* (Poon and Williams 1999) describes the use of a 2^{8-3} fractional factorial experiment to study the effect of eight factors on two responses; percentage volume matching (PVM) – the ratio of the actual printed solder paste volume to the designed volume; and nonconformities per unit (NPU) – the number of solder paste printing defects determined by visual inspection (20X magnification) after printing according to an industry workmanship standard. The test matrix and response data are in the following table.

Run order	A	B	C	D	E	F	G	H	PVM	NPU (%)
4	−	−	−	−	−	−	−	+	1.00	5
13	+	−	−	−	−	+	+	+	1.04	13
6	−	+	−	−	−	+	+	−	1.02	16
3	+	+	−	−	−	−	−	−	0.99	12
19	−	−	+	−	−	+	−	−	1.02	15
25	+	−	+	−	−	−	+	−	1.01	9
21	−	+	+	−	−	−	+	+	1.01	12
14	+	+	+	−	−	+	−	+	1.03	17
10	−	−	−	+	−	−	+	−	1.04	21
22	+	−	−	+	−	+	−	−	1.14	20
1	−	+	−	+	−	+	−	+	1.20	25
2	+	+	−	+	−	−	+	+	1.13	21
30	−	−	+	+	−	+	+	+	1.14	25
8	+	−	+	+	−	−	−	+	1.07	13
9	−	+	+	+	−	−	−	−	1.06	20
20	+	+	+	+	−	+	+	−	1.13	26
17	−	−	−	−	+	−	−	−	1.02	10
18	+	−	−	−	+	+	+	−	1.10	13
5	−	+	−	−	+	+	+	+	1.09	17
26	+	+	−	−	+	−	−	+	0.96	13
31	−	−	+	−	+	+	−	+	1.02	14
11	+	−	+	−	+	−	+	+	1.07	11
29	−	+	+	−	+	−	+	−	0.98	10
23	+	+	+	−	+	+	−	−	0.95	14
32	−	−	−	+	+	−	+	+	1.10	28
7	+	−	−	+	+	+	−	+	1.12	24
15	−	+	−	+	+	+	−	−	1.19	22
27	+	+	−	+	+	−	+	−	1.13	15
12	−	−	+	+	+	+	+	−	1.20	21
28	+	−	+	+	+	−	−	−	1.07	19
24	−	+	+	+	+	−	−	+	1.12	21
16	+	+	+	+	+	+	+	+	1.21	27

(a) Verify that the generators are $I = ABCF$, $I = ABDG$, and $I = BCDEH$ for this design.
(b) What are the aliases for the main effects and two-factor interactions? You can ignore all interactions of order three and higher.
(c) Analyze both PVM and NPU responses.
(d) Analyze the residual for both responses. Are there any problems with model adequacy?
(e) The ideal value of PVM is unity and the NPU response should be as small as possible. Recommend suitable operating conditions for the process based on the experimental results.

4.14 You need to design a screening experiment for six factors. Four of the factors are continuous but two of them are categorical with three levels. You only want to estimate the main effects of these factors.

(a) What is the smallest design that you can use, assuming that all interactions are negligible? Construct a D-optimal design with this number of runs.
(b) Find the alias relationships between the main effects and two-factor interactions for this design.
(c) Suppose that you can afford up to four additional runs. Find this design using the D criterion. Compare the alias relationships for the design to the minimal run design from part (a). What did you gain from the additional runs?

5

Further Principles of Experimental Design

5.1 Introduction

In Chapter 4, we presented an introduction to the general principles of experimental design and some basic analysis methods, including the analysis of variance. The primary emphasis was on factorial designs, the two-level design system (2^k designs) and fractional factorials (2^{k-p} designs). These designs are very useful for fitting models with only main effects or models with main effects and some low-order interactions. While these are extremely useful design and analysis methods, situations arise frequently where more complex models are necessary or where the response distribution is non-normal and designs that are more appropriate for those situations would be helpful. This chapter focuses on designs for these situations. We present designs for fitting second-order models, which often arise in the context of response surface methodology (RSM).

5.2 Response Surface Methods and Designs

RSM is a collection of mathematical and statistical techniques useful for the modeling and analysis of problems in which a response of interest is influenced by several variables and the objective is to optimize this response. For example, suppose that a reliability engineer wishes to find the settings of temperature (x_1) and vibration (x_2) that maximize the failure rate (y) of an electronic component. The failure rate is a function of the settings of temperature and vibration level, say,

$$y = f(x_1, x_2) + \epsilon,$$

where ϵ represents the noise or error observed in the response y. If we denote the expected response by $E(y) = f(x_1, x_2) = \eta$, then the surface represented by $\eta = f(x_1, x_2)$ is called a response surface.

We usually represent the response surface graphically, such as in Figure 5.1, where the response y is plotted versus the levels of x_1 and x_2. To help visualize the shape of a response surface, we often plot the contours of the response surface as shown in Figure 5.2. In the contour plot, lines of constant response are drawn in the x_1–x_2 plane. Each contour corresponds to a particular height of the response surface. Contour plots can be very useful, particularly in experiments where the number of design factors (the xs) is small. In most RSM problems, the form of the relationship between the response and the independent variables is unknown. Therefore, the first step in RSM is to find a suitable approximation for the true

Design of Experiments for Reliability Achievement, First Edition.
Steven E. Rigdon, Rong Pan, Douglas C. Montgomery, and Laura J. Freeman.
© 2022 John Wiley & Sons, Inc. Published 2022 by John Wiley & Sons, Inc.
Companion website: www.wiley.com/go/rigdon/designexperiments

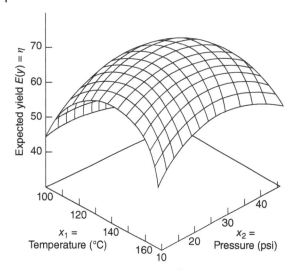

Figure 5.1 A three-dimensional response surface showing the expected yield as a function of temperature (x_1) and time (x_2).

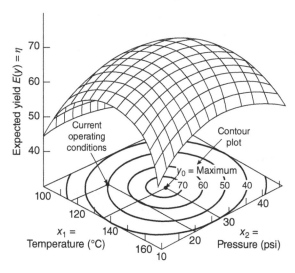

Figure 5.2 A three-dimensional response surface showing the expected yield as a function of temperature (x_1) and time (x_2) with a contour plot in the x_1–x_2 plane.

functional relationship between y and the independent variables. Usually, the experimenter employs a low-order polynomial in some region of the independent variables. If the response is well-modeled by a linear function of the independent variables, then the appropriate approximating function is the first-order model

$$y = \beta_0 + \beta_1 x_1 + \beta_2 x_2 + \cdots + \beta_k x_k + \epsilon. \tag{5.1}$$

If there is curvature in the system, then a polynomial of higher degree must be used, usually the second-order model

$$y = \beta_0 + \sum_{i=1}^{k} \beta_i x_i + \sum_{i=1}^{k} \beta_{ii} x_i^2 + \sum_{i<j} \sum \beta_{ij} x_i x_j + \epsilon. \tag{5.2}$$

Almost all RSM problems use one or both of these models. It is unlikely that a polynomial model will be a reasonable approximation of the true functional relationship over the entire space of the independent variables, but for a relatively small region, they are often adequate.

We use the method of least squares to estimate the parameters in the approximating polynomials. The response surface analysis is performed using the fitted surface. If the fitted surface is an adequate approximation of the true response function, then analysis of the fitted surface will be approximately equivalent to analysis of the actual system. The model parameters are estimated most effectively if good experimental designs are used to collect the data. We call designs for fitting response surface models response surface designs. We discuss these designs in Section 5.4.

Experimenters often deploy RSM as a sequential procedure. The first phase of an RSM study is often a factor screening experiment to identify the active factors that affect the response. This serves to reduce the number of factors used in the RSM deployment. In the screening experiment, we often discover that we are at a point on the response surface that is far from the optimum, such as the current operating conditions shown in Figure 5.3. When this occurs, we often observe that there is little curvature in the system and the first-order model will be appropriate. Our objective here is to lead the experimenter rapidly and efficiently along a path of improvement toward the general vicinity of the optimum. Once we have arrived in the region of the optimum, a more elaborate model, such as the second-order model, is typically employed, and an analysis is performed to locate the optimum. From Figure 5.3, we see that the analysis of a response surface can be thought of as "climbing a hill," where the top of the hill represents the point of maximum response. If the true optimum is a point of minimum response, then we may think of "descending into a valley."

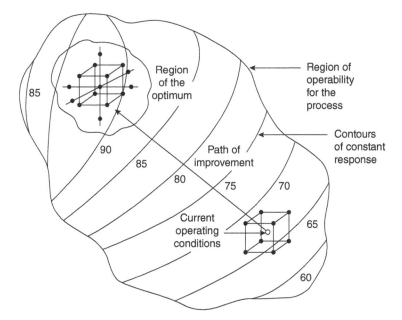

Figure 5.3 The sequential application of RSM.

The eventual objective of RSM is to determine the optimum operating conditions for the system or to determine a region of the factor space in which operating requirements are satisfied. More extensive presentations of RSM and many useful references are in Khuri and Cornell (1987), Myers et al. (2016), and Box and Draper (1987).

5.3 Optimization Techniques in Response Surface Methodology

Figure 5.3 illustrates the sequential deployment of RSM. The initial experiment is usually performed around either the current operating conditions for the system, some initial estimate of these conditions that are often obtained from some smaller scale operations such as a pilot plant or a development laboratory, or from other preliminary testing. If there is an indication of curvature in the fitted response surface, the experimenter usually proceeds to fit a second-order model and estimate the optimum operating conditions. However, sometimes we find that the response values obtained at this location are disappointing; for example, a failure rate of 15% or 20%, and we desire much better performance. In that case, the experimenter probably wants to shift the region of experimentation to an area where encountering better results are likely. This is the "path of improvement" shown in Figure 5.3.

Suppose that the experimenter has fit a first-order model in three variables, say,

$$\hat{y} = 25 + 12x_1 + 8x_2 - 4x_3$$

and the objective is to maximize the response y. If we want to move as rapidly as possible toward to optimum, the steps along this "path of improvement" should be in the gradient direction. In other words, the steps should be proportional to the model regression coefficients. In our example, this implies that we should move an amount equal to a $12/12 = 1$ coded unit step in the positive x_1 direction, $8/12 = 2/3 = 0.67$ coded units step in the positive x_2 direction, and $-4/12 = -1/3 = -0.33$ coded units step in the negative x_3 direction. Every step along this path will be proportional to these step sizes. This direction is usually called the **direction of steepest ascent**. If we were trying to minimize the response, then the steps would be in the negative direction (reversing the signs of the estimates of the model parameters) and this would be a direction of steepest descent. Operationally we take steps along this path and observe the response at each new point until the response changes direction. Sometimes a second round of steepest ascent is performed, resulting in a "mid-course correction" that leads to the region that likely contains the optimum. For detailed examples of the method of steepest ascent, see Myers et al. (2016). Adding points solely on the path of steepest ascent is like a one-factor experiment. If you are not a huge distance from the optimum it may be of value to choose points having more volume in the space of interest. In so doing, you may improve your model while simultaneously moving in the direction of the desired response. See Jones and Nachtsheim (2011b) for more details.

When the experimenter is relatively close to the optimum, a model that incorporates curvature is usually required to approximate the response. In most cases, the second-order model in Eq. (5.2) is adequate. We now show how to use this fitted model to find the optimum set of operating conditions for the xs and to characterize the nature of the response surface.

Suppose that $\hat{\boldsymbol{\beta}}$ is the least squares estimate of the vector of model parameters $\boldsymbol{\beta}$ for the model in Eq. (5.2). We can write the fitted second-order linear regression model in matrix form as

$$\hat{y} = \hat{\beta}_0 + \boldsymbol{x}'\boldsymbol{b} + \boldsymbol{x}'\boldsymbol{B}\boldsymbol{x}, \tag{5.3}$$

where

$$
x = \begin{bmatrix} x_1 \\ x_2 \\ \vdots \\ x_k \end{bmatrix}, \quad b = \begin{bmatrix} \hat{\beta}_1 \\ \hat{\beta}_2 \\ \vdots \\ \hat{\beta}_k \end{bmatrix}, \quad \text{and} \quad B = \begin{bmatrix} \hat{\beta}_{11} & \hat{\beta}_{12}/2 & \cdots & \hat{\beta}_{1k}/2 \\ & \hat{\beta}_{22} & \cdots & \hat{\beta}_{2k}/2 \\ & & \ddots & \vdots \\ & & & \hat{\beta}_{kk} \end{bmatrix}.
$$

That is, b is a $k \times 1$ vector of the estimates of the first-order model parameters, and B is a $k \times k$ symmetric matrix whose main diagonal elements are the estimates of the quadratic or second-order model parameters and whose off-diagonal elements are one-half the estimates of the interaction or mixed second-order model parameters. The optimum point on the fitted surface is the point at which all of the first derivatives of the fitted model with respect to x equal zero. This point is called the stationary point and denoted β_S. The derivative of $\hat{y}$ with respect to the elements of the vector x equated to 0 is

$$
\frac{\partial \hat{y}}{\partial x} = x + 2Bx = 0. \tag{5.4}
$$

The stationary point is the solution to Eq. (5.4), which is

$$
x_S = -\frac{1}{2}B^{-1}b \tag{5.5}
$$

assuming that the matrix B can be inverted. Furthermore, by substituting Eq. (5.5) into Eq. (5.3), we can find the predicted response at the stationary point as

$$
\hat{y}_S = \hat{\beta}_0 + \frac{1}{2}x'_S b. \tag{5.6}
$$

Once we have found the stationary point, it is usually necessary to characterize the response surface in the immediate vicinity of this point. By characterize, we mean determining whether the stationary point is a point of maximum or minimum response or a saddle point. We also usually want to study the relative sensitivity of the response to the variables $x_1, x_2, \ldots, x_k$. Figures 5.4–5.6 illustrate these three types of fitted response surfaces.

A relatively straightforward way to do this is to examine a contour plot of the fitted model as shown in Figures 5.4–5.6. If there are only two or three process variables (the xs), the construction and interpretation of this *contour plot* is relatively easy. However, even when there are relatively few variables, a more formal analysis, called the *canonical analysis*, can be useful.

To perform the canonical analysis we must first transform the model into a new coordinate system with the origin at the stationary point xs and then rotate the axes of this system until they are parallel to the principal axes of the fitted response surface. This transformation is illustrated in Figure 5.7. We can show that this results in the fitted model

$$
\hat{y} = \hat{y}_S + \lambda_1 w_1^2 + \lambda_2 w_2^2 + \cdots + \lambda_k w_k^2, \tag{5.7}
$$

where $w = [w_1, w_2, \ldots, w_k]'$ are the transformed independent variables (usually called the *canonical variables*) and $\lambda_1, \lambda_2, \ldots, \lambda_k$ are the eigenvalues (or characteristic roots) of the matrix B.

The nature of the response surface can be determined from the stationary point and the *signs* and *magnitudes* of the eigenvalues λ_i. First, suppose that the stationary point is within the region of exploration for fitting the second-order model. If the λ_i are all positive, the stationary point is a point of minimum response; if all of the λ_i are negative, the stationary point is a point of maximum response; and if the λ_i have different signs, the stationary point is a saddle point. Furthermore, the surface is steepest in

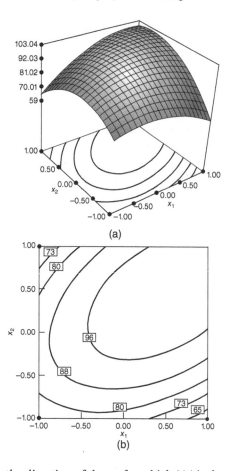

Figure 5.4 (a) Response surface and (b) contour plot illustrating a surface with a maximum.

the direction of the w_i for which $|\lambda_i|$ is the greatest. For example, Figure 5.4 depicts a system for which the stationary point is a maximum (λ_1 and λ_2 are negative) with $|\lambda_1| > |\lambda_2|$.

It is not unusual to encounter variations of the pure maximum, minimum, or saddle point response surfaces discussed in Section 5.2. Ridge systems, in particular, are fairly common. Consider the canonical form of the second-order model given previously in Eq. (5.7). Now suppose that the stationary point is within the region of experimentation; furthermore, let one or more of the eigenvalues be very small in magnitude (i.e., $\lambda_i \approx 0$). The response variable is then very insensitive to the variables w_i multiplied by the small eigenvalue λ_i.

A contour plot illustrating this situation is shown in Figure 5.7 for $k = 2$ variables with $\lambda_1 = 0$. (In practice, λ_1 would be close to but not exactly equal to zero.) The canonical model for this response surface is theoretically

$$\hat{y} = \hat{y}_s + \lambda_1 w_1^2$$

with λ_2 negative. Notice that the severe elongation in the λ_1 direction has resulted in a line of centers at $\hat{y} = 70$ and the optimum may be taken anywhere along that line. This type of response surface is called a *stationary ridge system*.

Figure 5.5 (a) Response surface and (b) contour plot illustrating a surface with a minimum.

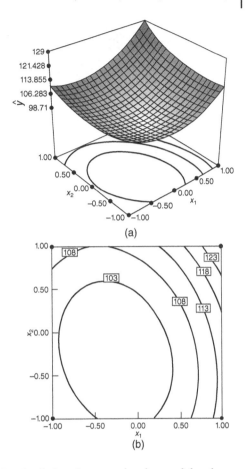

(a)

(b)

If the stationary point is far outside the region of exploration for fitting the second-order model and one (or more) λ_i is near zero, then the surface may be a rising ridge. Figure 5.8 illustrates a rising ridge for $k = 2$ variables with λ_1 near zero and λ_2 negative. In this type of ridge system, we cannot draw inferences about the true surface or the stationary point because x_s is outside the region where we have fit the model. However, further exploration is warranted in the λ_1 direction. If λ_2 had been positive, we would call this system a falling ridge.

Example 5.1 Box and Draper (1987) describe an experiment conducted to investigate the effects of three factors, $x_1 = $ length, $x_2 = $ amplitude, and $x_3 = $ load on the number of cycles to effective failure, y, of worsted yarn. The original authors describe the complete experiment describe more detail. Table 5.1 presents the 3^3 factorial experiment.

A complete second-order response surface model is fit to the cycles to failure data. Figure 5.9 shows the output for fitting this model with JMP. The overall model fit results in an R^2 of 0.94 and several of the model parameters are significant. However, the plot of the residuals versus the predicted values exhibits an indication that the variance of the response is increasing as the mean increases. We used the Box–Cox procedure implemented in JMP to determine whether a transformation on the response would

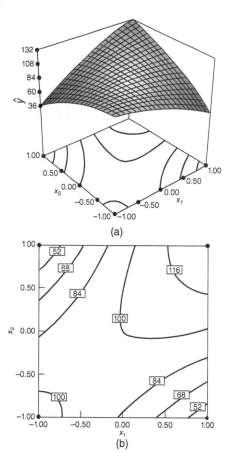

(a)

(b)

Figure 5.6 (a) Response surface and (b) contour plot illustrating a surface with a saddle point.

Figure 5.7 A stationary ridge system.

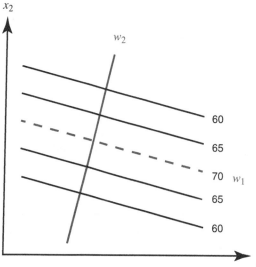

Figure 5.8 A rising ridge system.

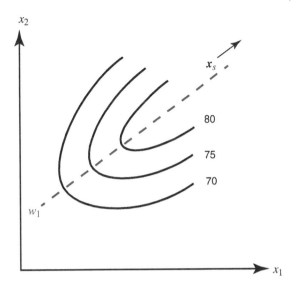

be helpful in resolving this problem. The procedure indicated that a transformation parameter of -0.203 or a transformed $y^* = y^{-0.203}$ is appropriate for variance stabilization. However, the value -0.203 is very close to 0, and 0 implies a log transformation, which is much more intuitive to explain, so we decided to use the transformation $y^* = \log(y)$.

After fitting the complete second-order model to the log response, we find that only the first-order model terms are significant. After dropping the second-order terms, the JMP output for the final fitted model is shown in Figure 5.10. The overall model is significant and the R^2 is 0.97. The residual plots are satisfactory so the log transformation has resolved the inequality of variance observed in the model that was fit to the original response. The prediction profiler indicates that in order to maximize the number of cycles to failure, x_1 should be set to the high level and the other two variables should be set to the low level. The final prediction equation for cycles to failure is

$$\hat{y} = \exp\left(6.33 + 0.833x_1 - 0.631x_2 - 0.392x_3\right). \qquad \blacksquare$$

5.4 Designs for Fitting Response Surfaces

5.4.1 Classical Response Surface Designs

Suppose that we want to fit the first-order model in k variables. There is a unique class of designs that minimize the variance of the model regression coefficients. These are the orthogonal first-order designs. A first-order design is orthogonal if the off-diagonal elements of the $(X'X)$ matrix are all zero. This implies that the cross products of the columns of the X matrix sum to zero. The class of orthogonal first-order designs includes the 2^k factorial and fractions of the 2^k series in which main effects are not aliased with each other, that is, designs of at least Resolution III. In using these designs, we assume that the low and high levels of the k factors are coded to the usual ± 1 levels.

Table 5.1 3^3 design and results for the worsted yarn example.

x_1	x_2	x_3	y
−1	−1	−1	674
0	−1	−1	1414
1	−1	−1	3636
−1	0	−1	338
0	0	−1	1022
1	0	−1	1568
−1	1	−1	170
0	1	−1	442
1	1	−1	1140
−1	−1	0	370
0	−1	0	1198
1	−1	0	3184
−1	0	0	266
0	0	0	620
1	0	0	1070
−1	1	0	118
0	1	0	332
1	1	0	884
−1	−1	1	292
0	−1	1	634
1	−1	1	2000
−1	0	1	210
0	0	1	438
1	0	1	566
−1	1	1	90
0	1	1	220
1	1	1	360

There are two designs that have been widely used in fitting second-order response surface models. The first of these is the *central composite design*, or the CCD. The CCD consists of a 2^k factorial (or fractional factorial of Resolution V) with n_F factorial runs, 2^k axial or star runs, and n_C center runs. Figure 5.11 shows the CCD for $k = 2$ and $k = 3$ factors.

The practical deployment of a CCD often arises through sequential experimentation. That is, a 2^k design was used to fit a first-order model, this model has exhibited lack of fit, and the axial runs are then added to allow the quadratic terms to be incorporated into the model. The CCD is an efficient design for fitting

Figure 5.9 The Worsted yarn experiment.

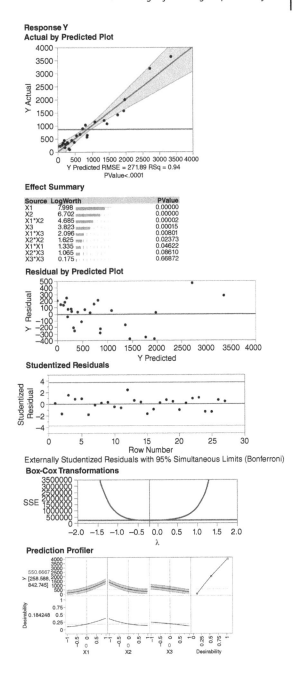

Response LnY
Actual by Predicted Plot

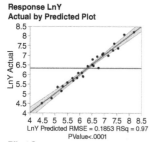

LnY Predicted RMSE = 0.1853 RSq = 0.97
PValue<.0001

Effect Summary

Source	LogWorth		PValue
X1	14.869		0.00000
X2	12.292		0.00000
X3	8.252		0.00000

Residual by Predicted Plot

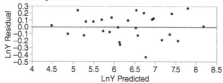

Studentized Residuals

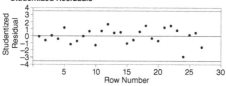

Externally Studentized Residuals with 95% Simultaneous Limits (Bonferroni)

Analysis of Variance

Source	DF	Sum of Squares	Mean Square	F Ratio
Model	3	22.425894	7.47530	217.6409
Error	23	0.789980	0.03435	**Prob > F**
C. Total	26	23.215874		<.0001*

Parameter Estimates

Term	Estimate	Std Error	t Ratio	Prob>|t|
Intercept	6.3348148	0.035667	177.61	<.0001*
X1	0.8333333	0.043683	19.08	<.0001*
X2	−0.630556	0.043683	−14.43	<.0001*
X3	−0.392222	0.043683	−8.98	<.0001*

Prediction Profiler

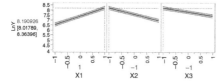

Figure 5.10 JMP output for the first-order model for the log of the cycles to failure response for the worsted yard data in Table 5.1.

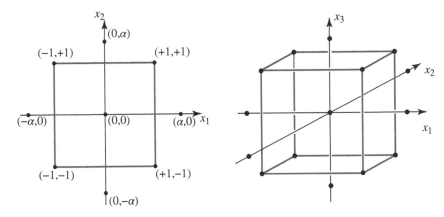

Figure 5.11 The central composite design for $k = 2$ and $k = 3$ factors.

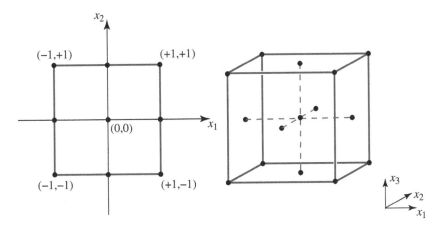

Figure 5.12 Face centered designs for $k = 2$ and $k = 3$.

the second-order model in this sequential experimentation approach. There are two parameters in the design that must be specified: the distance α of the axial runs from the design center and the number of center points n_C.

It is important for the second-order model to provide good predictions of the response throughout the region of interest. One way to define "good" is to require that the model should have a reasonably consistent and stable variance of the predicted response at points of potential interest. Box and Hunter (1957) suggested that a second-order response surface design should be rotatable. This means that the variance of the predicted response is the same at all of the points x in the region of interest that are the same distance, in coded units, from the design center. That is, the variance of predicted response is constant on spheres. A CCD is made rotatable by the choice of α. The value of α to achieve rotatability depends on the number of points in the factorial portion of the design; in fact, $\alpha = n_F^{1/4}$ yields a rotatable CCD where n_F is the number of points used in the factorial portion of the design. For example, if the

CCD had a 2^2 factorial design in the factorial portion, this design has four runs, so the rotatable value of $n_F^{1/4} = 1.414$.

Rotatability is a spherical property; that is, it makes the most sense as a design criterion when the region of interest is a sphere. However, it is not important to have exact rotatability to have a good design. For a spherical region of interest, a good choice of α from a prediction variance viewpoint for the CCD is to set $\alpha = \sqrt{k}$. This design, called a spherical CCD, puts all the factorial and axial design points on the surface of a sphere of radius $\sqrt{k}$. There is usually little practical difference between the rotatable and the spherical CCD. In fact, for $k = 2$ both designs are the same. For more discussion of this, see Myers et al. (2016).

In many situations, the region of interest is cuboidal rather than spherical. In these cases, a useful variation of the CCD is the face-centered CCD or the face-centered cube, in which $\alpha = 1$. This design locates the star or axial points on the centers of the faces of the cube, as shown in Figure 5.12 for $k = 3$. This variation of the CCD is also sometimes used because it requires only three levels of each factor, and in practice it is frequently difficult to change factor levels. However, note that face-centered CCD are not rotatable.

The choice of α in the CCD is dictated primarily by the region of interest. When this region is a sphere, the design must include center runs to provide reasonably stable variance of the predicted response. Generally, three to five center runs are usually adequate. The face-centered cube does not require any center runs but again in most situations at least two center runs are desirable if the experimenter wants a model-independent estimate of error. However, note that the face-center cube with no center runs is the 14-run D-optimal design for fitting a second-order model in three factors.

Another classical RSM design is the Box–Behnken design (Box and Behnken 1960). These are three-level designs for fitting response surfaces. These designs are formed by combining 2^k factorials with incomplete block designs. The resulting designs are usually very efficient in terms of the number of required runs, and they are either rotatable or nearly rotatable. Table 5.2 shows a three-variable Box–Behnken design. The design is shown geometrically in Figure 5.13. Notice that the three-factor Box–Behnken design is a spherical design, with all points lying on a sphere of radius $\sqrt{k}$. There are no points at the vertices of the cubic region created by the upper and lower limits for each variable. It can

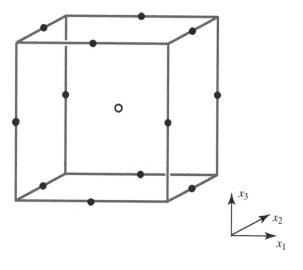

Figure 5.13 Box–Behnken design for $k = 3$.

Table 5.2 A three-variable Box–Behnken design with three runs at the center point.

Run	x_1	x_2	x_3
1	−1	−1	0
2	−1	1	0
3	1	−1	0
4	1	1	0
5	−1	0	−1
6	−1	0	1
7	1	0	−1
8	1	0	1
9	0	−1	−1
10	0	−1	1
11	0	1	−1
12	0	1	1
13	0	0	0
14	0	0	0
15	0	0	0

be shown that the Box–Behnken design with two center runs is an alias-optimal design if the model that is being protected against is the full cubic model. See Jones and Nachtsheim (2011b) for discussion of the alias-optimal design criterion.

5.4.2 Definitive Screening Designs

Jones and Nachtsheim (2011a) have introduced a useful class of response surface designs for quantitative factors whose construction can be facilitated by an optimal design algorithm. They are called *definitive screening designs* (DSDs) because they are small enough to allow efficient screening of potentially many factors yet they can accommodate second-order effects in many situations without additional runs. In that regard, we can think of these designs as "one-step RSM designs." Table 5.3 shows the general structure of these designs with m factors.

Notice that for m factors, there are only $2m + 1$ runs based on m fold-over pairs and an overall center point. Each run (excluding the center run) has exactly one factor level at its center point and all others levels at the extremes. These designs have the following desirable properties:

1. The number of required runs is only one more than twice the number of factors. Consequently, these are very small designs.
2. Unlike Resolution III designs, main effects are completely independent of two-factor interactions. As a result, estimates of main effects are not biased by the presence of active two-factor interactions, regardless of whether the interactions are included in the model.

Table 5.3 Box–Behnken design in $k = 3$ variables.

Run	x_1	x_2	x_3
1	−1	−1	0
2	−1	1	0
3	1	−1	0
4	1	1	0
5	−1	0	−1
6	−1	0	1
7	1	0	−1
8	1	0	1
9	0	−1	−1
10	0	−1	1
11	0	1	−1
12	0	1	1
13	0	0	0
14	0	0	0
15	0	0	0

3. Unlike Resolution IV designs, two-factor interactions are not completely confounded with other two-factor interactions, although they may be correlated.
4. Unlike Resolutions III, IV, and V designs with added center points, all quadratic effects can be estimated in models comprised of any number of linear and quadratic main effect terms.
5. Quadratic effects are orthogonal to main effects and not completely confounded (though correlated) with interaction effects.
6. With six or more factors, the designs are capable of estimating all possible full quadratic models involving three or fewer factors with very high levels of statistical efficiency.

These designs are an excellent compromise between Resolution III fractions for screening and small RSM designs. They also admit the possibility of moving directly from screening to optimization using the results of a single experiment. Jones and Nachtsheim found these designs using an optimization technique they had previously developed for finding minimum aliasing designs. Their algorithm minimizes the sum of the squares of the elements of the alias matrix subject to a constraint on the D-efficiency of the resulting design.

DSDs can also be constructed from *conference matrices* (see Xiao et al., 2012). A conference matrix C is an $n \times n$ matrix that has diagonal elements equal to zero and all off-diagonal elements equal to ±1. They have the property that $C'C$ is a multiple of the identity matrix. For the $n \times n$ conference matrix C, $C'C = (n - 1)I$. Conference matrices first arose in connection with a problem in telephony. They were used in constructing ideal telephone conference networks from ideal transformers. These networks were represented by conference matrices. There are other applications as well.

The conference matrix of order 6 is

$$\begin{bmatrix} 0 & +1 & +1 & +1 & +1 & +1 \\ +1 & 0 & +1 & -1 & -1 & +1 \\ +1 & +1 & 0 & +1 & -1 & -1 \\ +1 & -1 & +1 & 0 & +1 & -1 \\ +1 & -1 & -1 & +1 & 0 & +1 \\ +1 & +1 & -1 & -1 & +1 & 0 \end{bmatrix}.$$

The 13-run 6-factor DSD can be found by folding over each row of this conference matrix and adding a row of zeros at the bottom. In general, if C is the conference matrix of order n, the m-factor DSD matrix can be found as follows:

$$D = \begin{bmatrix} C \\ -C \\ 0 \end{bmatrix}.$$

DSDs are intended for use with continuous factors. However, DSDs can be modified to include two-level categorical variables. The process is straightforward. Begin by constructing the usual DSD for continuous factors, except adding two rows of zeros instead of one. Change the zeros in the columns for the categorical factors to either +1 or −1. If the zeros are the added rows of zeros, make all the categorical factors −1 for the first row and +1 for the second row. If the zeros are from the conference matrix and its fold over, make the factor −1 for the first row and +1 for the second row.

As an example, consider a DSD for four continuous factors and two categorical factors. The DSD for six factors with the two added zeros is shown in Table 5.4.

Now the zeros in the last two columns of Table 5.4 are changed to ±1 in order to produce the categorical variables E and F. The design is shown in Table 5.5.

When two-level categorical variables are added to a DSD as shown, the main effects of the categorical factors have some correlation with other factors, but the correlations are small.

It is also possible to construct DSDs in orthogonal blocks. The procedure is as follows:

1. Start by repeating the steps for creating a design for continuous factors except adding only as many rows of zeros as there are blocks. Arrange the design in standard order.
2. Assign the first fold-over pair to the first block, the second fold-over pair to the second block, and so on until you get to the last block, then start over again assigning the next fold-over pair to the first block.
3. Assign each center run to a separate block.

As an example, consider a DSD for six factors in two blocks. The six-factor DSD in standard order with two added zeros. Now arrange the blocks as described earlier, with the first fold-over pair forming the first block, the second fold-over pair forming the second block, and so on, and then finish by assigning the zeros to each block. The final design is shown in Table 5.6.

Table 5.4 Definitive screening design for $m = 6$ factors.

A	B	C	D	E	F
0	+1	+1	+1	+1	+1
+1	0	+1	−1	−1	+1
+1	+1	0	+1	−1	−1
+1	−1	+1	0	+1	−1
+1	−1	−1	+1	0	+1
+1	+1	−1	−1	+1	0
0	−1	−1	−1	−1	−1
−1	0	−1	+1	+1	−1
−1	−1	0	−1	+1	+1
−1	+1	−1	0	−1	+1
−1	+1	+1	−1	0	−1
−1	−1	+1	+1	−1	0
0	0	0	0	0	0
0	0	0	0	0	0

Table 5.5 Definitive screening design for $m = 6$ factors with categorical variables in columns *E* and *F*.

A	B	C	D	E	F
0	+1	+1	+1	+1	+1
+1	0	+1	−1	−1	+1
+1	+1	0	+1	−1	−1
+1	−1	+1	0	+1	−1
+1	−1	−1	+1	−1	+1
+1	+1	−1	−1	+1	−1
0	−1	−1	−1	−1	−1
−1	0	−1	+1	+1	−1
−1	−1	0	−1	+1	+1
−1	+1	−1	0	−1	+1
−1	+1	+1	−1	+1	−1
−1	−1	+1	+1	−1	+1

Table 5.6 Definitive screening design in two blocks.

Block	A	B	C	D	E	F
1	0	+1	+1	+1	+1	+1
1	0	−1	−1	−1	−1	−1
2	+1	0	−1	+1	+1	−1
2	−1	0	+1	−1	−1	+1
1	+1	−1	0	−1	+1	+1
1	−1	+1	0	+1	−1	−1
2	+1	+1	−1	0	−1	+1
2	−1	−1	+1	0	+1	−1
1	+1	+1	+1	−1	0	−1
1	−1	−1	−1	+1	0	+1
2	+1	−1	+1	+1	−1	0
2	−1	+1	−1	−1	+1	0
1	0	0	0	0	0	0
2	0	0	0	0	0	0

5.4.3 Optimal Designs in RSM

The standard response surface designs discussed in the previous sections, such as the CCD, the Box–Behnken design, and their variations (such as the face-centered cube), are widely used in RSM applications. If the experimental region is either a cube or a sphere, often a standard response surface design will be applicable to the problem. However, quite frequently an experimenter encounters a situation where a standard response surface design may not be the obvious choice. Optimal designs are an alternative to use in these cases. As we have noted before, there are several very common situations where an optimal design is appropriate. For example, if a non-standard model is required (that is something other than the usual first- or second-order polynomial), a non-standard sample size or there is some unusual blocking requirement, or when the experimental region is not the usual sphere or cube but is impacted by constraints. We recommend that I-optimal designs be used in these situations. In fact, we recommend considering I-optimal designs first because they are often superior to even very good standard designs.

For example, suppose that we are going to fit a second-order model in $k = 4$ factors and that the face-centered cube is being considered. If we use two center runs in this design, it will have 26 total runs. Consider an alternative design, an I-optimal design with only 24 runs. Figure 5.14 shows the fraction of design space plot comparing these two designs. In Figure 5.14 the CCD is represented by the lower curve

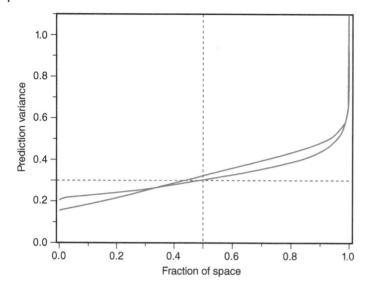

Figure 5.14 Fraction of design space plot comparing a face centered cube and an I-Optimal design.

at the left-hand side of the graph, indicating that the CCD has better prediction performance over the initial part of the design space, but the I-optimal design begins to outperform it fairly quickly and has lower prediction variance over more than 50% of the design space. The average variance of prediction for the face-centered CCD is $0.324947\sigma^2$ and the average prediction variance for the I-optimal design is $0.315392\sigma^2$ despite the fact that it has two fewer runs.

Problems

5.1 The data shown in Table 5.7 were collected in an experiment to optimize crystal growth as a function of three variables x_1, x_2, and x_3. Large values of y (yield in grams) are desirable. Fit a second-order model and analyze the fitted surface. Under what set of conditions is maximum growth achieved?

5.2 The data in Table 5.8 were collected by a chemical engineer. The response y is filtration time, x_1 is temperature, and x_2 is pressure. Fit a second-order model.
   (a) What operating conditions would you recommend if the objective is to minimize the filtration time?
   (b) What operating conditions would you recommend if the objective is to operate the process at a mean filtration rate very close to 46?

Table 5.7 Data for Problem 5.1.

x_1	x_2	x_3	y
−1	−1	−1	66
−1	−1	1	70
−1	1	−1	78
−1	1	1	60
1	−1	−1	80
1	−1	1	70
1	1	−1	100
1	1	1	75
−1.682	0	0	100
1.682	0	0	80
0	−1.682	0	68
0	1.682	0	63
0	0	−1.682	65
0	0	1.682	82
0	0	0	113
0	0	0	100
0	0	0	118
0	0	0	88
0	0	0	100
0	0	0	85

Table 5.8 Data for Problem 5.2.

x_1	x_2	y
−1	−1	54
−1	1	45
−1	−1	32
−1	1	47
−1.414	0	50
1.414	0	53
0	−1.414	47
0	1.414	51
0	0	41
0	0	39
0	0	44
0	0	42
0	0	40

Table 5.9 Data for Problem 5.3.

x_1	x_2	y
1	0	68
0.5	$\sqrt{0.75}$	74
−0.5	$\sqrt{0.75}$	65
−1	0	60
−0.5	$-\sqrt{0.75}$	63
0.5	$-\sqrt{0.75}$	70
0	0	58
0	0	60
0	0	57
0	0	55
0	0	69

5.3 The design in Table 5.9 is used in an experiment that has the objective of fitting a second-order model:

(a) Fit the second-order model.

(b) Perform the canonical analysis. What type of surface has been found?

(c) What operating conditions on x_1 and x_2 lead to the stationary point?

(d) Where would you run this process if the objective is to obtain a response that is as close to 65 as possible?

(e) Plot the design points. What kind of arrangement do they form? Is this design rotatable?

5.4 Verify that an orthogonal first-order design is also first-order rotatable.

5.5 Show that augmenting a 2^k design with an arbitrary number of center points does not affect the estimates of the model regression coefficients but that the estimate of the intercept is the average of all design runs.

5.6 Suppose that you need to design an experiment to fit a quadratic model over the region $-1 \leq x_i \leq 1$, $i = 1,2$ subject to the constraint $x_1 + x_2 \leq 1$. If the constraint is violated, the process will not work properly. You can afford to make no more than $n = 12$ runs. What design would you recommend?

5.7 An experimenter has run a Box–Behnken design and obtained the results as shown in Table 5.10, where the response variable is the viscosity of a polymer.

(a) Fit the second-order model.

(b) Perform the canonical analysis. What type of surface has been found?

(c) What operating conditions on x_1, x_2, and x_3 lead to the stationary point?

(d) What operating conditions would you recommend if it is important to obtain a viscosity that is as close to 600 as possible?

Table 5.10 Data for Problem 5.7.

Level	Temp	Agitation rate	Pressure	x_1	x_2	x_3
High	200	100	25	+1	+1	+1
Middle	175	7.5	20	0	0	0
Low	150	5.0	15	−1	−1	−1

Run	x_1	x_2	x_3	y
1	−1	−1	0	535
2	+1	−1	0	580
3	−1	+1	0	596
4	+1	+1	0	563
5	−1	0	−1	645
6	+1	0	−1	458
7	−1	0	+1	350
8	+1	0	+1	600
9	0	−1	−1	595
10	0	+1	−1	648
11	0	−1	+1	532
12	0	+1	+1	656
13	0	0	0	653
14	0	0	0	599
15	0	0	0	620

5.8 Suppose that you need to design an experiment to fit a quadratic model over the region $-1 \leq x_i \leq 1$, $i = 1,2$. You can afford 12 runs. Construct both an I-optimal and a D-optimal design. Compare the performance of both designs in terms of the precision with which the model parameters are estimated and the ability to accurately predict the mean response. Which design would you prefer?

5.9 An article in the *Journal of Chromatography A* (Morris et al. 1997) describes an experiment to optimize the production of ranitidine, a compound that is the primary active ingredient of Zantac, a pharmaceutical product used to treat ulcers, gastroesophageal reflux disease (a condition in which backward flow of acid from the stomach causes heartburn and injury of the esophagus), and other conditions where the stomach produces too much acid, such as Zollinger–Ellison syndrome. The authors used three factors (x_1 = pH of the buffer solution, x_2 = the electrophoresis voltage, and x_3 = the concentration of one component of the buffer solution) in a central composite design. The response is chromatographic exponential function (CEF), which should be minimized. The design is shown in Table 5.11.

Table 5.11 The ranitidine separation experiment.

Standard order	x_1	x_2	x_3	CEF
1	−1	−1	−1	17.3
2	1	−1	−1	45.5
3	−1	1	−1	10.3
4	1	1	−1	11,751.1
5	−1	−1	1	16.942
6	1	−1	1	25.4
7	−1	1	1	31,697.2
8	1	1	1	12,039.2
9	−1.68	0	0	7.5
10	1.68	0	0	6.3
11	0	−1.68	0	11.1
12	0	1.68	0	6.664
13	0	0	−1.68	16,548.7
14	0	0	1.68	26,351.8
15	0	0	0	9.9
16	0	0	0	9.6
17	0	0	0	8.9
18	0	0	0	8.8
19	0	0	0	8.013
20	0	0	0	8.059

(a) Fit a second-order model to the CEF response. Analyze the residuals from this model. Does it seem that all model terms are necessary?

(b) Reduce the model from part (a) as necessary. Did model reduction improve the fit?

(c) Does transformation of the CEF response seem like a useful idea? What aspect of either the data or the residual analysis suggests that transformation would be helpful?

(d) Fit a second-order model to the transformed CEF response. Analyze the residuals from this model. Does it seem that all model terms are necessary? What would you choose as the final model?

(e) Suppose that you had some information that suggests that the separation process malfunctioned during run 7. Delete this run and analyze the data from this experiment again.

(f) What conditions would you recommend to minimize CEF?

5.10 An article in *Quality Progress* (Johnson and Burrows 2011) describes using a central composite design to improve the packaging of one-pound coffee (Table 5.12). The objective is to produce an airtight seal that is easy to open without damaging the top of the coffee bag. The

Table 5.12 Starbucks' packaging of one-pound coffee.

Run	Viscosity	Pressure	Plate gap	Tear	Leakage
Center	350	180	0	0	0.15
Axial	250	170	0	0	0.5
Factorial	319	186	1.8	0.45	0.15
Factorial	380	174	1.8	0.85	0.05
Center	350	180	0	0.35	0.15
Axial	300	180	0	0.3	0.45
Axial	400	180	0	0.7	0.25
Axial	350	190	0	1.9	0
Center	350	180	0	0.25	0.05
Factorial	319	186	−1.8	0.1	0.35
Factorial	380	186	−1.8	0.15	0.4
Axial	350	180	3	3.9	0
Factorial	380	174	−1.8	0	0.45
Center	350	180	0	0.55	0.2
Axial	350	180	−3	0	1
Factorial	319	174	−1.8	0.05	0.2
Factorial	319	174	1.8	0.4	0.25
Factorial	380	186	1.8	4.3	0.05
Center	350	180	0	0	0

experimenters studied three factors: x_1 = plastic viscosity (300–400 centipoise), x_2 = clamp pressure (170–190 psi), and x_3 = plate gap (−3, +3 mm) and two responses: y_1 = tear and y_2 = leakage. The design is shown in the following table. The tear response was measure on a scale from 0–9 (good to bad) and leakage was proportion failing. Each run used a sample of 20 bags for response measurement.

(a) Build a second-order model for the tear response.
(b) Build a second-order model for the leakage response.
(c) Analyze the residuals for both models. Do transformations seem necessary for either response? If so, refit the models in the transformed metric.
(d) Construct response surface plots and contour plots for both responses. Provide interpretations for the fitted surfaces.
(e) What conditions would you recommend for process operation to minimize leakage and keep tear below 0.75?

5.11 Table 5.13 shows a six-variable definitive screening design from Jones and Nachtsheim (2011b). Analyze the response data from this experiment.

Table 5.13 Definitive screening design from Jones and Nachtsheim (2011b).

Run(i)	$x_{i,1}$	$x_{i,2}$	$x_{i,3}$	$x_{i,4}$	$x_{i,5}$	$x_{i,6}$	y_i
1	0	1	−1	−1	−1	−1	21.04
2	0	−1	1	1	1	1	10.48
3	1	0	−1	1	1	−1	17.89
4	−1	0	1	−1	−1	1	10.07
5	−1	−1	0	1	−1	−1	7.74
6	1	1	0	−1	1	1	21.01
7	−1	1	1	0	1	−1	16.53
8	1	−1	−1	0	−1	1	20.38
9	1	−1	1	−1	0	−1	8.62
10	−1	1	−1	1	0	1	7.80
11	1	1	1	1	−1	0	23.56
12	−1	−1	−1	−1	1	0	15.25
13	0	0	0	0	0	0	19.91

Source: Modified from Jones and Nachtsheim (2011b).

5.12 Suppose that you want to fit a second-order response surface model in a situation where there are $k = 4$ factors; however, one of the factors is categorical with two levels. What model should you consider for this experiment? Suggest an appropriate design for this situation.

5.13 Suppose that you want to fit a second-order model in $k = 5$ factors. You cannot afford more than 25 runs. Construct both a D-optimal and I-optimal design for this situation. Compare the prediction variance properties of the designs. Which design would you prefer?

5.14 The design shown in Table 5.14 was used in an experiment that has the objective of fitting a second-order model.
(a) Fit the second-order model.
(b) What type of surface has been found?
(c) What operating conditions on the process variables lead to the stationary point?
(d) Where would you run this process if the objective is to obtain a response that is as close to 65 as possible?
(e) Is the design rotatable?

Table 5.14 Experimental data for Problem 5.15.

X_1	X_2	Y
1	0	68
0.5	$\sqrt{0.75}$	74
−0.5	$\sqrt{0.75}$	65
−1	0	60
−0.5	$-\sqrt{0.75}$	63
0.5	$-\sqrt{0.75}$	70
0	0	58
0	0	60
0	0	57
0	0	55
0	0	69

Part III

Regression Models for Reliability Studies

6

Parametric Regression Models

6.1 Introduction to Failure-Time Regression

The chapters in Part I of this text focused on analyzing a single data set of units produced and tested under the same conditions. Often, however, the distribution of lifetimes will depend on the values of one or more predictor variables. These predictor variables could be:

- **product design characteristics** – such as the wall thickness of the casing, the type of metal used in a rod, the fan blade thickness, etc.,
- **process variables** – that is, characteristics of the process that produced the items, for example, temperature and extrusion rate for a plastic molding process,
- **uncontrollable environmental factors** – such as the ambient temperature and humidity, the amount of particulates in the air, etc., and
- **accelerating variables** – that is, variables designed to speed up the failure process, such as temperature, vibration, or voltage.

These variables can be categorical (e.g. type of metal, or supplier for raw material), continuous (temperature, wire diameter), or discrete (number of surface treatments). In Part II, we saw how Design of Experiments is used to systematically vary these predictor variables or factors in a purposeful series of tests.

The question for this chapter is "How do the predictor variables affect the lifetime?" In Part II, linear regression models that assumed a normal distribution were used to analyze the data. The worsted yarn example showed how transformations can be useful for non-normal distributions. This chapter provides a more complete review of analysis methods available for lifetime data.

For reliability studies, linear regression is frequently inadequate for a number of reasons. First, note that a linear regression has the form

$$y_i = \beta_0 + \beta_1 x_{i1} + \cdots + \beta_k x_{ki} + \epsilon_i,$$

where $\epsilon \sim N(0, \sigma^2)$, so that it is possible for a predicted value for y_i to be negative, which is not possible in the context of life testing. Also, the assumption that the ϵ_i are i.i.d. $N(\mu, \sigma^2)$ means that the responses (lifetimes in this case) are normally distributed; that is,

$$y_i \sim N\left(\beta_0 + \beta_1 x_{i1} + \cdots + \beta_k x_{ki}, \sigma^2\right).$$

Design of Experiments for Reliability Achievement, First Edition.
Steven E. Rigdon, Rong Pan, Douglas C. Montgomery, and Laura J. Freeman.
© 2022 John Wiley & Sons, Inc. Published 2022 by John Wiley & Sons, Inc.
Companion website: www.wiley.com/go/rigdon/designexperiments

The normal distribution is usually a poor model for lifetimes. Product failure times are often best represented by distributions that allow for skewness. For example, products with design flaws leading to infant mortality failure modes are best represented by the Weibull distribution with a shape parameter less than 1, or product wear out by a Weibull distribution with a shape parameter greater than 1.

Instead of the usual normal-theory linear regression, we focus instead on how an item ages and how this aging depends on the predictor variables. There are several parametric approaches for modeling failure times. The ones we will cover in this chapter include transformations of the lifetime response coupled with linear regression, generalized linear models (GLMs), location-scale models, and Weibull regression. We will also discuss the scale-accelerated failure time (SAFT) model, which has been widely used in the engineering community and closely connects with location-scale models for the Weibull distribution.

6.2 Regression Models with Transformations

Transformations provide a simple tool for working with data that do not meet the assumptions of linear regression models. Power functions and log transformations can be used to transform skewed distributions into symmetric or normal distributions. Transformations are also useful for stabilizing variance to meet assumptions of homogeneity of variance. Common transformations include log,[1] square root, and power functions. In this section we will focus on the log transform due to its frequent applicability to lifetime data. However, the methods provided work for any transformation.

While the normal distribution tends to provide a poor fit to lifetime data, the lognormal distribution is often an appropriate model for failure times. The relationship between the normal and lognormal distribution, whose log is normally distributed, provides simple modeling methodology for failure time data. That is, if failure times, $t_i \sim \text{LOGN}(\mu, \sigma^2)$ then $\log(t_i) \sim N(\mu, \sigma^2)$. This relationship allows us to use linear regression models to include response surface models for failure times if the lognormal distribution is a good fit.

In Chapter 5 we saw how using the Box–Cox or a log transformation can be used to transform the response variable such that the assumptions of normality are satisfied. Figure 6.1 illustrates this concept

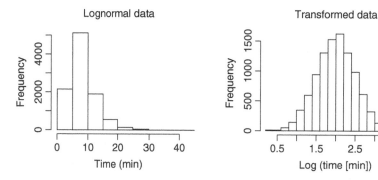

Figure 6.1 Simulated failure time data.

1 Throughout, we will use log to indicate logarithm to the base *e*, that is, the natural logarithm.

visually. In the figure, notional failure time data, t_i, on the left follow a lognormal distribution with parameters $\mu = 2$ and $\sigma = 0.5$, applying a log transformation the data on the right, $\log(t_i)$, follow a normal distribution with the parameters $\mu = 2$ and $\sigma = 0.5$.

Because the transformed data follow a normal distribution, we can now use linear models including response surface models to make the appropriate inferences as shown in Part II. However, a challenge with transformations is that all predicted lifetimes will be in the transformed units. We must transform back to the original time units. The estimates and confidence interval for quantities in the original time scale need to be constructed.

6.2.1 Estimation and Confidence Intervals for Transformed Data

For maximum likelihood estimates (MLEs), the process for reversing transformations and calculating interval estimates is fairly straightforward:

1. Calculate the statistic of interest leveraging the invariance property of MLE.
2. Calculate the standard error via the Fisher information matrix.
3. Select the appropriate form of the Wald statistic and estimate the interval.

In this textbook we invert the Wald statistic to construct confidence intervals. However, numerous other methods are available for calculating confidence intervals, including inverting the likelihood ratio test statistic and bootstrapping. Wald confidence intervals are based on the asymptotic distribution of the MLE and are therefore most accurate for larger samples. We use them here because of their wide applicability, their integration into software packages including JMP and R, and the fact that they provide closed form solutions. This approach uses the multivariate delta method (i.e. propagation of error) to estimate the variance.

Step 1: Calculate the statistic of interest. In general, let θ be a vector of coefficients estimated using maximum likelihood estimation resulting in the vector of estimates $\hat{\theta}$. The invariance property of MLE tells us that if $\hat{\theta}$ is the MLE of θ and $g(\theta)$ is a function of θ, then $g(\hat{\theta})$ is the MLE of $g(\theta)$. Because of the invariance property it is straightforward to calculate the statistics of interest.

Examples of statistics include the mean, median, or percentile of the failure time distribution. For our lognormal example these might include

$$E(T) = \exp(\mu + \sigma^2/2) \tag{6.1}$$

and

$$V(T) = \exp(2\mu + \sigma^2)\left(e^{\sigma^2} - 1\right). \tag{6.2}$$

An additional function of interest in reliability data is the percentile function. For the lognormal distribution the percentile function is

$$t_p = \exp\left(\mu + \Phi_N^{-1}(p)\sigma\right), \tag{6.3}$$

where $\Phi_N^{-1}(p)$ is the inverse cumulative distribution function (CDF) of the standard normal distribution. Additionally, note that plugging in $p = 0.5$ to the percentile functions yields the median: $\text{med}(T) = \exp(\mu)$.

In the lognormal case, if $\hat{\mu}$ and $\hat{\sigma}$ are the MLEs of μ and σ, then the MLE of any function of these statistics can be obtained by simply plugging in the values of $\hat{\mu}$ and $\hat{\sigma}$. However, their confidence intervals are more complex because we have to account for the contribution to the variability due to both $\hat{\mu}$ and $\hat{\sigma}$.

Step 2: Calculate the standard error. To calculate the standard error we need first to calculated the standard errors of the MLEs, then propagate those variances into the new function of interest using the multivariate delta method. To derive the covariance matrix for the MLE $\hat{\theta}$, we leverage the Fisher information matrix presented in Chapter 3. Recall,

$$I(\hat{\theta}) = -E\left[\frac{\partial^2 \log L(\theta)}{\partial \theta_i \, \partial \theta_j}\right]\Bigg|_{\theta=\hat{\theta}}. \tag{6.4}$$

When the expectation is not tractable, we can obtain the *observed Fisher information matrix*, which is obtained by substuting the MLE $\hat{\theta}$ for θ in the second-derivative matrix. The observed Fisher information matrix is

$$\hat{I}(\hat{\theta}) = \left[-\frac{\partial^2 \log L(\theta)}{\partial \theta_i \, \partial \theta_j}\right]_{\theta=\hat{\theta}}.$$

The estimated covariance matrix for $\hat{\theta}$ is then

$$\hat{V}(\hat{\theta}) = \hat{\Sigma}_{\hat{\theta}} = [\hat{I}(\hat{\theta})]^{-1} = \left[-\frac{\partial^2 \log L(\theta)}{\partial \theta_i \, \partial \theta_j}\Bigg|_{\theta=\hat{\theta}}\right]^{-1}.$$

Then the covariance matrix for $\hat{\theta}$ is:

$$V(\hat{\theta}) = \hat{\Sigma}_{\hat{\theta}} = [I(\hat{\theta})]^{-1} = \left[-\frac{\partial^2 \log L(\hat{\theta})}{\partial \theta_i \, \partial \theta_j}\right]^{-1}.$$

The standard error is then simply the square root of the diagonal element of the resulting matrix that corresponds to the parameter of interest.

Multivariate delta method for functions of MLEs. The multivariate delta method is often referred to by its purpose, propagation of error. It is useful for making inferences on functions of estimated parameters (e.g. the percentile function, or estimates of quantities in pre-transformed units). Let $g(\theta)$ represent a function of interest. Then, by the invariance property of MLEs, $\widehat{g(\theta)} = g(\hat{\theta})$. In sufficiently large samples, $g(\hat{\theta})$ is approximately normally distributed with mean $g(\theta)$ and covariance matrix

$$\Sigma_{\hat{g}} = \left[\frac{\partial g(\theta)}{\partial \theta}\right]^T \Sigma_{\hat{\theta}} \left[\frac{\partial g(\theta)}{\partial \theta}\right]. \tag{6.5}$$

An estimate of the covariance matrix, evaluated at $\theta = \hat{\theta}$, is

$$\hat{\Sigma}_{\hat{g}} = \left[\frac{\partial g(\theta)}{\partial \theta}\right]^T \hat{\Sigma}_{\hat{\theta}} \left[\frac{\partial g(\theta)}{\partial \theta}\right]. \tag{6.6}$$

Notice that if $\theta = \theta$, which is a scalar, then the expression reduces to the widely used expression:

$$\hat{V}(g(\hat{\theta})) = \left[\frac{\partial g(\theta)}{\partial \theta}\right]^2 \hat{V}(\hat{\theta}).$$

Let's return to our lognormal example and select the percentile function as our function of interest:

$$g(\mu, \sigma) = t_p = \exp(\mu + \Phi_N^{-1}(p)\sigma).$$

For this function we can calculate the variance of g as

$$
\widehat{\Sigma}_{\hat{g}} = \begin{bmatrix} \dfrac{\partial g}{\partial \mu} \\[4pt] \dfrac{\partial g}{\partial \sigma} \end{bmatrix}^T \Sigma_{\hat{\mu},\hat{\sigma}} \begin{bmatrix} \dfrac{\partial g}{\partial \mu} \\[4pt] \dfrac{\partial g}{\partial \sigma} \end{bmatrix}
$$

$$
= \begin{bmatrix} \hat{g} \\[4pt] \hat{g}\Phi_N^{-1}(p) \end{bmatrix}^T \begin{bmatrix} \widehat{V}(\hat{\mu}) & 0 \\[4pt] 0 & \widehat{V}(\hat{\sigma}) \end{bmatrix} \begin{bmatrix} \hat{g} \\[4pt] \hat{g}\Phi_N^{-1}(p) \end{bmatrix}
$$

$$
= \hat{g}^2 \widehat{V}(\hat{\mu}) + (\Phi_N^{-1}(p))^2 \hat{g}^2 \times \widehat{V}(\hat{\sigma}).
$$

The standard error for $\hat{t}_p$ is

$$
\text{s.e.}(\hat{t}_p) = \text{s.e.}(g(\hat{\mu},\hat{\sigma})) = \hat{t}_p \sqrt{V(\hat{\mu}) + (\Phi_N^{-1}(p))^2 V(\hat{\sigma})}. \tag{6.7}
$$

With this expression one can calculate confidence intervals on any failure time percentile. Consider, for example, the median: here $\Phi_N^{-1}(p) = \Phi_N^{-1}(0.5) = 0$. In this case the expression simplifies substantially and the variance on $t_{0.5}$ can be calculated as $\widehat{\text{s.e.}}(t_{0.5}) = \exp(\hat{\mu})\sqrt{\widehat{V}(\hat{\mu})} = \exp(\hat{\mu})\hat{\sigma}/\sqrt{n}$.

Step 3: Select the Wald statistic and estimate interval. The Wald statistic is specified by

$$
W(\theta) = (\hat{\theta} - \theta)^T [\Sigma_{\hat{\theta}}]^{-1} (\hat{\theta} - \theta). \tag{6.8}
$$

has an asymptotic chi-squared distribution with p degrees of freedom, where p is the number of parameters in θ. This is based on the large sample normal approximation of the MLE. Similarly, we can construct confidence regions for $g(\theta)$ using

$$
W(g(\theta)) = (g(\hat{\theta}) - g(\theta))^T \left[\Sigma_{g(\hat{\theta})}\right]^{-1} (g(\hat{\theta}) - g(\theta)). \tag{6.9}
$$

When g is a scalar, this nicely reduces to the easily constructed confidence interval:

$$
[g_{\text{lower}}, g_{\text{upper}}] = \widehat{g(\theta)} \pm z_{1-\alpha/2}\widehat{\text{s.e.}}(\hat{g}). \tag{6.10}
$$

This Wald, or normal, approximation interval assumes that $g(\hat{\theta})$ is normally distributed with mean $g(\theta)$ and corresponding standard error (s.e.). However, depending on the function, this may not be a reasonable assumption. For lifetime data we still have to deal with the fact that lifetimes are positively bounded and skewed. Therefore, for functions like the mean, median, and percentiles of the lognormal distribution it may be more reasonable to assume

$$
\frac{\log\left(\widehat{g(\theta)}\right) - \log(g(\theta))}{\widehat{\text{s.e.}} \, \log\left(\widehat{g(\theta)}\right)} \sim N(0,1).
$$

It can then be shown using another execution of the multivariate delta method assuming the new function of interest is $q = \log(g)$, that a $(1 - \alpha)100\%$ confidence interval on $g(\theta)$ is

$$
[g_{\text{lower}}, g_{\text{upper}}] = \left[\frac{\widehat{g(\theta)}}{w}, w \times \widehat{g(\theta)}\right], \tag{6.11}
$$

where

$$
w = \exp\left(z_{1-\alpha/2}\frac{\widehat{\text{s.e.}} \, \log\left(\widehat{g(\theta)}\right)}{\widehat{g(\theta)}}\right).
$$

Another common statistic of interest for failure time data is for a given time t_p the percentage of the population expected to fail. In this case it may be best to assume the logit of percentile function is normally distributed.

Example 6.1   Table 6.1 shows 28 failures times in minutes from the simulated distribution we showed histograms for in Figure 6.1. We would like to calculate the mean and 50th percentile (median) with corresponding confidence intervals based on this sample of data.

Table 6.1 Twenty eight failure times (min) from the LOGN(2, 0.5) distribution.

5.132	4.480	8.783	12.845
3.067	5.288	3.934	4.636
7.847	16.604	8.913	24.503
20.142	6.957	8.763	9.922
18.936	11.728	4.445	6.109
6.766	18.477	7.060	7.382
9.510	12.801	7.782	13.638

Solution:

For this data set, we have a vector, **t**, of failures times and a vector, **y**, of log transformed failure times. We assume the failure times are lognormal.

First we will use maximum likelihood estimation for the normal distribution to obtain the estimates of μ and σ. Note that these values are the same for both the normal distribution and the lognormal distribution. For the normal distribution:

$$\hat{\mu} = \frac{1}{n}\sum_{i=1}^{n} \log t_i = \frac{1}{n}\sum_{i=1}^{n} y_i = 2.149$$

and

$$\hat{\sigma} = \sqrt{\frac{1}{n}\sum_{i=1}^{n}(y_i - \bar{y})^2} = 0.529.$$

Step 1: Calculate the statistic of interest. We are interested in the expected value of the lognormal, $E(T)$, and the median, $t_{0.5}$:

$$E(T) = \exp(\mu + \sigma^2/2),$$
$$t_{0.5} = \exp(\mu + \Phi_N^{-1}(0.5)\sigma) = \exp(\mu).$$

The estimated mean and median of the lognormal distribution are, respectively,

$$\bar{t} = \exp(\hat{\mu} + \hat{\sigma}^2/2) = \exp(2.149 + 0.539^2/2) = 9.864$$

and

$$\hat{t}_{0.5} = \exp(\hat{\mu}) = \exp(2.149) = 8.575.$$

Note, that these values are slightly different than if you were to directly compute the mean and median of the 28 data points; this is because we fit the data to a normal distribution first and then transform back into the measurement scale. For large samples one can show these values converge to one another.

Step 2: Calculate the standard error for MLE. To construct a confidence interval about the mean and median of the lognormal distribution, we use the Fisher information matrix for the MLEs of the normal distribution. Recall that if $y = \log t \sim N(\mu, \sigma^2)$, then the probability density function (PDF) is

$$f(y_i | \mu, \sigma) = \frac{1}{\sqrt{2\pi\sigma^2}} \exp\left(-\frac{1}{2} \frac{(y_i - \mu)^2}{\sigma^2} \right).$$

The likelihood for this single sample is

$$L(\mu, \sigma | y) = \left(\frac{1}{\sqrt{2\pi\sigma^2}} \right)^n \exp\left(-\frac{1}{2\sigma^2} \sum_{i=1}^{n} (y_i - \mu)^2 \right).$$

Note, for simplicity we are using $\log L = \ell$ to denote the log-likelihood. This is

$$\ell(\mu, \sigma | y) = -\frac{n}{2} \log(2\pi) - n \log(\sigma^2) - \frac{1}{2\sigma^2} \sum_{i=1}^{n} (y_i - \mu)^2.$$

For the one-sample case when $\theta = (\mu, \sigma)^T$ the variance–covariance matrix is

$$
\begin{aligned}
\hat{\Sigma}_{\hat{\theta}} &=
\begin{bmatrix}
\hat{V}(\hat{\mu}) & \widehat{\text{cov}}(\hat{\mu}, \hat{\sigma}) \\
\widehat{\text{cov}}(\hat{\mu}, \hat{\sigma}) & \hat{V}(\hat{\sigma})
\end{bmatrix} \\
&=
\begin{bmatrix}
-\dfrac{\partial^2 \ell(\mu, \sigma)}{\partial \mu^2} & -\dfrac{\partial^2 \ell(\mu, \sigma)}{\partial \mu\, \partial \sigma} \\
-\dfrac{\partial^2 \ell(\mu, \sigma)}{\partial \mu\, \partial \sigma} & -\dfrac{\partial^2 \ell(\mu, \sigma)}{\partial \sigma^2}
\end{bmatrix}^{-1} \Bigg|_{(\hat{\mu}, \hat{\sigma})} \\
&=
\begin{bmatrix}
\dfrac{\hat{\sigma}^2}{n} & 0 \\
0 & \dfrac{\hat{\sigma}^2}{2n}
\end{bmatrix} \\
&=
\begin{bmatrix}
0.01000283 & 0 \\
0 & 0.005001416
\end{bmatrix}.
\end{aligned}
$$

For the mean we will use the function

$$g_1(\mu, \sigma) = \exp\left(\mu + \frac{\sigma^2}{2} \right)$$

for which

$$
\hat{\Sigma}_{\hat{g}_1} =
\begin{bmatrix}
\dfrac{\partial g_1}{\partial \mu} \\
\dfrac{\partial g_1}{\partial \sigma}
\end{bmatrix}^T
\Sigma_{\hat{\mu}, \hat{\sigma}}
\begin{bmatrix}
\dfrac{\partial g_1}{\partial \mu} \\
\dfrac{\partial g_1}{\partial \sigma}
\end{bmatrix}
$$

$$
= \begin{bmatrix} \hat{g}_1 \\ \hat{g}_1 \sigma \end{bmatrix}^T \begin{bmatrix} \hat{V}(\hat{\mu}) & 0 \\ 0 & \hat{V}(\hat{\sigma}) \end{bmatrix} \begin{bmatrix} \hat{g}_1 \\ \hat{g}_1 \sigma \end{bmatrix}
$$

$$
= \hat{g}_1^2 \hat{V}(\hat{\mu}) + \sigma^2 \hat{g}_1^2 \times \hat{V}(\hat{\sigma})
$$

and

$$
\widehat{s.e.}(g_1) = \hat{g}_1 \sqrt{\hat{V}(\hat{\mu}) + \sigma^2 \hat{V}(\hat{\sigma})} = 1.053342.
$$

For the median we will use the function g_2, where

$$
g_2(\mu, \sigma) = \exp(\mu).
$$

We previously showed in Eq. (6.7) the standard error for a percentile reduces to the following for the median:

$$
\widehat{s.e.}(g_2) \doteq \widehat{s.e.}(t_{.5}) = \exp(\hat{\mu}) \sqrt{\hat{V}(\hat{\mu})} = \exp(\hat{\mu}) \hat{\sigma} / \sqrt{n} = 0.8576121.
$$

Step 3: Select the Wald statistic and estimate interval. For large samples the estimator $\hat{g}$ is approximately normally distributed, so calculating the Wald confidence interval is straightforward:

$$
[g_{\text{lower}}, g_{\text{upper}}] = \widehat{g(\theta)} \pm z_{1-\alpha/2} \widehat{s.e.}(\hat{g}).
$$

Using this formulation of Wald's interval our 95% confidence interval on the mean (expected value) of the lognormal distribution works out to [7.799, 11.928], and the 95% confidence interval on median (50th percentile) of the lognormal distribution works out to: [6.894, 10.256]. However, it is unlikely that the assumption that the distributions of the estimators for the mean and median follow a normal distribution. For functions like the mean and median it is more reasonable that

$$
\frac{\log\left(\widehat{g(\theta)}\right) - \log(g(\theta))}{\widehat{s.e.} \, \log\left(\widehat{g(\theta)}\right)} \overset{\text{approx}}{\sim} N(0,1).
$$

Then a $(1 - \alpha)100\%$ confidence interval on $g(\theta)$ is

$$
[g_{\text{lower}}, g_{\text{upper}}] = \left[\frac{\widehat{g(\theta)}}{w}, w \times \widehat{g(\theta)} \right],
$$

where

$$
w = \exp\left(z_{\alpha/2} \frac{\widehat{s.e.} \, \log\left(\widehat{g(\theta)}\right)}{\widehat{g(\theta)}} \right).
$$

The value of $w(g_1)$ and the corresponding 95% confidence interval on the expected value are then

$$
w(g_1) = \exp\left(z_{0.025} \frac{\widehat{s.e.}(g_1)}{\hat{g}_1} \right) = 1.23282
$$

and

$$
[g_{1.\text{lower}}, g_{1.\text{upper}}] = [8.001, 12.160].
$$

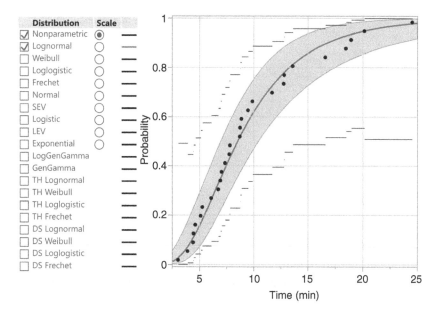

Figure 6.2 Parametric lifetime distribution fit to the failure time data.

The value of $w(g_2)$ and the corresponding 95% confidence interval on the median are:

$$w(g_1) = \exp\left(z_{0.025} \frac{\widehat{\text{s.e.}}(g_2)}{\hat{g}_2} \right) = 1.216561$$

and

$$\left[g_{2.\text{lower}}, g_{2.\text{upper}} \right] = [7.048, 10.432] \,.$$

The Fit Life Distribution application in JMP is useful for confirming our analysis. Use the Analyze $\rightarrow$ Reliability and Survival $\rightarrow$ Life Distribution path to get distribution fitting options. Note, there is an option to select the Confidence Interval Method, which defaults to Wald, but Likelihood (short for Likelihood Ratio) is another option. Once in the platform the user can select the Lognormal distribution. Figure 6.2 shows a lognormal distribution fit to the data set.

The covariance matrix output can be used to confirm the standard error estimates (note: variances are shown in covariance matrix). The Profilers allow for the exploration of different value of the distribution. Figure 6.3 shows that plugging into the quantile profiler $p = 0.5$ confirms the Wald interval for the median as [7.048, 10.432]. Therefore, JMP is assuming that the estimator of $\log(t_p)$ is distributed normally. ∎

Example 6.2 Recall the worsted yarn example from Chapter 5, where we considered the effect of three factors, x_1 = length, x_2 = amplitude, and x_3 = load on the number of cycles to effective failure, y, of worsted yarn. In Chapter 5, we showed that using a transformation $y^* = \log t$ we can model the data with a linear model for y^*.

Use a log transformation on the number of cycles to failure and fit a regression model to the corresponding log failure cycles. Calculate the 50th percentile lifetime in the original time scale at the configuration

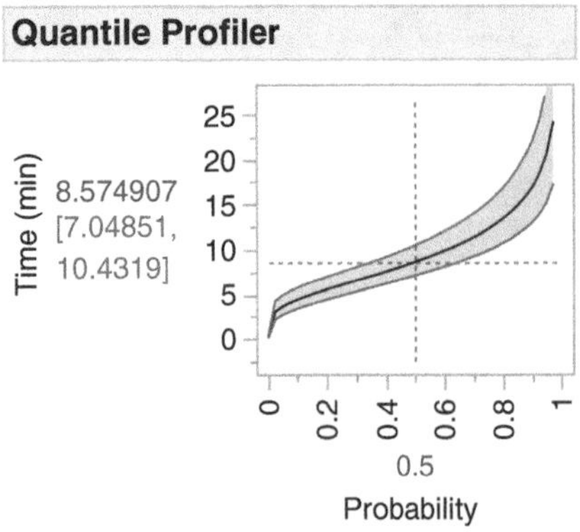

Quantile Profiler

Figure 6.3 Quantile profiler showing the median estimate and 95% confidence bounds.

of length, amplitude, and load that maximizes lifetime. Provide corresponding confidence intervals in the original time scale.

Solution:
Recall that we are considering the impact of three factors, x_1 = length, x_2 = amplitude, and x_3 = load on the number of cycles to effective failure, y, of worsted yarn. The final models from Chapter 5 are

$$y^* = 6.335 + 0.832x_1 - 0.631x_2 - 0.392x_3$$

and

$$y = \exp(6.335 + 0.832x_1 - 0.631x_2 - 0.392x_3).$$

A similar process can be used to make predictions and provide associated confidence intervals.

Step 1: Calculate the statistic of interest. Recall, the median, $t_{0.50}$, is given by

$$t_{0.5} = \exp(\mu + \Phi_N^{-1}(0.5)\sigma)$$
$$= \exp(\mu)$$

which is estimated by

$$\hat{t}_{0.5} = \exp\left(6.335 + 0.832x_1 - 0.631x_2 - 0.392x_3\right).$$

Therefore, the prediction expression given earlier is actually for the 50% percentile (median) of the lognormal. Consider the values of x_1 = length, x_2 = amplitude, and x_3 = load within the feasible region that maximize the number of cycles until failure: $(1, -1, -1)$, respectively. So maximizing length, minimizing amplitude, and minimizing load gives the maximum predicted number of cycles to failure. Specifically,

$$\hat{t}_{0.5} = \exp(\hat{\mu})$$
$$= \exp\left(6.335 + 0.832 \times (1) - 0.631 \times (-1) - 0.392(-1)\right)$$
$$= 3606.6.$$

Step 2: Calculate the standard error for MLE. To construct a confidence interval about the predicted median of the lognormal distribution, we use the Fisher information matrix for the MLEs of the normal distribution. We will use JMP to get the final intervals for this problem, following the same process as the

Table 6.2 Regression parameter estimates and standard errors.

Parameter Estimates				
Term	Estimate	Std Error	Lower 95%	Upper 95%
Intercept	6.33466433	0.0329717	6.270041	6.3992876
x1	0.83238416	0.0403819	0.7532371	0.9115312
x2	−0.6309916	0.0403819	−0.710139	−0.551845
x3	−0.392494	0.0403819	−0.471641	−0.313347
σ	0.17132582	0.0233145	0.1256303	0.2170214
Confidence Intervals are Wald				

previous example, but with our new g function. Table 6.2 shows the parameter estimates and the standard errors for the MLE.

These standard errors are based on the parameter's Fisher information matrix. For the linear regression case $\theta = (\beta_0, \beta_1, \beta_2, \beta_3, \sigma)^T$ and the information matrix is:

$$\hat{\Sigma}_{\hat{\theta}} = \begin{bmatrix} \hat{V}(\hat{\beta}_0) & \widehat{\text{cov}}(\hat{\beta}_0, \hat{\beta}_1) & \widehat{\text{cov}}(\hat{\beta}_0, \hat{\beta}_2) & \widehat{\text{cov}}(\hat{\beta}_0, \hat{\beta}_3) & \widehat{\text{cov}}(\hat{\beta}_0, \hat{\sigma}) \\ \widehat{\text{cov}}(\hat{\beta}_0, \hat{\beta}_1) & \hat{V}(\hat{\beta}_1) & \widehat{\text{cov}}(\hat{\beta}_1, \hat{\beta}_2) & \widehat{\text{cov}}(\hat{\beta}_1, \hat{\beta}_3) & \widehat{\text{cov}}(\hat{\beta}_1, \hat{\sigma}) \\ \widehat{\text{cov}}(\hat{\beta}_0, \hat{\beta}_2) & \widehat{\text{cov}}(\hat{\beta}_1, \hat{\beta}_2) & \hat{V}(\hat{\beta}_2) & \widehat{\text{cov}}(\hat{\beta}_2, \hat{\beta}_3) & \widehat{\text{cov}}(\hat{\beta}_2, \hat{\sigma}) \\ \widehat{\text{cov}}(\hat{\beta}_0, \hat{\beta}_3) & \widehat{\text{cov}}(\hat{\beta}_1, \hat{\beta}_3) & \widehat{\text{cov}}(\hat{\beta}_2, \hat{\beta}_3) & \hat{V}(\hat{\beta}_3) & \widehat{\text{cov}}(\hat{\beta}_3, \hat{\sigma}) \\ \widehat{\text{cov}}(\hat{\beta}_0, \hat{\sigma}) & \widehat{\text{cov}}(\hat{\beta}_1, \hat{\sigma}) & \widehat{\text{cov}}(\hat{\beta}_2, \hat{\sigma}) & \widehat{\text{cov}}(\hat{\beta}_3, \hat{\sigma}) & \hat{V}(\hat{\sigma}) \end{bmatrix}$$

$$= \begin{bmatrix} -\frac{\partial^2 \ell(\beta_0,\beta_1,\beta_2,\beta_3,\sigma)}{\partial\beta_0^2} & -\frac{\partial^2 \ell(\beta_0,\beta_1,\beta_2,\beta_3,\sigma)}{\partial\beta_0\partial\beta_1} & -\frac{\partial^2 \ell(\beta_0,\beta_1,\beta_2,\beta_3,\sigma)}{\partial\beta_0\partial\beta_2} & -\frac{\partial^2 \ell(\beta_0,\beta_1,\beta_2,\beta_3,\sigma)}{\partial\beta_0\partial\beta_3} & -\frac{\partial^2 \ell(\beta_0,\beta_1,\beta_2,\beta_3,\sigma)}{\partial\beta_0\partial\sigma} \\ -\frac{\partial^2 \ell(\beta_0,\beta_1,\beta_2,\beta_3,\sigma)}{\partial\beta_0\partial\beta_1} & -\frac{\partial^2 \ell(\beta_0,\beta_1,\beta_2,\beta_3,\sigma)}{\partial\beta_1^2} & -\frac{\partial^2 \ell(\beta_0,\beta_1,\beta_2,\beta_3,\sigma)}{\partial\beta_1\partial\beta_2} & -\frac{\partial^2 \ell(\beta_0,\beta_1,\beta_2,\beta_3,\sigma)}{\partial\beta_1\partial\beta_3} & -\frac{\partial^2 \ell(\beta_0,\beta_1,\beta_2,\beta_3,\sigma)}{\partial\beta_1\partial\sigma} \\ -\frac{\partial^2 \ell(\beta_0,\beta_1,\beta_2,\beta_3,\sigma)}{\partial\beta_0\partial\beta_2} & -\frac{\partial^2 \ell(\beta_0,\beta_1,\beta_2,\beta_3,\sigma)}{\partial\beta_1\partial\beta_2} & -\frac{\partial^2 \ell(\beta_0,\beta_1,\beta_2,\beta_3,\sigma)}{\partial\beta_2^2} & -\frac{\partial^2 \ell(\beta_0,\beta_1,\beta_2,\beta_3,\sigma)}{\partial\beta_2\partial\beta_3} & -\frac{\partial^2 \ell(\beta_0,\beta_1,\beta_2,\beta_3,\sigma)}{\partial\beta_2\partial\sigma} \\ -\frac{\partial^2 \ell(\beta_0,\beta_1,\beta_2,\beta_3,\sigma)}{\partial\beta_0\partial\beta_3} & -\frac{\partial^2 \ell(\beta_0,\beta_1,\beta_2,\beta_3,\sigma)}{\partial\beta_1\partial\beta_3} & -\frac{\partial^2 \ell(\beta_0,\beta_1,\beta_2,\beta_3,\sigma)}{\partial\beta_2\partial\beta_3} & -\frac{\partial^2 \ell(\beta_0,\beta_1,\beta_2,\beta_3,\sigma)}{\partial\beta_3^2} & -\frac{\partial^2 \ell(\beta_0,\beta_1,\beta_2,\beta_3,\sigma)}{\partial\beta_3\partial\sigma} \\ -\frac{\partial^2 \ell(\beta_0,\beta_1,\beta_2,\beta_3,\sigma)}{\partial\beta_0\partial\sigma} & -\frac{\partial^2 \ell(\beta_0,\beta_1,\beta_2,\beta_3,\sigma)}{\partial\beta_1\partial\sigma} & -\frac{\partial^2 \ell(\beta_0,\beta_1,\beta_2,\beta_3,\sigma)}{\partial\beta_2\partial\sigma} & -\frac{\partial^2 \ell(\beta_0,\beta_1,\beta_2,\beta_3,\sigma)}{\partial\beta_3\partial\sigma} & -\frac{\partial^2 \ell(\beta_0,\beta_1,\beta_2,\beta_3,\sigma)}{\partial\sigma^2} \end{bmatrix}.$$

Wald's method can again be used to calculate confidence intervals on the model parameters. For the estimated median we will use the function g_{reg} defined by

$$g_{\text{reg}}(\beta_0, \beta_1, \beta_2, \beta_3, \sigma) = \exp\left(\beta_0 + \beta_1 x_1 + \beta_2 x_2 + \beta_3 x_3\right).$$

In this case $\theta = \left(\beta_0 + \beta_1 x_1 + \beta_2 x_2 + \beta_3 x_3\right)^T$. Just as before, the standard error for the function g_{reg} can be calculated using:

$$\hat{\Sigma}_{\hat{g}} = \left[\frac{\partial \mathbf{g}(\theta)}{\partial \theta}\right]^T \hat{\Sigma}_{\hat{\theta}} \left[\frac{\partial \mathbf{g}(\theta)}{\partial \theta}\right].$$

Step 3: Select the Wald statistic and estimate interval. As you can see, the math can quickly grow cumbersome, so we will rely on JMP to get the final estimates. The Fit Parametric Survival Function under

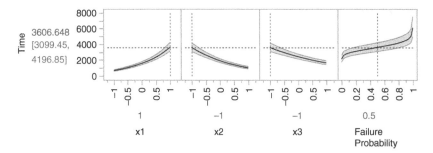

Figure 6.4 Quantile profiler from JMP® survival fit showing factor settings that lead to maximum median failure cycles.

Analyze →Reliability and Survival → Fit Parametric Survival brings up an analysis box similar to the traditional fit model. However, now we can choose from distributions more appropriate for failure time data. Selecting the lognormal distribution yields the results captured in Figure 6.4. Using the quantile profiler you can determine that the predicted time for 50% of the failures to occur and the corresponding 95% confidence interval at $x_1 = 1$, $x_2 = -1$, and $x_3 = -1$ are 3606.6 cycles [3099.5, 4196.9]. ∎

6.3 Generalized Linear Models

Data transformations are often an effective way to deal with the problem of non-normal responses and the associated inequality of variance. As we saw in Example 5.1, the Box–Cox method is an easy and effective way to select the form of the transformation. However, the use of a data transformation poses some problems. One problem is that the experimenter may be uncomfortable in working with the response in the transformed scale. That is, he or she is interested in the number of defects, not the square root of the number of defects, or in the number of cycles to failure instead of the logarithm of the number of cycles to failure. On the other hand, if a transformation is successful and improves the analysis and the associated model for the response, experimenters will usually quickly adopt the new metric.

A more serious problem is that a transformation can result in a *nonsensical value* for the response variable over some portion of the design factor space that is of interest to the experimenter. For example, suppose that we have used the square root transformation in an experiment involving the number of defects observed on semiconductor wafers, and for some portion of the region of interest the predicted square root of the count of defects is negative. This is likely to occur for situations where the actual number of defects observed is small. Consequently, the model for the experiment has produced an obviously unreliable prediction in the very region where we would like this model to have good predictive performance. Finally, we often use transformations to accomplish several objectives; stabilize variance, induce normality, and model simplification. There is no assurance that a transformation will effectively accomplish all of these objectives simultaneously.

An alternative to the typical approach of data transformation followed by standard least squares analysis of the transformed response is to use the generalized linear model (GLM). This is an approach developed by Nelder and Wedderburn (1972) that essentially unifies linear and nonlinear models with both normal and non-normal responses. McCullagh and Nelder (1989) and Myers et al. (2012) give comprehensive presentations of GLMs, and Myers and Montgomery (1997) provide a tutorial.

Like all regression models, a GLM is made up of a random component (what we have usually called the error term) and a function of the design factors (the xs) and some unknown parameters (the βs). In a standard normal-theory linear regression model, we write

$$y = \beta_0 + \beta_1 x_1 + \beta_2 x_2 + \cdots + \beta_k x_k + \epsilon, \tag{6.12}$$

where the error term ϵ is assumed to have a normal distribution with mean zero and constant variance, and the mean of the response variable y is

$$E(y) = \beta_0 + \beta_1 x_1 + \beta_2 x_2 + \cdots + \beta_k x_k = \boldsymbol{x}' \boldsymbol{\beta}. \tag{6.13}$$

The portion $\boldsymbol{x}' \boldsymbol{\beta}$ of Eq. (6.13) is called the *linear predictor*. The GLM contains Eq. (6.12) as a special case. In a GLM, the response variable can have any distribution that is a member of the *exponential family*. This family includes the normal, Poisson, binomial, exponential, and gamma distributions, so the exponential family is a very rich and flexible collection of distributions applicable to many experimental situations, including several frequently encountered in reliability engineering. Also, the relationship between the response mean μ and the linear predictor $\boldsymbol{x}' \boldsymbol{\beta}$ is determined by a link function

$$g(\mu) = \boldsymbol{x}' \boldsymbol{\beta}. \tag{6.14}$$

The regression model that represents the mean response is then given by

$$\mu = g^{-1}(\boldsymbol{x}' \boldsymbol{\beta}).$$

For example, the link function leading to the ordinary linear regression model in Eq. (6.12) is called the *identity link* because $\mu = g^{-1}(\boldsymbol{x}' \boldsymbol{\beta}) = \boldsymbol{x}' \boldsymbol{\beta}$. As another example, the *log link*

$$\log \mu = \boldsymbol{x}' \boldsymbol{\beta} \tag{6.15}$$

produces the model where

$$\mu = \exp(\boldsymbol{x}' \boldsymbol{\beta}). \tag{6.16}$$

The log link is often used with count data (Poisson response) and with continuous responses that have a distribution that has a long tail to the right (the exponential or gamma distribution). Another important link function used with binomial data is the *logit link*

$$\log \frac{\mu}{1 - \mu} = \boldsymbol{x}' \boldsymbol{\beta}. \tag{6.17}$$

This choice of link function leads to the model

$$\mu = \frac{1}{1 + \exp(-\boldsymbol{x}' \boldsymbol{\beta})}. \tag{6.18}$$

Many choices of link function are possible, but it must always be monotonic and differentiable. Also, note that in a GLM, the variance of the response variable does not have to be a constant; it can be a function of the mean (and the predictor variables through the link function). For example, if the response is Poisson, the variance of the response is exactly equal to the mean.

To use a GLM in practice, the experimenter must specify a response distribution and a link function. Then the model fitting or parameter estimation is done by the method of maximum likelihood, which

Table 6.3 Canonical links for generalized linear models.

Distribution	Type	Typical reliability use	Canonical link function	Link name
Normal	Continuous	Failure times (with transformation)	$x'\beta = \mu$	Identity
Poisson	Discrete	Number of defects in a fixed time/length	$x'\beta = \log(\mu)$	Log
Binomial	Discrete	Number of independent failures	$x'\beta = \log(\frac{\mu}{1-\mu})$	Logit
Gamma	Continuous	Failure times, cycles until failure	$x'\beta = -\mu^{-1}$	Negative inverse

for the exponential family can be achieved by an iterative version of weighted least squares. For ordinary linear regression or experimental design models with a normal response variable, this reduces to standard least squares. Using an approach that is analogous to the analysis of variance for normal-theory data, we can perform inference and diagnostic checking for a GLM. Refer to Myers and Montgomery (1997) and Myers et al. (2012) for the details and examples. Software packages that support the GLMs include SAS (PROC GENMOD), R (glm()), JMP, and Minitab.

Table 6.3 lists several distributions in the exponential family, their typical uses in a reliability context, and the canonical link function. Notice, by inverting the link function we can find the expression for the mean (μ) function.

Logistic regression is widely used and may be the most common application of the GLM. While much of this textbook focuses on failure times and lifetimes, logistic regression assumes a binary outcome. As discussed in Chapter 2, the binomial distribution is the discrete probability distribution of the number of successes (or failures) in a sequence of n independent experiments. Logistic regression uses a logistic function to model a binary dependent variable. It finds wide application in the biomedical field with dose-response studies (where the design factor is the dose of a particular therapeutic treatment and the response is whether the patient responds successfully to the treatment). Many reliability engineering experiments involve binary (success–failure) data, such as when units of products or components are subjected to a stress or load and the response is whether the unit fails.

Example 6.3 A consumer products company is studying the factors that impact the chances that a customer will redeem a coupon for one of its personal care products. A 2^3 factorial experiment was conducted to investigate the following variables: A = coupon value (low, high), B = length of time for which the coupon is valid, and C = ease of use (easy, hard). A total of 1000 customers were randomly selected for each of the eight cells of the 2^3 design, and the response is the number of coupons redeemed. The experimental results are shown in Table 6.4. Use an appropriate GLM to determine the factors that affect coupon redemption.

Solution:
We can think of the response as the number of successes out of 1000 Bernoulli trials in each cell of the design, so a reasonable model for the response is a GLM with a binomial response distribution and a logit link. This particular form of the GLM is usually called *logistic regression*.

The experimenters decided to fit a model involving only the main effects and two-factor interactions. In JMP, Analyze →Fit Model → Personality: Generalized Linear Model, Distribution: Binomial, Link

Table 6.4 Number of coupons redeemed for each of eight treatments.

A	B	C	y
−	−	−	200
+	−	−	250
−	+	−	265
+	+	−	347
−	−	+	210
+	−	+	286
−	+	+	272
+	+	+	326

Function: Logit. Note, to fully specify the model in any software package to the data table previously you have to include an additional data column for frequency with all values set to 1000. For the logistic regression model, the response is the total number of successes divided by N, where $N = 1000$ for each row. In JMP the "Freq" Role allows for the inclusion of the total sample size column. Table 6.5 presents a portion of the JMP output for the data in Table 6.4 for the coupon redemption experiment. The table shows the model term effect summary for the full model involving all three main effects and the three two-factor interactions. Notice that currently only two of the effects appear to be statistically significant, as shown by the vertical line in the chart and the small P-values. Table 6.6 shows the model term estimates and standardized errors. Additionally, notice that the output contains a likelihood ratio (L-R) Chi-Square and P-value (Prob>Chi-Sq). The likelihood ratio test is the standard method for testing model parameter significance in a GLM. It is an alternative to the Wald test described previously. In the likelihood ratio test two models are compared based on their contribution to the likelihood as measured through the deviance (-2 log-likelihood). The test compares the model with the term under consideration to the null model (intercept only).

Tables 6.5–6.7 present the analysis for a reduction of the model to the final model containing the three main effects and the BC interaction (factor C was included to maintain model hierarchy). This first step in model selection uses backwards selection to remove the two least significant terms. Arguably, the model could be reduced further by eliminating factor C and the $B * C$ interaction. However, they were

Table 6.5 JMP® output for full model in the coupon redemption experiment.

Source	LogWorth		PValue
B: Time	10.666		0.00000
A: Coupon Value	10.473		0.00000
B: Time*C: Ease of Use	0.952		0.11180
C: Ease of Use	0.459		0.34725
A: Coupon Value*B: Time	0.105		0.78567
A: Coupon Value*C: Ease of Use	0.050		0.89160

Table 6.6 JMP® parameter estimates for the full model.

Term	Estimate	Std Error	L-R ChiSquare	Prob>ChiSq	Lower CL	Upper CL
Intercept	−1.010907	0.0255113	1785.9437	<.0001*	−1.061115	−0.961107
A: Coupon Value	0.1685814	0.0255056	43.952526	<.0001*	0.1186469	0.2186364
B: Time	0.1702662	0.0255113	44.824093	<.0001*	0.1203217	0.2203335
C: Ease of Use	0.0239715	0.0255061	0.8834647	0.3473	−0.026012	0.07398
A: Coupon Value*B: Time	−0.006936	0.0255084	0.0739534	0.7857	−0.056955	0.0430454
A: Coupon Value*C: Ease of Use	−0.003465	0.025428	0.0185717	0.8916	−0.053312	0.0463734
B: Time*C: Ease of Use	−0.04043	0.0254307	2.5285329	0.1118	−0.090294	0.0094015

Table 6.7 JMP® output for reduced model in the coupon redemption experiment.

Term	Estimate	Std Error	L-R ChiSquare	Prob>ChiSq	Lower CL	Upper CL
Intercept	−1.010765	0.0255035	1785.9057	<.0001*	−1.060957	−0.960979
A: Coupon Value	0.1679851	0.025419	43.927436	<.0001*	0.1182174	0.2178666
B: Time	0.1697089	0.0254287	44.808277	<.0001*	0.1199234	0.2196109
C: Ease of Use	0.0236817	0.0254272	0.867583	0.3516	−0.026147	0.0735345
B: Time*C: Ease of Use	−0.040369	0.0254273	2.5216104	0.1123	−0.090226	0.0094556

maintained to illustrate the relative impact of these not highly significant terms. The fitted model is

$$\hat{y} = \frac{1}{1 + \exp(-1.01 + 0.168x_1 + 0.170x_2 + 0.023x_3 - 0.040x_2x_3)}.$$

This gives the predicted probability of a success (redeeming a coupon) as a function of the predictor variables.

Figure 6.5 shows the percent of coupon redemption for a high coupon value, a long validity time, and low ease of use. Notice the slope on ease of use is relatively flat. The coefficients table (Table 6.7) suggests this is the case, as the 95% confidence intervals include the value 0. ∎

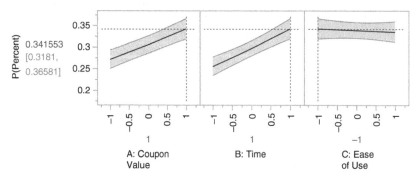

Figure 6.5 Prediction profiler for the percent coupon redemption.

Example 6.4 Return to the worsted yarn example from Chapter 5 one more time to illustrate the application of a GLM to the data. Use an appropriate GLM to analyze this data set, and compare your results with the log transform approach.

Solution:

Recall the response is the number of cycles to failure. The data were analyzed originally using the standard (least squares) approach, and data transformation was necessary to stabilize the variance. The natural log of the cycles to failure data was found to yield an adequate model in terms of overall model fit and satisfactory residual plots. The model in the original response units of number of cycles to failure is

$$\hat{y} = \exp(6.33 + 0.832x_1 - 0.631x_2 - 0.392x_3).$$

Now we analyze this experiment using the GLM and selecting the gamma response distribution and the log link. In R we used the `glm()` function and specified Gamma(link = "log"). Notice that this is not the canonical link shown in Table 6.3; we use the log-link here for a comparable model form to the previously demonstrated log-transformation approach. We used exactly the same model form found by least squares analysis of the log-transformed response. The summary.glm() function gives the following result.

```
Call:
glm(formula = y ~ x1 + x2 + x3,
    family = Gamma(link = "log"),
    data = WorstedYarnData)

Deviance Residuals:
Min        1Q       Median     3Q        Max
-0.43391   -0.11553  -0.00922   0.10260   0.25342

Coefficients:
Estimate Std. Error t value Pr(>|t|)
(Intercept)   6.34891      0.03422 185.511  < 2e-16  ***
x1            0.84251      0.04192  20.100 4.34e-16  ***
x2           -0.63132      0.04192 -15.062 2.10e-13  ***
x3           -0.38513      0.04192  -9.188 3.68e-09  ***
---
Signif. codes:  0 '***' 0.001 '**' 0.01 '*' 0.05 '.' 0.1 ' ' 1

(Dispersion parameter for Gamma family taken to be 0.0316243)

Null deviance: 22.88613  on 26   degrees of freedom
Residual deviance:  0.76939  on 23   degrees of freedom
AIC: 332.76

Number of Fisher Scoring iterations: 5
```

The resulting prediction expression is

$$\hat{y} = \exp(6.35 + 0.84x_1 - 0.63x_2 - 0.39x_3).$$

Table 6.8 Comparing worsted yarn reliability results.

| Obs | Least squares methods with log data | | | | | | Length of 95% confidence interval | |
| | Transformed | | Untransformed | | Using a GLM | | | |
	Estimate of mean	95% confidence interval	Estimate of mean	95% confidence interval	Estimate of mean	95% confidence interval	Least squares	GLM
1	2.83	(2.76, 2.91)	682.50	(573.85, 811.52)	680.52	(583.83, 793.22)	237.67	209.39
2	2.66	(2.60, 2.73)	460.26	(397.01, 533.46)	463.00	(407.05, 526.64)	136.45	119.59
3	2.49	(2.42, 2.57)	310.38	(260.98, 369.06)	315.01	(271.49, 365.49)	108.09	94.00
4	2.56	(2.50, 2.62)	363.25	(313.33, 421.11)	361.96	(317.75, 412.33)	107.79	94.58
5	2.39	(2.34, 2.44)	244.96	(217.92, 275.30)	246.26	(222.55, 272.51)	57.37	49.96
6	2.22	(2.15, 2.28)	165.20	(142.50, 191.47)	167.55	(147.67, 190.10)	48.97	42.42
7	2.29	(2.21, 2.36)	193.33	(162.55, 229.93)	192.52	(165.69, 226,370)	67.38	58.01
8	2.12	(2.05, 2.18)	130.38	(112.46, 151.15)	130.98	(115.43, 148.64)	38.69	33.22
9	1.94	(1.87, 2.02)	87.92	(73.93, 104.54)	89.12	(76.87, 103.32)	30.62	26.45
10	3.20	(3.13, 3.26)	1569.28	(135,394, 1819.28)	1580.00	(1390.00, 1797.00)	465.34	407.00
11	3.02	(2.97, 3.08)	1058.28	(941.67, 1189.60)	1072.00	(972.52, 1189.00)	247.92	216.48
12	2.85	(2.79, 2.92)	713.67	(615.60, 827.37)	731.50	(644.35, 830.44)	211.97	186.09
13	2.92	(2.87, 2.97)	835.41	(743.19, 938.86)	840.54	(459.65, 930.04)	195.67	170.39
14	2.75	(2.72, 2.78)	563.25	(523.24, 606.46)	571.87	(536.67, 609.38)	83.22	72.70
15	2.58	(2.53, 2.63)	379.84	(337.99, 426.97)	389.08	(351.64, 430.51)	88.99	78.87
16	2.65	(2.58, 2.71)	444.63	(383.53, 515.35)	447.07	(393.81, 507.54)	131.82	113.74
17	2.48	(2.43, 2.53)	299.85	(266.75, 336.98)	304.17	(275.13, 336.28)	70.23	61.15
18	2.31	(2.24, 2.37)	202.16	(174.42, 234.37)	206.95	(182.03, 235.27)	59.95	53.23
19	3.56	(3.48, 3.63)	3609.11	(3034.59, 4292.40)	3670.00	(3165.00, 4254.00)	1257.81	1089.00
20	3.39	(3.32, 3.45)	2433.88	(2099.42, 2821.63)	2497.00	(2200.00, 2833.00)	722.21	633.00
21	3.22	(3.14, 3.29)	1641.35	(1380.07, 1951.64)	1699.00	(1462.00, 1974.00)	571.57	512.00
22	3.28	(3.22, 3.35)	1920.88	(1656.91, 2226.90)	1952.00	(1720.00, 2215.00)	569.98	495.00
23	3.11	(3.06, 3.16)	1295.39	(1152.66, 1455.79)	1328.00	(1200.00, 1470.00)	303.14	270.00
24	2.94	(2.88, 3.01)	873.57	(753.53, 1012.74)	903.51	(793.15, 1028.00)	259.22	235.85
25	3.01	(2.93, 3.08)	1022.35	(859.81, 1215.91)	1038.00	(894.79, 1205.00)	356.10	310.21
26	2.84	(2.77, 2.90)	689.45	(597.70, 799.28)	706.34	(620.99, 803.43)	204.58	182.44
27	2.67	(2.59, 2.74)	464.94	(390.93, 552.97)	480.57	(412.29, 560.15)	162.04	147.86

This is nearly identical to the model found using least squares. Table 6.8 presents the predicted values from the least squares model and the GLM, along with 95% confidence intervals on the mean response at each of the 27 points in the design. A comparison of the lengths of the confidence intervals reveals that the GLM is likely to be a more precise predictor than the least squares model. ∎

6.4 Incorporating Censoring in Regression Models

As we saw in Chapter 3, censoring data is easily incorporated into the model through the likelihood function. The likelihood function measures the goodness of fit of a statistical model with unknown parameters to a given sample of data. It is used to develop estimates of the parameters via maximization of the likelihood function (maximum likelihood estimation) given a set of data. The likelihood contribution essentially integrates over that area to include the censored information in the estimation of model parameters. This general framework provides the methods for incorporating censoring in any analysis that leverages likelihood functions for assessing the fit of a model to a set of data.

Recall, that the total likelihood can be expressed as

$$L(\theta, t) = c \prod_{i=1}^{n} (L_i(\theta, t_i)) \tag{6.19}$$

$$= L(\theta, t) c \prod_{i=1}^{n} (F(t_i)^{l_i} (f_i(\theta, t_i)^{\delta}_i)(1 - F(t_i)^{r_i}, \tag{6.20}$$

where $\ell_i = 1$ if observation i is left censored at time t_i, and 0 otherwise, $r_i = 1$ if observation i is right censored at time t_i and 0 otherwise, and $\delta_i = 1 - \ell_i - r_i$ (i.e. δ_i is one for failure times, and 0 for censored times). Here c is a constant of proportionality that is independent of the parameters.

All parametric regression models can be estimated via maximum likelihood estimation, making it straightforward to included censoring for any distribution. However, when using commercial software not all packages have incorporated censoring, it is often necessary to develop code to incorporate censoring in the analysis. For example, the GLM packages in R and JMP do not include censoring options, only the survival analysis packages allow censoring, so if we wanted to use the gamma distribution with censoring, we would need to derive the likelihood, estimates, and inferences and code them into software. Here we show a single predictor variable regression example of that process for the location scale and log-location scale distributions.

6.4.1 Parameter Estimation for Location Scale and Log-Location Scale Models

The most common location scale distributions in reliability data analysis include the normal distribution, the smallest extreme value (SEV) distribution, and the logistic distribution. The likelihood function for location scale distributions if there is no censoring present is

$$L(\mu, \sigma; \mathbf{y}) = \prod_{i=1}^{n} f(y_i; \mu, \sigma) = \prod_{i=1}^{n} \frac{1}{\sigma} \phi(z_i), \tag{6.21}$$

where $z_i = (y_i - \mu)/\sigma$ and ϕ is the PDF of the standard distribution (i.e. the distribution with location parameter $\mu = 0$ and scale parameter $\sigma = 1$).

If right censoring is present in the data because not all of the units have failed, the likelihood is

$$L(\mu, \sigma | \mathbf{y}) = c \prod_{i=1}^{n} \left[\frac{1}{\sigma} \phi\left(z_i\right) \right]^{\delta_i} \left[1 - \Phi\left(z_i\right) \right]^{1-\delta_i}, \tag{6.22}$$

where

$$\delta_i = \begin{cases} 1, & \text{if the observation is a failure time} \\ 0, & \text{if the observation is censored} \end{cases}$$

and c is a constant which varies based on the censoring type (Type I or Type II) and $z_i = (y_i - \mu)/\sigma$. The constant does not affect the MLEs and therefore is generally taken as $c = 1$ for simplicity. These likelihood expressions are general expressions for all location-scale distributions. One must substitute the appropriate PDF and CDF to obtain the likelihood for a specific distribution. For example, for the normal distribution one would use $\phi = \phi_{\text{Norm}}$ (the PDF for the standardized normal distribution) and $\Phi = \Phi_{\text{Norm}}$ (the CDF for the standardized normal distribution). The likelihood can easily be adapted to accommodate left censoring or interval censoring by multiplying the likelihood by additional terms if these types of censoring are present. The right censored likelihood is presented here because it is the most common type of censoring in reliability data analysis.

To find the MLEs, the likelihood is then maximized with respect to the model parameters μ and σ. Typically, we maximize the log-likelihood function for simplicity of calculations. Numerical methods are used to maximize the likelihood expression with respect to μ and σ except in cases where a closed form solution exists. The resulting estimates for μ and σ are the MLEs denoted by $\hat{\mu}$ and $\hat{\sigma}$.

6.4.2 Maximum Likelihood Method for Log-Location Scale Distributions

Common log-location scale distributions used in reliability data analysis are the lognormal distribution, the Weibull distribution, and the log-logistic distribution. The likelihood function for these distributions if there is no censoring present is

$$L(\mu, \sigma; \mathbf{t}) = \prod_{i=1}^{n} f(t_i; \mu, \beta) = \prod_{i=1}^{n} \left\{ \frac{1}{\sigma t_i} \phi\left(\frac{\log t_i - \mu}{\sigma} \right) \right\}. \tag{6.23}$$

If right censoring is present in the data because not all of the units have failed, the likelihood is

$$L(\mu, \sigma; \mathbf{t}) = c \prod_{i=1}^{n} \left\{ \frac{1}{\sigma t_i} \phi\left(\frac{\log t_i - \mu}{\sigma} \right) \right\}^{\delta_i} \left\{ 1 - \Phi\left(\frac{\log t_i - \mu}{\sigma} \right) \right\}^{1-\delta_i}. \tag{6.24}$$

Again, we can assume $c = 1$ for simplicity for obtaining MLEs for both Type I and Type II censoring.

Common choices for ϕ and Φ include the following:

1. $\Phi_{\text{SEV}}(z) = 1 - \exp(-\exp(z))$. This is the SEV distribution for Y and the Weibull distribution for T.
2. $\Phi_N(z) = \int_{-\infty}^{z} \frac{1}{\sqrt{2\pi}} e^{-z^2/2} \, dz$. This is the CDF of the standard normal distribution for Y and the log-normal distribution for T.
3. $\Phi_{\text{Logistic}}(z) = \frac{e^z}{1 + e^z}$. This is the logistic distribution for Y and leads to the log-logistic distribution for T.

The log likelihood function must be maximized with respect to μ and σ using numerical methods to obtain the MLEs $\hat{\mu}$ and $\hat{\sigma}$ except in cases where closed form solutions exist.

For the SEV distribution with right censoring, the log-likelihood reduces to

$$\ell(\mu, \sigma) = -r \log \sigma + \sum_{i=1}^{n} \left[\delta_i z_i - \exp(z_i) \right],$$ (6.25)

where r is the total number of observed failures and $z_i = (\log t_i - \mu)/\hat{\sigma}$.

We can solve one of the two likelihood equations (i.e. the equations obtained by taking the first partial derivative with respect to each of the parameters) for μ in terms of σ. This yields

$$\hat{\mu} = \hat{\sigma} \log \left(\frac{1}{r} \sum_{i=1}^{n} e^{(\log t_i)/\hat{\sigma}} \right).$$ (6.26)

We can then substitute this into the other likelihood equation, which must then be solved using numerical methods. Once this solution $\hat{\sigma}$ is obtained, we can substitute it back into (6.26) to obtain the MLE $\hat{\mu}$.

6.4.3 Inference for Location Scale and Log-Location Scale Models

Wald's method can be used to find confidence intervals on μ and σ for the location scale and log-location scale distributions. It can be shown that the estimator $\hat{\mu}$ is asymptotically normal. Because it is already on the log scale, this approximation is often good. However, because σ is a positive parameter, it is common practice to use a log transformation and apply the delta method. The $100\,(1-\alpha)\%$ confidence intervals for μ and σ are respectively:

$$\left[\hat{\mu} - z_{\alpha/2} \widehat{\text{s.e.}}(\hat{\mu}), \quad \hat{\mu} + z_{\alpha/2} \widehat{\text{s.e.}}(\hat{\mu}) \right]$$ (6.27)

and

$$\left[\frac{\hat{\sigma}}{w}, \quad w \hat{\sigma} \right],$$ (6.28)

where $w = z_{\alpha/2} \widehat{\text{s.e.}}(\hat{\sigma})/\hat{\sigma}$. The standard error of the estimates comes from the inverse of the estimation of the parameter's Fisher's information matrix. The information matrix for the location scale and log-location scale parameters is

$$
\hat{\Sigma} = \begin{bmatrix} \hat{V}(\hat{\mu}) & \widehat{\text{cov}}(\hat{\mu}, \hat{\sigma}) \\ \widehat{\text{cov}}(\hat{\sigma}, \hat{\mu}) & \hat{V}(\hat{\sigma}) \end{bmatrix}
$$ (6.29)

$$
= \begin{bmatrix} -\dfrac{\partial^2 \ell(\mu, \sigma)}{\partial \mu^2} & -\dfrac{\partial^2 \ell(\mu, \sigma)}{\partial \mu \partial \sigma} \\ -\dfrac{\partial^2 \ell(\mu, \sigma)}{\partial \mu \partial \sigma} & -\dfrac{\partial^2 \ell(\mu, \sigma)}{\partial \sigma^2} \end{bmatrix}^{-1} \Bigg|_{(\hat{\mu}, \hat{\sigma})} .
$$

where the partial derivatives of the log-likelihood are evaluated at the MLEs $\hat{\mu}$ and $\hat{\sigma}$. It is important to note that for the Weibull distribution, a significant correlation between the two parameters exists. The standard error of each of the parameters is then given by the square-root of its variance estimate, which is one of the diagonal elements in $\hat{\Sigma}$. Alternative methods for computing confidence intervals include using the likelihood ratio method of computing confidence intervals and Monte-Carlo simulation.

6.4.4 Location Scale and Log-Location Scale Regression Models

For location scale and log-location scale models it is common to include predictor variables in the location parameter, μ. We assume that the location parameter for the distribution $Y = \log T$ satisfies

$$\mu = \mu(\mathbf{x}) = \beta_0 + \beta_1 x_1 + \cdots + \beta_k x_k.$$

A similar process to the earlier can be used to derive the MLEs and their standard errors. Now, though, instead of the parameter vector being $\theta = (\mu, \sigma)^T$, we have $\theta = (\beta_0, \beta_1, \ldots, \beta_k, \sigma)^T$. Then the delta method can be implemented to calculate standard errors for functions of $\hat{\mu}$ and $\hat{\sigma}$ making inference on additional quantities possible. Additionally, standard errors for other statistics of interest, for example, $\hat{t}_p$ can be calculated as before.

6.5 Weibull Regression

Probably the most common distribution for modeling failure times is the Weibull distribution. Note that in the log-location scale family the lognormal relationship to the normal matches the relationship between the SEV and the Weibull distribution. Therefore, one approach to developing regression models for lifetimes is analogous to the lognormal transformation approach in regression models when normally distributed errors are assumed.

A common parameterization for the distribution of the lifetime T_i for a unit having corresponding predictor $\mathbf{x}_i$ is

$$T_i | \mathbf{x}_i \sim \text{WEIB}\left(\theta = e^{\mathbf{x}_i' \beta}, \kappa\right),$$

where $\kappa = 1/\sigma$ from the log-location scale parameterization and $\theta = e^{\mu}$. This model is called Weibull regression and is equivalent to the SAFT model in Section 6.8. Note, in this widely used parameterization, that the scale parameter θ is related to the predictor variables through the β parameter only; that is,

$$\theta = e^{\mathbf{x}_i' \beta}.$$

The scale parameter of the Weibull distribution is constrained to be positive, and taking the exponential of $\mathbf{x}_i' \beta$ assures this. The shape parameter κ is often assumed to be constant (not dependent on the values of the predictor variable $\mathbf{x}_i$), although as previously discussed this assumption can be relaxed.

There are several properties of Weibull regression that are important to understand.

1. **Relation to the SEV distribution**. In Chapter 1, we showed that if $T_i \sim \text{WEIB}(\theta_i, \kappa)$, then $\log T_i \sim \text{SEV}(\mu_i = \log \theta_i, \sigma = 1/\kappa)$. Thus, if

$$T_i \sim \text{WEIB}\left(\theta_i = e^{\mathbf{x}_i' \beta}, \kappa\right)$$

then

$$\log T_i \sim \text{SEV}\left(\mu_i = \log \theta = \mathbf{x}_i' \beta, \sigma = 1/\kappa\right).$$

Recall that the SEV distribution is a location-scale distribution with parameters μ (which is not the mean) and σ (which is not the standard deviation). Assuming that the shape parameter κ is constant means that the scale parameter of the SEV distribution is constant. This is akin in normal theory regression or ANOVA, to assuming that the standard deviation (i.e. the scale parameter) is constant.

2. **The special case of** $\kappa = 1$. When $\kappa = 1$, the Weibull distribution reduces to the exponential distribution. In the context of Weibull regression, this means that

$$T_i \sim \text{WEIB}\left(\theta = e^{x_i'\beta}, 1\right) = \text{EXP}\left(\theta = e^{x_i'\beta}\right).$$

This is often called **exponential regression**. The logarithms of the failure times therefore have distribution

$$\log T_i \sim \text{SEV}\left(\mu = x_i'\beta, \sigma = 1\right).$$

This is akin, in normal theory regression or ANOVA, to assuming that the standard deviation (i.e. the scale parameter) is known to be 1.

3. **The proportional hazards property**. The hazard function for the Weibull(θ, κ) distribution is

$$h(t) = \frac{\kappa}{\theta}\left(\frac{t}{\theta}\right)^{\kappa-1} = \kappa\theta^{-\kappa}t^{\kappa-1}, \ t > 0.$$

For Weibull regression, where $T_i \sim \text{WEIB}\left(\theta = e^{x_i'\beta}, \kappa\right)$, hazard function is

$$h\left(t|\mathbf{x}_i\right) = \kappa\left(e^{x_i'\beta}\right)^{-\kappa}t^{\kappa-1} = \kappa e^{-\kappa x_i'\beta}t^{\kappa-1}, \ t > 0.$$

Now, let us compare the hazard functions for two different levels of the predictor variables, say, $\mathbf{x}_i$ and $\mathbf{x}_j$. The ratio of these hazards is

$$\frac{h\left(t|\mathbf{x}_i\right)}{h\left(t|\mathbf{x}_j\right)} = \frac{\kappa e^{-\kappa x_i'\beta}t^{\kappa-1}}{\kappa e^{-\kappa x_j'\beta}t^{\kappa-1}} = \frac{e^{-\kappa x_i'\beta}}{e^{-\kappa x_j'\beta}} = \exp\left(-\kappa\left(\mathbf{x}_i - \mathbf{x}_j\right)'\beta\right)$$

which does not depend on time t. This means that the hazard functions for different values of $\mathbf{x}$ are always proportional to the baseline hazard, which we take to be $h(t|\mathbf{x} = \mathbf{0})$. Figure 6.6 shows the consequence of this assumption; that all hazard functions have basically the same shape and are scaled up or down according to the levels of the predictor $\mathbf{x}$.

4. **The accelerated failure time property**. In an accelerated failure time model, predictor variables have the effect of scaling time; in other words, the aging process is thought to speed up or slow down depending on the predictor variables. The assumption is that there is a function $A(\mathbf{x})$ with the property that

$$S(t|\mathbf{x}) = S_0(A(\mathbf{x})t),$$

where S_0 is the survival function at some baseline, usually $\mathbf{x}_0 = \mathbf{0}$. For example, if $A(\mathbf{x}_1) = 2$ for some level $\mathbf{x}_1$ of $\mathbf{x}$, then

$$S(t|\mathbf{x}) = S_0(2t).$$

Thus, the probability of surviving 100 hours is equal to the probability that a baseline unit (a unit whose predictor is $\mathbf{x}_0 = \mathbf{x} = \mathbf{0}'$) will survive 200 hours. The aging process for the unit with predictor $\mathbf{x}_1$ has essentially doubled; it is as if time were passing by at twice the rate for the $\mathbf{x}_1$ unit compared to the baseline ($\mathbf{x}_0$) unit. In Weibull regression, we often choose the baseline survival function to be

$$S_0(t) = \exp\left(-\left(\frac{t}{e^{\beta_0}}\right)^{\kappa}\right), \ t > 0,$$

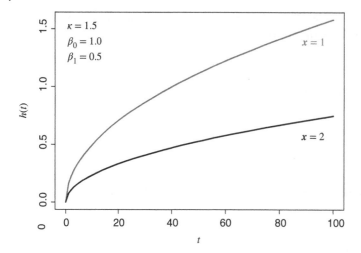

Figure 6.6 illustration of the proportional hazards property for Weibull regression with a constant κ. The upper curve is approximately 2.117 times the lower one for all $t > 0$.

and the acceleration factor to be

$$A(\mathbf{x}) = e^{-\beta_1 x_1 - \beta_2 x_2 - \cdots - \beta_p x_p}.$$

This means that the survival function will be

$$
\begin{aligned}
S(t|\mathbf{x}) &= S_0(A(\mathbf{x})t) \\
&= \exp\left(-\left(\frac{t\, e^{-\beta_1 x_1 - \beta_2 x_2 - \cdots - \beta_p x_p}}{e^{\beta_0}}\right)\right) \\
&= \exp\left(-\left(\frac{t}{e^{\mathbf{x}'\beta}}\right)^{\kappa}\right), \qquad t > 0
\end{aligned}
$$

which is the survival function for the Weibull distribution with parameters $\theta = e^{\mathbf{x}'\beta}$ and κ.

5. **The Weibull distribution is the only continuous distribution having both the proportional hazards property and the accelerated time property.** Of course, this property also holds for the exponential distribution because the exponential is a special case of the Weibull distribution.

Example 6.5 Table 6.9 gives the failure times of rolling ball bearings for a 2^3 unreplicated factorial design. These data were first provided by Hellstrand (1989). Apply Weibull regression to fit a first-order model.

Solution:
The predictor variables outer ring osculation, inner ring heat treatment, and cage design are nominal and coded -1 for the low level, and 1 for the high level. Let $x_1, x_2,$ and x_3 denote these coded values and let t denote the lifetime. There was no censoring.

The model is

$$T_i \sim \text{WEIB}(\theta_i, \kappa),$$

Table 6.9 Lifetimes of rolling ball bearings in a 2^3 experiment.

Outer ring Osculation	Inner ring Heat treatment	Cage design	t
−1	−1	−1	17
1	−1	−1	25
−1	1	−1	26
1	1	−1	85
−1	−1	1	19
1	−1	1	21
−1	1	1	16
1	1	1	128

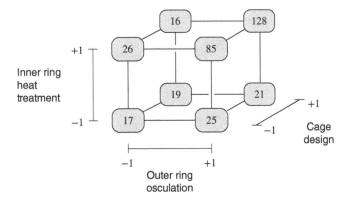

Figure 6.7 Display of lifetimes of rolling ball bearings from 2^3 experiment.

where

$$\log \theta_i = \beta_0 + \beta_1 x_{i1} + \beta_2 x_{i2} + \beta_3 x_{i3}.$$

The data from the 2^3 experiment are shown graphically in Figure 6.7.
    The output from the surveg function in R's survival package is

```
               Value Std. Error      z        p
(Intercept)   3.6080     0.1128  31.98  < 2e-16
x1            0.4454     0.1649   2.70   0.0069
x2            0.4412     0.1637   2.70   0.0070
x3            0.0995     0.1114   0.89   0.3717

Log(scale)   -1.1924     0.2996  -3.98 6.9e-05

Scale= 0.303
```

These results tell us that there is evidence that outer ring osculation (x_1) and inner ring heat treatment (x_2) affect the lifetime; there is no evidence that cage design (x_3) affects the lifetime. All slope coefficients are positive, indicating that larger values lead to larger lifetimes. The characteristic life is estimated to be

$$\hat{\theta} = \exp\left(3.6080 + 0.4454x_1 + 0.4412x_2 + 0.0995x_3\right)$$

and the common shape parameter estimate is $\hat{\kappa} = 1/0.303 = 3.30$. The optimal choice of predictors would be $(x_1, x_2, x_3) = (+1, +1, +1)$, although, because the coefficient for x_3 is not significant, we are unsure of the last entry in this vector.

With JMP, one can choose the "Fit Parametric Survival" function in the "Analyze" -> "Reliability and Survival" group. The model dialog window is as Figure 6.8, where, in the model effect field, one can specify the factors that may affect the location parameter of a log-location-scale lifetime distribution model. For Weibull regression, this means the natural logarithm of the characteristic life. Note that, if needed, a regression function for the scale parameter can be defined too. In such a case, the scale parameter will no longer be a constant. The output from JMP for this example is given in Figure 6.9, which matches the result from R.

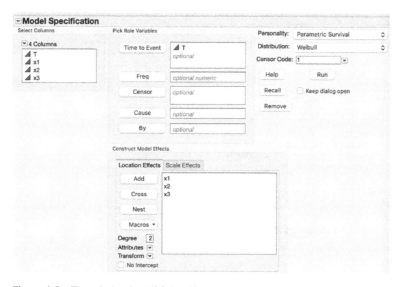

Figure 6.8 The window in JMP® for lifetime regression model specification. Source: SAS Institute Inc.

The next example illustrates what happens if we apply a model with only the significant variables.

Example 6.6 We will now apply a first-order model with interaction to the data from Example 6.5, ignoring the cage design variable.

Solution:
If we collapse the variable $x_3 =$ cage design, we obtain the situation illustrated in Figure 6.10. Here we see that there is a clear interaction between the variables x_1 and x_2, outer ring osculation and inner ring heat treatment. When x_1 is at the low level, the effect of moving x_2 from low to high is +3.0, whereas if

▼ ⊟**Parametric Survival Fit**

▶ **Effect Summary**

Time to event: T	AICc	101.5220	Observation Used	8
Distribution: Weibull	BIC	71.9192	Uncensored Values	8
	-2*LogLikelihood	61.5220		

▶ **Whole Model Test**

▼ **Parameter Estimates**

Term	Estimate	Std Error	Lower 95%	Upper 95%
Intercept	3.6079749	0.1128175	3.3868567	3.8290931
x1	0.44539502	0.1649252	0.1221475	0.7686425
x2	0.44122582	0.1637047	0.1203704	0.7620812
x3	0.09947156	0.1113633	-0.118797	0.3177397
δ	0.30348712	0.0909385	0.125251	0.4817233

Figure 6.9 The parameter estimation table provided by JMP's® Weibull regression. Source: SAS Institute Inc.

Figure 6.10 Rolling ball bearing data with cage design ignored.

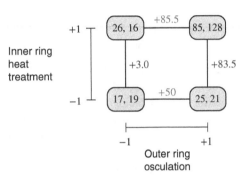

x_1 is at the high level, the effect of moving x_2 from low to high is +83.5. Clearly, the effect of x_2 depends on what the level of x_1 is, and the difference between the effects (83.5 − 3.0 = 80.5) is a measure of the interaction. Another way to look at this is to consider the effect of x_1 when x_2 is at the low or high setting. When x_2 is at the low setting, the effect of x_1 is only +5.0, whereas when x_2 is at the high setting the effect of x_1 is +85.5. Note that the difference in effects is 85.5 − 5.0 = 80.5, just as before.

If we were to run the Weibull regression in R using a first-order model with interaction, we would obtain the following results:

```
             Value Std. Error      z        p
(Intercept)  3.5013     0.0462  75.79 < 2e-16
x1           0.4638     0.0441  10.52 < 2e-16
x2           0.4709     0.0449  10.50 < 2e-16
x1:x2        0.3342     0.0441   7.58 3.6e-14

Log(scale)  -2.0818     0.3117  -6.68 2.4e-11

Scale= 0.125
```

Both factors x_1 and x_2 along with the interaction are significant. The same result can be obtained from JMP.

The positive interaction indicates that there is a synergistic effect between x_1 and x_2. Increasing x_1 from low to high increases the lifetime and increasing x_2 from low to high also increases the lifetime. But if we increase *both* x_1 and x_2 from the (low, low) level to the (high, high) level, we increase the lifetime by more than the sum of the main effects. This is a lot of information from a simple 2^3 factorial design. ∎

The next example presents a larger experiment that involves censoring.

Example 6.7 This example involves a life testing experiment on the shafts of an automatic transmission. The data were first presented by Davis (1995) and given as an exercise in Wu and Hamada (2000). The experiment involved the following seven factors:

(A) spline end profile (spherical = −1 and grooved = +1)
(B) annealing amount (zero, one, or four hours)
(C) shaft diameter (16.1, 17.7, 18.8 mm)
(D) shot intensity (3, 6, or 9 Almen)
(E) shot coverage (200%, 400%, or 600%)
(F) tempering temperature (140, 160, or 180 °C)
(G) shot blasting (without = −1 and with = +1).

The outcome was the lifetime in thousands of cycles, where the lifetime is defined to be the time until a crack appears in the shaft. The design and the responses are shown in Table 6.10. The design is a fractional factorial with two factors (*A* and *G*) at two levels, and the remaining five factors at three levels. There are two replicates. The design is not balanced with respect to the factor *G*; there are 24 runs with this factor at the low level and 12 runs with this factor at the high level. The design is orthogonal for the main effects but there is partial aliasing between the main effects and the two-factor interactions (Table 6.11).

Factors *A* and *C* are characteristics of the product itself, and factors *B*, *D*, *E*, *F*, and *G* are characteristics of the manufacturing process. There are no accelerating variables. Before the experiment, the process was running at the following levels:

$A = -1$ (spherical)

$B = $ zero hours

$C = 17.7$ mm

$D = 6$ Almen

$E = 400\%$

$F = 160\ °C$

$G = -1$ (without shot blasting).

Plot the data and use a Weibull regression model with censoring to analyze the data, determine the significant factors, and make recommendations to the process settings.

Table 6.10 Transmission experimental design and responses.

A	B	C	D	E	F	G	Lifetime	Cens
−1	0	17.7	6	400	160	−1	322	1
−1	0	17.7	6	400	160	−1	2000	0
−1	1	17.7	3	200	140	1	95	1
−1	1	17.7	3	200	140	1	95.4	1
−1	4	17.7	9	600	180	−1	2000	0
−1	4	17.7	9	600	180	−1	125	1
−1	0	16.1	6	200	180	−1	747	1
−1	0	16.1	6	200	180	−1	414	1
−1	1	16.1	3	600	160	−1	821	1
−1	1	16.1	3	600	160	−1	192	1
−1	4	16.1	9	400	140	1	2000	0
−1	4	16.1	9	400	140	1	2000	0
−1	0	18.8	3	400	140	−1	972	1
−1	0	18.8	3	400	140	−1	2000	0
−1	1	18.8	9	200	180	−1	2000	0
−1	1	18.8	9	200	180	−1	1920	1
−1	4	18.8	6	600	160	1	2000	0
−1	4	18.8	6	600	160	1	2000	0
1	0	17.7	9	600	140	−1	739	1
1	0	17.7	9	600	140	−1	285	1
1	1	17.7	6	400	180	1	1080	1
1	1	17.7	6	400	180	1	634	1
1	4	17.7	3	200	160	−1	2000	0
1	4	17.7	3	200	160	−1	1940	1
1	0	16.1	3	600	180	1	2000	0
1	0	16.1	3	600	180	1	1790	1
1	1	16.1	9	400	160	−1	2000	0
1	1	16.1	9	400	160	−1	617	1
1	4	16.1	6	200	140	−1	2000	0
1	4	16.1	6	200	140	−1	2000	0
1	0	18.8	9	200	160	1	1380	1
1	0	18.8	9	200	160	1	1110	1
1	1	18.8	6	600	140	−1	2000	0
1	1	18.8	6	600	140	−1	2000	0
1	4	18.8	3	400	180	−1	2000	0
1	4	18.8	3	400	180	−1	2000	0

Table 6.11 Correlation between estimators of main effects and two-factor interactions for transmission data.

	A	B	C	D	E	F	G
A	1.00						
B	0.00	1.00					
C	0.00	0.00	1.00				
D	0.00	0.00	0.00	1.00			
E	0.00	0.00	0.00	0.00	1.00		
F	0.00	0.00	0.00	0.00	0.00	1.00	
G	0.00	0.00	0.00	0.00	0.00	0.00	1.00
AB	0.19	0.00	0.00	0.55	0.55	0.00	0.54
AC	0.07	0.00	0.00	0.00	0.00	0.00	0.00
AD	0.00	0.56	0.00	0.00	0.00	0.50	0.00
AE	0.00	0.56	0.00	0.00	0.00	0.00	0.00
AF	0.00	0.00	0.00	0.50	0.00	0.00	0.58
AG	0.33	0.52	0.00	0.00	0.00	0.54	0.00
BC	0.00	0.07	0.19	0.34	0.33	0.61	0.02
BD	0.55	0.00	0.34	0.19	0.34	0.10	0.25
BE	0.55	0.00	0.33	0.34	0.19	0.34	0.25
BF	0.00	0.00	0.61	0.10	0.34	0.19	0.33
BG	0.51	0.33	0.02	0.24	0.24	0.31	0.18
CD	0.00	0.35	0.00	0.07	0.07	0.32	0.14
CE	0.00	0.33	0.00	0.07	0.07	0.28	0.21
CF	0.00	0.62	0.00	0.32	0.28	0.07	0.00
CG	0.00	0.02	0.33	0.13	0.20	0.00	0.07
DE	0.00	0.34	0.08	0.00	0.00	0.31	0.18
DF	0.50	0.10	0.32	0.00	0.31	0.00	0.18
DG	0.00	0.24	0.14	0.33	0.17	0.17	0.00
EF	0.00	0.34	0.29	0.31	0.00	0.00	0.35
EG	0.00	0.24	0.20	0.17	0.33	0.33	0.00
FG	0.54	0.32	0.00	0.17	0.33	0.33	0.00

Solution:

Figure 6.11 shows scatter plots of the lifetimes against each of the predictor variables. A small amount of jitter has been added to the factor to distinguish multiple points at the same location; this is particularly important for the censored values that all occur at 2000. This plot suggests that none of the factors has a large effect on the lifetime, with the possible exception of factor *B*, annealing time.

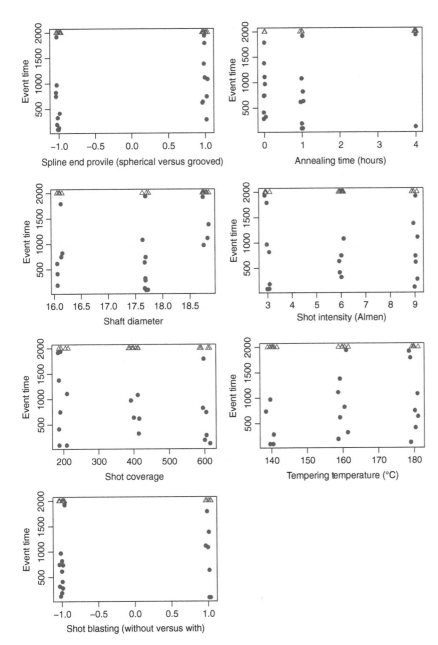

Figure 6.11 Scatterplot of transmission shaft lifetimes versus factors *A* through *G*. Triangles indicate censoring.

We assumed a Weibull regression model for which the log of the characteristic life is linear; that is

$$\log \theta = \beta_0 + \beta_1 A + \beta_2 B + \beta_3 C + \beta_4 D + \beta_5 E + \beta_6 F + \beta_7 G.$$

The results of a first-order Weibull regression, found using R's `survreg` function in the `survival` package, are as follows:

```
              Value Std. Error      z      p
(Intercept) 0.40184    5.31022   0.08 0.9397
A           0.21801    0.24591   0.89 0.3753
B           0.49049    0.18918   2.59 0.0095
C           0.26462    0.20209   1.31 0.1904
D           0.01902    0.10238   0.19 0.8526
E           0.00189    0.00171   1.10 0.2705
F           0.00740    0.01527   0.48 0.6280
G          -0.18061    0.25749  -0.70 0.4830
Log(scale) -0.08237    0.20103  -0.41 0.6820
Scale= 0.921
```

Here we see that the only factor that is significant is B = annealing amount. All other factors have considerably larger P-values, except possibly C = shaft diameter which has a P-value of 0.1904. We created a contour plot of the estimated characteristic life, holding all variables except B and C constant at levels A = "spherical," $D = 6.0$, $E = 400$, $F = 160$, and G = "without." This plot, shown in Figure 6.12, indicates that the most reliable transmission shafts would occur when B and C are both at the high end, that is at $A = 4$ hours and $C = 18.8$ mm. If further experiments were feasible, they should be conducted at or beyond the upper right corner of the original design region Alternatively, runs could be made along the path of steepest ascent in an initial attempt at locating where the most reliable transmission shafts might be built.

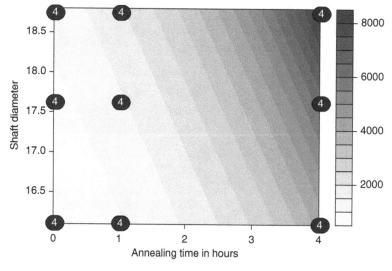

Figure 6.12 Contour plot for characteristic life θ as a function of factors B and C, assuming a first-order model. The numbers indicate the number of runs at each of these nine design points.

We then fit a full second-order model with factors B and C while not accounting for the other variables. The results of the Weibull regression are shown in the following text:

```
                Value Std. Error      z       p
(Intercept) 218.9415    76.2302   2.87  0.0041
B            -0.1068     3.2853  -0.03  0.9741
C           -24.6438     8.8248  -2.79  0.0052
B2            0.2070     0.1397   1.48  0.1386
C2            0.7146     0.2548   2.80  0.0050
BC           -0.0154     0.1866  -0.08  0.9344
Log(scale)   -0.1425     0.1931  -0.74  0.4604

Scale= 0.867
```

A contour plot of the second-order model, shown in Figure 6.13, indicates a fairly flat characteristic life surface throughout much of the middle region of the design space. The most reliable transmission shafts would be those produced at the highest settings for factors B and C; that is, $A = 4$ hours and $C = 18.8$ mm.

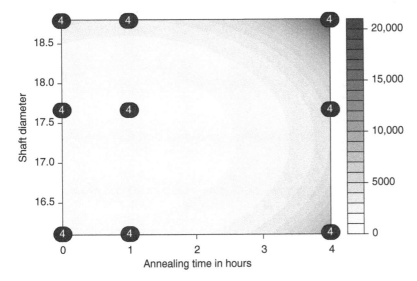

Figure 6.13 Contour plot for characteristic life θ as a function of factors B and C, assuming a second-order model. The numbers indicate the number of runs at each of these nine design points.

Under the linear model, the estimated characteristic lifetime at the current operating condition, as stated previously, is

$$\hat{\theta} = 1213 \, (1000 \, \text{cycles})$$

which is in agreement with the two observations at this level of the experiment: 322, and 2000+. The estimate for the setting where $B = 4$ hours and $C = 18.8$ mm is

$$\hat{\theta} = 11,546 \, (1000 \, \text{cycles}).$$

Under the quadratic model in just factors B and C, the estimated characteristic lifetime at the current operating condition is

$$\hat{\theta} = 753 \,(1000 \,\text{cycles}),$$

and under the conditions $B = 4$ hours and $C = 18.8$ mm, the estimate is

$$\hat{\theta} = 20,671 \,(1000 \,\text{cycles}).$$

Regardless of which model is selected, changing factors B and C to $B = 4$ hours and $C = 18.8$ mm, leads to a significant improvement in the reliability of the transmission shaft. Further experiments above and to the right of the region in Figure 6.12 or 6.13 may uncover settings that improve the reliability even more. ∎

JMP is also capable of performing Weibull regression. The next example illustrates this.

Example 6.8 Reanalyze the transmission data from the previous example using JMP.

Solution:

The results of the first-order model are shown in Figure 6.14. The results here are nearly identical to the ones obtained by R.

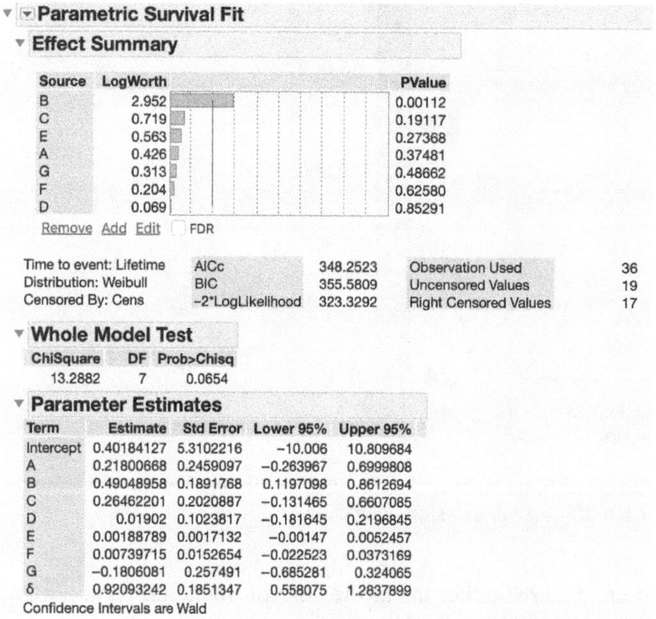

▼ ▽ **Parametric Survival Fit**

  ▼ **Effect Summary**

Source	LogWorth		PValue
B	2.952		0.00112
C	0.719		0.19117
E	0.563		0.27368
A	0.426		0.37481
G	0.313		0.48662
F	0.204		0.62580
D	0.069		0.85291

  Remove Add Edit   ○ FDR

Time to event: Lifetime	AICc	348.2523	Observation Used	36
Distribution: Weibull	BIC	355.5809	Uncensored Values	19
Censored By: Cens	−2*LogLikelihood	323.3292	Right Censored Values	17

  ▼ **Whole Model Test**

ChiSquare	DF	Prob>Chisq
13.2882	7	0.0654

  ▼ **Parameter Estimates**

Term	Estimate	Std Error	Lower 95%	Upper 95%
Intercept	0.40184127	5.3102216	−10.006	10.809684
A	0.21800668	0.2459097	−0.263967	0.6999808
B	0.49048958	0.1891768	0.1197098	0.8612694
C	0.26462201	0.2020887	−0.131465	0.6607085
D	0.01902	0.1023817	−0.181645	0.2196845
E	0.00188789	0.0017132	−0.00147	0.0052457
F	0.00739715	0.0152654	−0.022523	0.0373169
G	−0.1806081	0.257491	−0.685281	0.324065
δ	0.92093242	0.1851347	0.558075	1.2837899

  Confidence Intervals are Wald

Figure 6.14 Output from JMP® for the first-order model applied to the transmission data. Source: SAS Institute Inc.

When JMP performs Weibull regression on a second-order model, the system automatically centers the quadratic terms. The model in factors B and C is therefore

$$\mu = \log\theta = \gamma_0 + \gamma_1 B + \gamma_2 C + \gamma_{12}(B - 1.66667)(C - 17.53333)$$
$$+ \gamma_{11}(B - 1.66667)^2 + \gamma_{22}(C - 17.53333)^2.$$

$$= \gamma_0 + 1.66667 \times 17.53333\gamma_{12} + 1.66667^2\gamma_{11} + 17.53333^2\gamma_{22}$$
$$+ \left(\gamma_1 - 17.53333\gamma_{12} - 2 \times 1.6667\gamma_{11}\right) B$$
$$+ \left(\gamma_2 - 1.66667\gamma_{12} - 2 \times 17.53333\gamma_{22}\right) C$$
$$+ \gamma_{12}BC + \gamma_{11}B^2 + \gamma_{22}C^2.$$

Thus, the parameters $\beta_0, \beta_1, \beta_2, \beta_{11}, \beta_{22}$, and β_{12} from R and $\gamma_0, \gamma_1, \gamma_2, \gamma_{11}, \gamma_{22}$, and γ_{12} from JMP are related as follows:

$$\beta_0 = \gamma_0 + \gamma_1 B + \gamma_2 C + \gamma_{12}(B - 1.66667)(C - 17.53333),$$
$$\beta_1 = \gamma_1 - 17.53333\gamma_{12} - 2 \times 1.6667\gamma_{11},$$
$$\beta_2 = \gamma_2 - 1.66667\gamma_{12} - 2 \times 17.53333\gamma_{22},$$
$$\beta_{11} = \gamma_{11},$$
$$\beta_{22} = \gamma_{22},$$
$$\beta_{12} = \gamma_{12}.$$

Thus, estimates for the quadratic terms are the same in the two formulations of Weibull regressions, but the intercept and the first-order terms are different. If we take the estimated parameters from JMP's output and substitute these into the previous formulas, we see that they match R's output very closely. The intercept differs slightly, but the first-order terms are nearly identical (Figure 6.15).

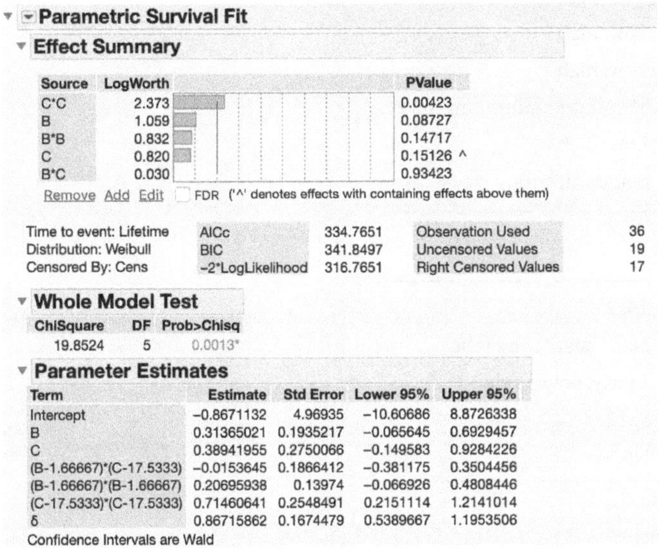

▼ ⊟ Parametric Survival Fit

▼ Effect Summary

Source	LogWorth		PValue
C*C	2.373		0.00423
B	1.059		0.08727
B*B	0.832		0.14717
C	0.820		0.15126 ^
B*C	0.030		0.93423

Remove Add Edit ☐ FDR ('^' denotes effects with containing effects above them)

Time to event: Lifetime	AICc	334.7651	Observation Used	36
Distribution: Weibull	BIC	341.8497	Uncensored Values	19
Censored By: Cens	−2*LogLikelihood	316.7651	Right Censored Values	17

▼ Whole Model Test

ChiSquare	DF	Prob>Chisq
19.8524	5	0.0013*

▼ Parameter Estimates

Term	Estimate	Std Error	Lower 95%	Upper 95%
Intercept	−0.8671132	4.96935	−10.60686	8.8726338
B	0.31365021	0.1935217	−0.065645	0.6929457
C	0.38941955	0.2750066	−0.149583	0.9284226
(B-1.66667)*(C-17.5333)	−0.0153645	0.1866412	−0.381175	0.3504456
(B-1.66667)*(B-1.66667)	0.20695938	0.13974	−0.066926	0.4808446
(C-17.5333)*(C-17.5333)	0.71460641	0.2548491	0.2151114	1.2141014
δ	0.86715862	0.1674479	0.5389667	1.1953506

Confidence Intervals are Wald

Figure 6.15 Output from JMP® for the second-order model applied to the transmission data. Source: SAS Institute Inc. ∎

Example 6.9 Our next example involves the lifetimes of drill bits that are used to drill holes in circuit boards. Lifetimes are measured in cycles until failure, i.e. the drill bit breaks. Since the holes are designed to be small, the drill bit is thin which leads to frequent failures.

The experiment involved 11 factors (A, B, C, D, E, F, G, H, I, J, and L) that were related to the design of the drill bit. The raw data are given in Table D.4. The experimenters believed that these could affect the lifetime of the drill bit. Factor A had four levels, labeled 0, 1, 2, and 3, while all other factors had two levels, labeled "−" and "+." Factors such as these that are characteristics of the product itself are called *control factors*. The investigators also identified five factors (M, N, O, P, and Q) that involve the environment in which the drill bit was operated. Such factors are called *noise factors* (Table 6.12).

Table 6.12 The 11 control factors and 5 noise factors and their levels for the drill bit experiment.

Factor	Description	Levels
	Control factors	
A	Carbide cobalt	$A1$, $A2$, $A3$, $A4$
B	Body length	Minimum, minimum + 30%
C	Web thickness	Low, high
D	Web taper	Low, high
E	Moment of inertia	Low, high
F	Radial rake	Low, high
G	Helix angle	Low, high
H	Axial rake	Low, high
I	Flute length	Minimum, minimum + 50%
J	Point angle	Low, high
L	Point style	Standard, strong
	Noise factors	
M	Feed rate (in/min)	10, 20
N	Backup material	Hard board, phenolic
O	pcb material	Epoxy, polyamide
P	Number of layers	8, 12
Q	Four 2-oz layers	No, yes

The design for the 11 design factors was a Resolution III design with 16 runs. Suppose we create a 16-run design in four factors and write it in standard order, that is, the first column contains alternating − and +, the second column contains alternating pairs of − and +, etc. If we label the columns in this standard design as **1**, **2**, **3**, and **4**, then the control factors in the drill bit experiment can be written in terms of these. The first factor, A, has four levels, which can be represented by the first two columns of the design in standard form; that is,

$$A = \begin{cases} 1, & \mathbf{1} = - \text{ and } \mathbf{2} = -, \\ 2, & \mathbf{1} = + \text{ and } \mathbf{2} = -, \\ 3, & \mathbf{1} = - \text{ and } \mathbf{2} = +, \\ 4, & \mathbf{1} = + \text{ and } \mathbf{2} = +. \end{cases}$$

Factors $B, C, D, E, F, G, H, I, J,$ and L are then defined as

$$B = -13 \quad C = -23 \quad D = 3 \quad E = 124 \quad F = 123,$$
$$G = 4 \quad H = -14 \quad I = -24 \quad J = -34 \quad L = 234.$$

The noise array for factors $M, N, O, P,$ and Q is a 2_{III}^{5-2} design. Each row of the array for the design factors was run at each of the eight sets of levels for the noise array. This resulted in a total of $16 \times 8 = 128$ runs. Each drill bit was run until it failed or until it reached 3000 cycles. Thus, we have time censoring at $T = 3000$ where time is measured in cycles. Analyze this data using Weibull regression.

Solution:

We apply the `survreg` function in R's `survival` package. A first-order model in all variables yields the following output from R:

```
Call:
survreg(formula = Surv(y, cens) ~ A.factor + B + C + D +
        E + F + G + H + I + J + L + M + N + O + P + Q)
              Value  Std. Error    z       p
(Intercept)   5.3883   0.1882   28.63  < 2e-16
A.factor2     1.3413   0.2747    4.88  1.1e-06
A.factor3     1.3393   0.2764    4.85  1.3e-06
A.factor4     1.5035   0.2798    5.37  7.7e-08
B             0.0171   0.0994    0.17    0.864
C             0.2211   0.0990    2.23    0.026
D             0.3104   0.1003    3.09    0.002
E            -0.0715   0.1016   -0.70    0.481
F            -0.1252   0.0991   -1.26    0.207
G            -0.0517   0.1006   -0.51    0.607
H            -0.1630   0.0995   -1.64    0.102
I             0.1533   0.1024    1.50    0.134
J             0.2497   0.0991    2.52    0.012
L             0.1785   0.1031    1.73    0.083
M            -0.0417   0.1075   -0.39    0.698
N            -0.1778   0.1061   -1.68    0.094
O            -0.9354   0.1099   -8.51  < 2e-16
P            -0.6985   0.1083   -6.45  1.1e-10
Q             0.1347   0.1079    1.25    0.212
Log(scale)    0.0266   0.0763    0.35    0.728

Scale= 1.03

Weibull distribution
Loglik(model)= -824.9   Loglik(intercept only)= -882.5
Chisq= 115.27 on 18 degrees of freedom, p= 3.3e-16
Number of Newton-Raphson Iterations: 6
n= 128
```

Factors *A* (carbide cobalt), *C* (web thickness), *D* (web taper), *J* (point angle), *O* (pcb material), and *P* (number of layers) are significant factors. Level 4 of factor *A*, along with the high levels of *C*, *D*, and *J*, lead to the best reliability. Noise factors are usually not under the control of the investigators, but the highest reliability occurs when factors *O* and *P* are at the low levels.

A (mostly) two-level design like this is often used as a screening device to see which factors truly affect the lifetime. Once the set of factors that have a significant effect on reliability is determined, further experiments can be conducted based on this knowledge. A follow-up study with a second-order model in the control factors *A*, *C*, *D*, and *J* would help to identify the levels that optimize reliability.

Finally, note that last line of the list of parameter estimates is for `Log(scale)`. This provides inference for the parameter $\log \sigma = -\log \kappa$. The point estimate for this is $\log \hat{\sigma} = 0.0266$, which leads to $\hat{\sigma} = \exp(0.0266) \approx 1.03$. (This estimate is also provided in R's output.) The estimated standard error is $\widehat{s.e.} = 0.0763$ and the *P*-value for the test of whether this is zero is $P = 0.728$. Thus, it is possible that the scale parameter of the distribution of $\log T$ is one, which would lead to the exponential distribution. ■

Example 6.10   Consider the data shown in Table 6.13, giving survival times for a nickel super alloy specimen. This data set was described in Meeker and Escobar (1998a), p. 638 and first presented in Nelson (1984). The data set gives the lifetimes of items made of a nickel super alloy when placed under various

Table 6.13   Lifetimes of nickel superalloy specimens measured in kilocycles are in the columns labeled `kCycles`.

Pseudostress	kCycles	Cens	Pseudostress	kCycles	Cens
80.3	211.626	1	91.3	112.002	1
80.6	200.027	1	99.8	43.331	1
80.8	57.923	0	100.1	12.076	1
84.3	155.000	1	100.5	13.181	1
85.2	13.949	1	113.0	18.067	1
85.6	112.968	0	114.8	21.300	1
85.8	152.680	1	116.4	15.616	1
86.4	156.725	1	118.0	13.03	1
86.7	138.114	0	118.4	8.489	1
87.2	56.723	1	118.6	12.434	1
87.3	121.075	1	120.4	9.750	1
89.7	122.372	0	142.5	11.865	1
91.3	112.002	1	144.5	6.705	1
99.8	43.331	1	145.9	5.733	1

The predictor variable is the `Pseudostress` applied. The `Cens` variable is 1 for a failed specimen and 0 for one that was censored.

levels of stress as given in the variable `Pseudostress`. The lifetime is measured in kilocycles and is contained in the variable `kCycles`. The `Cens` variable is 1 for observations that were failures, and 0 for those that were censored. Apply Weibull regression models and assess the effect of `Pseudostress` of the lifetime.

Solution:
The data are plotted in Figure 6.16 along with some of the results of four models for the data.

Let's first consider the first-order model in `Pseudostress`:

$$\log \theta = \beta_0 + \beta_1 \text{Pseudostress}.$$

The code for performing this is developed and explained in Appendix A. The output is from R's `survreg` function is

```
Call:
survreg(formula = kCyclesSurv ~ Pseudostress,
        dist = "weibull")

                 Value Std. Error       z         p
(Intercept)    9.45774    0.45980   20.57  < 2e-16
Pseudostress  -0.05348    0.00426  -12.54  < 2e-16
Log(scale)    -0.68306    0.17633   -3.87  0.00011

Scale= 0.505

Weibull distribution
Loglik(model)= -99.5    Loglik(intercept only)= -118.4
  Chisq= 37.76 on 1 degrees of freedom, p= 8e-10
Number of Newton-Raphson Iterations: 7
n= 26
```

From this we conclude that the `Pseudostress` has a significant effect on the lifetime. This result is expected since a higher stress will usually shorten the specimen's lifetime. The estimated characteristic life is

$$\hat{\theta} = \exp(9.45774 - 0.05348\text{Pseudostress})$$

and the estimated mean function is

$$\hat{\mu} = \Gamma(1 + 1/\hat{\kappa}) \exp(\hat{\beta}_0 + \hat{\beta}_1 \text{Pseudostress})$$

$$= \Gamma(1 + 0.505) \exp(9.45774 - 0.05348\text{Pseudostress}).$$

This mean function is the smooth curve in the top figure of Figure 6.16. Since the y-axis is logarithmic, the graph of μ is a straight line. We can see that the cluster of points with `Pseudostress` between 110 and 122 lies entirely below the line and the points with `Pseudostress` above 140 lie entirely above the line. This suggests that a log linear model is not sufficient; the relationship between `Pseudostress` and `kCycles` seems to be concave up.

Next, we consider a second-order model with the one predictor `Pseudostress`; this model assumes

$$\theta = \exp(\beta_0 + \beta_1 \text{Pseudostress} + \beta_2 \text{Pseudostress}^2).$$

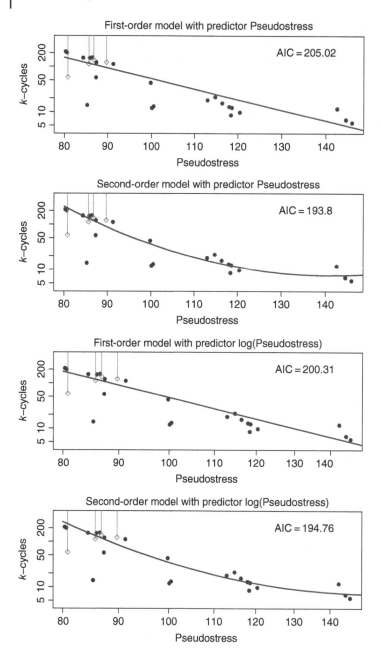

Figure 6.16 Scatter plots and estimated mean functions for four models using the nickel superalloy data. Note that the vertical axis of all four graphs is on a log scale, and the horizontal axis for the bottom two plots is also on a log scale.

The R output for this model is:

```
Call:
survreg(formula = Surv(kCycles, Cens) ~ Pseudostress +
                                        Pseudostress2,
                   dist = "weibull")
                  Value Std. Error      z        p
(Intercept)   20.805347   2.607281   7.98  1.5e-15
Pseudostress  -0.264811   0.047931  -5.52  3.3e-08
Pseudostress2  0.000939   0.000213   4.40  1.1e-05
Log(scale)    -1.015148   0.181386  -5.60  2.2e-08

Scale= 0.362

Weibull distribution
Loglik(model)= -92.9   Loglik(intercept only)= -118.4
        Chisq= 50.98 on 2 degrees of freedom, p= 8.5e-12
Number of Newton-Raphson Iterations: 8
n= 26
```

From R's output, we see that both `Pseudostress` and `Pseudostress`^{2} are significant. The second plot in Figure 6.16, gives the estimated mean curve superimposed on the scatterplot. The fit seems to be better than the first-order model that we applied earlier.

Meeker and Escobar (1998a) applied the log transformation to the `Pseudostress` variable before applying a first- or second-degree model for `kCycles`. Next, let's run these first- and second-degree models in R.

Once we define `logPS = log(Pseudostress)` we can run `survreg` in R. The output for the first-order model is this:

```
Call:
survreg(formula = Surv(kCycles, Cens) ~ logPS,
         dist = "weibull")
              Value Std. Error      z        p
(Intercept)  31.432      2.008   15.6  < 2e-16
logPS        -5.960      0.433  -13.8  < 2e-16
Log(scale)   -0.793      0.176   -4.5  6.7e-06

Scale= 0.452

Weibull distribution
Loglik(model)= -97.2   Loglik(intercept only)= -118.4
        Chisq= 42.46 on 1 degrees of freedom, p= 7.2e-11
Number of Newton-Raphson Iterations: 6
n= 26
```

The second-order model gives:

```
Call:
survreg(formula = Surv(kCycles, Cens) ~ logPS + logPS2,
         dist = "weibull")
              Value Std. Error      z        p
(Intercept) 217.611     62.132   3.50  0.00046
logPS       -85.522     26.546  -3.22  0.00127
logPS2        8.483      2.831   3.00  0.00273
```

Table 6.14 AIC (Akaike information criterion) for the four models for lifetimes of nickel superalloy specimens.

Predictors	Order	AIC
Pseudostress	1	205.0171
Pseudostress, Pseudostress^2	2	193.7985
log(Pseudostress)	1	200.3095
log(Pseudostress), log(Pseudostress)^2	2	194.7638

```
    Log(scale)    -0.982      0.179 -5.48 4.2e-08

    Scale= 0.375

    Weibull distribution
    Loglik(model)= -93.4    Loglik(intercept only)= -118.4
      Chisq= 50.01 on 2 degrees of freedom, p= 1.4e-11
    Number of Newton-Raphson Iterations: 8
    n= 26
```

■

From the results of the second-order model based on `log(Pseudostress)`, we see that the second-order term is significant, and from the bottom two graphs we can see that the second-order model seems to fit better than the first-order model.

The second-order models that are based on either `Pseudostress` or `log(Pseudostress)` seem to fit better than the corresponding first-order models. The Akaike information criterion (AIC), described in Section 3.6.3, can be used to compare models fit with the same data set. R can give the AIC for any fitted model using the `extractAIC` command, which takes as its argument the fitted model. For example,

```
mysurvreg1 = survreg( Surv(kCycles,Cens) ~ Pseudostress,
                      dist="Weibull")
summary( mysurvreg1)
AIC1 = extractAIC( mysurvreg1)
```

will perform the Weibull regression, summarize the results to the R console, and assign to the variable `AIC1` the value of the AIC for the object `mysurvreg1`. From the results of the AIC computations that are summarized in Table 6.14, we see that the second-order models have a lower (better) AIC, although the difference between the second-order models based on `Pseudostress` and `log(Pseudostress)` is negligible. Thus, either second-order model is likely to be adequate for modeling the relationship between `Pseudostress` and lifetime measured in `kCycles`.

6.6 Nonconstant Shape Parameter

So far, we have assumed that the scale parameter θ of the Weibull distribution is a function of the predictors, but the shape parameter κ is constant across all levels of the predictors. Equivalently, the location parameter $\mu = \log\theta$ of the distribution of the log lifetime is a function of the predictors, but the scale parameter $\sigma = 1/\kappa$ is assumed constant. This is similar to the assumption made in normal-theory regression and ANOVA where the variance is assumed constant.

When accelerated variables are used in an experiment to induce early failures (the topic of Section 8.2), higher levels of stress may create modes of failure that are not present at lower stress levels. The differences in failure modes at low and high stress levels could cause variation in the shape parameter κ. For example, Li et al. (1989) found that the shape parameter for dielectrics varies as the stress imposed by an electric field changes. Hellstrand (1989) found a similar result for dielectrics when temperature was the predictor.

Notice that in the scatter plots of Figure 6.16 the variability seems to be greater when the stress is lower, and the lifetimes are longer. Since κ and σ are inversely related, these figures suggest that κ is smaller for small values of Pseudostress. We can fit models where both the characteristic life (and scale parameter) θ and the shape parameter are functions of Pseudostress. We know from Example 6.10 that a second-order model in either Pseudostress or log(Pseudostress) provides a reasonable fit to the data. The variability of the data seems to decrease monotonically over Pseudostress or log(Pseudostress), so a first-order model for κ seems appropriate. Meeker and Escobar (1998a) analyzed this data set using a second-order model for θ using log(Pseudostress) and a first-order model for κ, also using log(Pseudostress). This model assumes that $T_i \sim \mathrm{WEIB}(\theta, \kappa)$, where

$$\log \theta = \beta_0 + \beta_1 \log\mathrm{PS} + \beta_2 \log\mathrm{PS}, \tag{6.30}$$

$$\log \kappa = \gamma_0 + \gamma_1 \log\mathrm{PS}, \tag{6.31}$$

or equivalently

$$\log T_i \sim \mathrm{SEV}(\mu, \sigma),$$
$$\mu = \beta_0 + \beta_1 \log\mathrm{PS} + \beta_2 \log\mathrm{PS},$$
$$\log \sigma = -\gamma_0 - \gamma_1 \log\mathrm{PS}.$$

We will present a Bayesian analysis for this model. One advantage of this approach is that with Markov chain Monte Carlo (MCMC), the posterior distribution can be obtained, regardless of its shape. The classical approach uses the asymptotic *normal* distribution of the parameter estimators; for small to moderate sample sizes, the normal approximation may not be reasonable.

Example 6.11   Perform a Bayesian analysis of the nickel superalloy data, assuming the model in (6.30) and (6.31).

Solution:
We follow the presentation given in Mueller and Rigdon (2015). There are six parameters to estimate: β_0, β_1, β_2, γ_0, and γ_1. We assume the relatively noninformative priors

$$\beta_0 \sim N(20, 1000),$$
$$\beta_1 \sim N(-100, 1000),$$
$$\beta_2 \sim N(10, 1000),$$
$$\gamma_0 \sim N(4, 10),$$
$$\gamma_1 \sim N(-1, 10).$$

The WinBUGS code for performing the MCMC can be found in Mueller and Rigdon (2015). (Note, however, that WinBUGS does not allow an index to be zero, so the parameters were relabeled `beta1`, `beta2`, `beta3`, `gamma1` and `gamma2`.) Since there was significant autocorrelation in the simulated values in the Markov chain, we took every 100th simulation (and discarded those in between). This is called *thinning*. We ran 5,000,000 burn-in iterations, followed by 10,000,000 simulations, of which 1 out of a 100 (100,000) were kept for analysis. Figure 6.17 shows the trace plots and histograms for the five parameters β_0, β_1, β_2, γ_0, and γ_1 for three separate Markov chains. Point estimates of the parameters can be obtained by taking the posterior means; this yields

$$\hat{\beta}_0 = 217.6,$$
$$\hat{\beta}_1 = -85.52,$$
$$\hat{\beta}_2 = 8.484,$$
$$\hat{\gamma}_0 = -4.73,$$
$$\hat{\gamma}_1 = 1.20.$$

Notice that the coefficient for `Pseudostress` in the model for κ is 1.20; being positive, this implies that the scale parameter κ increases as `Pseudostress` increases. Thus, the parameter $\sigma = 1/\kappa$ (the scale parameter of the distribution for $\log T_i$) is decreasing as `Pseudostress` increases. This agrees

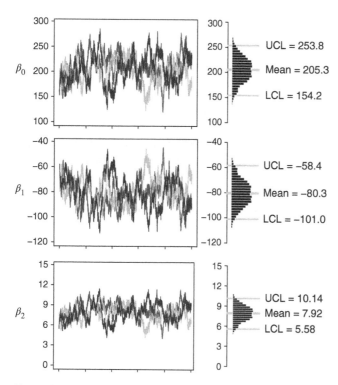

Figure 6.17 Trace plots and histograms for parameters β_0, β_1, β_2, γ_0, and γ_1.

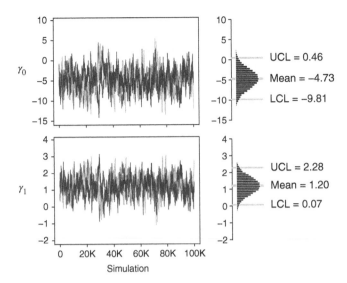

Figure 6.17 (*Continued*)

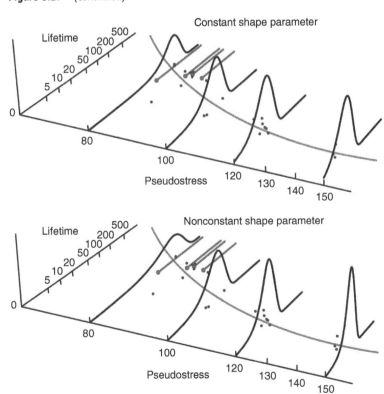

Figure 6.18 The effect on the nonconstant shape parameter of the lifetime distribution. Top, a constant κ was assumed. Bottom, the shape parameter was modeled as $\kappa = \exp(\gamma_0 + \gamma_1 \log(\text{Pseudostress}))$.

with the assessment made earlier that the variability seemed larger when the value of `Pseudostress` is small.

For comparison, we also ran a Bayesian analysis assuming a constant κ. We omit the details here, but further information can be found in Mueller and Rigdon (2015). The parameter estimates (posterior means) are

$$\hat{\beta}_0 = 167.2,$$
$$\hat{\beta}_1 = 6.18,$$
$$\hat{\beta}_2 = -63.94,$$
$$\hat{\kappa} = 2.28.$$

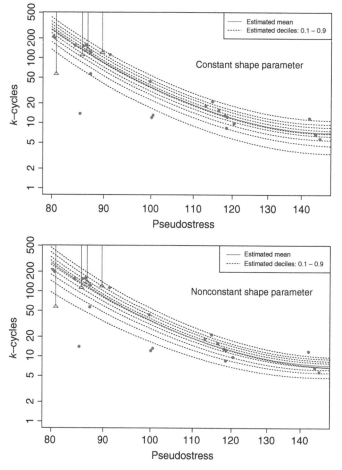

Figure 6.19 The shape of the estimated survival densities. Top, a constant κ was assumed. Bottom, the shape parameter was modeled as $\kappa = \exp(\gamma_0 + \gamma_1 \log(\text{Pseudostress}))$.

Figure 6.18 shows the estimated PDF for various values of `Pseudostress` for both the constant κ parameter case (top) and nonconstant κ. In the bottom plot, the higher variability is apparent for smaller values of `Pseudostress`. Figure 6.19 shows a scatterplot of the data along with the estimated mean response curves and the deciles (the 0.1, 0.2, ..., 0.9 quantiles) of the posterior distribution for the lifetime distribution. Once again, in the lower plot, the higher variability is evident for smaller values of `Pseudostress`. ∎

6.7 Exponential Regression

We saw in Section 2.2 that when $\kappa = 1$, or equivalently $\log \sigma = 0$, the Weibull reduces to the exponential distribution. We can also develop a regression model where the lifetimes are assumed to follow an exponential distribution. The model is

$$T_i \sim EXP\left(\exp(\beta_0 + \beta_1 x_1 + \cdots + \beta_p x_p)\right). \tag{6.32}$$

This is equivalent to the model where

$$Y_i = \log T_i \sim \text{SEV}\left(\beta_0 + \beta_1 x_1 + \cdots + \beta_p x_p, 1\right).$$

R's `survreg` in the `survival` package can be used to estimate parameters. In the `survreg` command, we simply tell R to use the `"exponential"` distribution.

Example 6.12 Apply exponential regression to the drill bit experiment presented in Example 6.9.

Solution:
The following R commands will run the exponential regression and summarize the results:

```
model0 = survreg( Surv(y,cens) ~ A.factor + B + C +
                  D + E + F + G + H + I + J + L +
                  M + N + O + P + Q,
                  dist="exponential" )
summary( model0 )
```

The output is the following:

```
Call:
survreg(formula = Surv(y, cens) ~ A.factor + B + C + D + E +
    F + G + H + I + J + L + M + N + O + P + Q, dist = "exponential")
             Value Std. Error     z       p
(Intercept)  5.4004    0.1804  29.94  < 2e-16
A.factor2    1.3382    0.2678   5.00  5.8e-07
A.factor3    1.3309    0.2686   4.95  7.2e-07
A.factor4    1.4928    0.2711   5.51  3.7e-08
B            0.0163    0.0969   0.17  0.8666
C            0.2209    0.0965   2.29  0.0221
D            0.3100    0.0978   3.17  0.0015
E           -0.0713    0.0991  -0.72  0.4721
F           -0.1237    0.0966  -1.28  0.2002
G           -0.0500    0.0979  -0.51  0.6097
H           -0.1608    0.0969  -1.66  0.0969
```

```
I                  0.1506      0.0996  1.51   0.1305
J                  0.2470      0.0963  2.57   0.0103
L                  0.1770      0.1006  1.76   0.0784
M                 -0.0399      0.1051 -0.38   0.7038
N                 -0.1776      0.1038 -1.71   0.0871
O                 -0.9333      0.1073 -8.69 < 2e-16
P                 -0.6946      0.1054 -6.59 4.4e-11
Q                  0.1365      0.1055  1.29   0.1957

Scale fixed at 1

Exponential distribution
Loglik(model)= -824.9   Loglik(intercept only)= -905.8
        Chisq= 161.77 on 18 degrees of freedom, p= 3.8e-25
Number of Newton-Raphson Iterations: 5
n= 128
```

Note that the estimates of the β parameters are very close to those obtained in Example 6.9. Also note that the R output says `Scale fixed at 1` to indicate that κ was set to one, which is a result of performing exponential regression as opposed to Weibull regression. ∎

The exponential assumptions carries some strong assumptions, including a constant hazard function and the memoryless property. Caution should be taken before applying exponential regression in a practical problem.

6.8 The Scale-Accelerated Failure-Time Model

The idea behind the SAFT model is that the predictor variable(s) have the effect of "scaling" time; that is, speeding up time, or slowing down time. Let $\mathbf{x}_0 = [x_{01}, x_{02}, \ldots, x_{0k}]'$ denote the baseline condition for the predictor variables. The SAFT model asserts that for other levels of the predictors, say, $\mathbf{x}_1 = [x_1, x_2, \ldots, x_k]'$ the time axis is appropriately scaled. If the level $\mathbf{x}_1$ leads to shorter lifetimes, then time will be scaled to go faster, thereby accelerating the failure time. To make this more precise, we say that

$$T(\mathbf{x}) = \frac{T(\mathbf{x}_0)}{AF(\mathbf{x})}. \tag{6.33}$$

Here, $T(\mathbf{x})$ denotes time it takes for the item to reach a fixed amount of deterioration, and $AF(\mathbf{x})$ is the acceleration factor. The acceleration factor has the properties that

- $AF(\mathbf{x}) > 0$ for every condition $\mathbf{x}$, and
- $AF(\mathbf{x}_0) = 1$.

In other words, time is scaled by a factor of 1 (i.e. not scaled at all) for the baseline. Thus, the magnitude of time acceleration is relative to the baseline condition $\mathbf{x}_0$. To illustrate this, suppose that $\mathbf{x}$ is a condition for which $AF(\mathbf{x}) = 2$. This means that at this condition the item will age at twice the rate that an item would age at the baseline condition $\mathbf{x}_0$. An item at baseline that is 100 hours old will be in exactly the same condition (i.e. have the same reliability) as a 50-hour old unit at condition $\mathbf{x}$. Here $T(\mathbf{x}) = 50$, $T(\mathbf{x}_0) = 100$, and $AF(\mathbf{x}) = 2$. These values clearly satisfy the condition in (6.33).

If, as is normally the case, that $AF(\mathbf{x}) > 1$, then the condition accelerates time, leading to generally shorter lifetimes relative to the baseline $\mathbf{x}_0$. If $AF(\mathbf{x}) < 1$, then condition $\mathbf{x}$ actually decelerates time relative to the baseline.

The acceleration model in (6.33) implies that

$$P(T \le t | \mathbf{x}_0) = P\left(T \le \frac{t}{AF(\mathbf{x})} | \mathbf{x}\right).$$

To see this, substitute the concrete values $t = 100$ and $AF = 2$ described previously. If time accelerates by a factor of $AF = 2$ then the unit at level $\mathbf{x}$ that is $100/2 = 50$ hours old will be in the same condition as a baseline unit at age 100 hours. Thus, the probability of a baseline unit failing before 100 hours is the same as a unit at $\mathbf{x}$ failing before 50 hours. Thus,

$$F(t | \mathbf{x}_0) = F\left(\frac{t}{AF(\mathbf{x})} | \mathbf{x}\right)$$

or equivalently

$$F(t | \mathbf{x}) = F(AF(\mathbf{x})t | \mathbf{x}_0).$$

If we let $t_p(\mathbf{x})$ denote the pth percentile for a unit at level $\mathbf{x}$, then since the percentiles will scale by the factor $AF(\mathbf{x})$, we have

$$t_p(\mathbf{x}) = \frac{t_p(\mathbf{x}_0)}{AF(\mathbf{x})}. \tag{6.34}$$

The accelerating factor cannot be negative, so any model involving an accelerating factor must assure that it is greater than zero. The most common assumption regarding how the level of the variables $\mathbf{x}$ affect the acceleration factor is that

$$AF(\mathbf{x}) = \exp(-\beta_1 x_1 - \beta_2 x_2 - \cdots - \beta_k x_k)$$
$$= \exp(-\mathbf{x}'\boldsymbol{\beta}), \tag{6.35}$$

where $\boldsymbol{\beta} = [\beta_1, \beta_2, \ldots, \beta_k]$. Here it is assumed that the baseline condition is coded so that $\mathbf{x} = \mathbf{0} = [0, 0, \ldots, 0]'$. Note that with this convention, $AF(\mathbf{x}_0) = \exp(0) = 1$, as required. Note also that there is no intercept term. The lifetime at the baseline is accounted for in the baseline CDF $F(t | \mathbf{x}_0)$.

Even with the stipulation that $AF(\mathbf{x}) = \exp(-\mathbf{x}'\boldsymbol{\beta})$ the SAFT model does not fully describe the failure process, only how the aging process changes relative to the baseline. To get a complete specification of the failure process, we could give a specific form for the distribution at baseline, and combine that with the specific form of aging given by the SAFT model (Figure 6.20). For example, if we specify that at the baseline $\mathbf{x}_0$, the lifetime distribution is WEIB(θ, κ); that is

$$F(t | \mathbf{x}_0) = 1 - \exp\left(-\left(\frac{t}{\theta}\right)^\kappa\right), \qquad t > 0.$$

Then the CDF for level $\mathbf{x}$ is

$$F(t | \mathbf{x}) = F(AF(\mathbf{x})t | \mathbf{x}_0) = 1 - \exp\left(-\left(\frac{AF(\mathbf{x})t}{\theta}\right)^\kappa\right). \tag{6.36}$$

If we make the additional assumption that the acceleration factor is as in (6.35), then

$$F(t | \mathbf{x}) = 1 - \exp\left(-\left(\frac{t}{\theta \exp(\mathbf{x}'\boldsymbol{\beta})}\right)^\kappa\right).$$

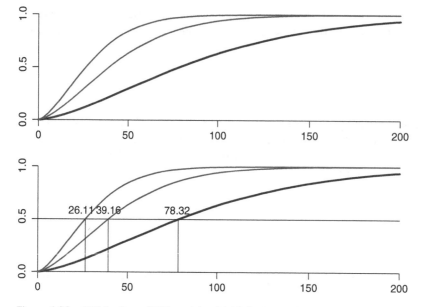

Figure 6.20 CDF for three SAFT models with Weibull baseline. The bottom curve is the baseline, and the upper curves correspond to AF = 2 (middle) and AF = 3 (top). In the bottom plot, the medians are marked; note that the medians for the two blue curves are equal to one half and one third times the baseline median.

If we let $\beta_0 = \log \theta$, or equivalently $\theta = \exp(\beta_0)$, then the model becomes

$$F(t|\mathbf{x}) = 1 - \exp\left(-\left(\frac{t}{\exp(\beta_0 + \beta_1 x_1 + \cdots + \beta_k x_k)}\right)\right), \qquad t > 0.$$

In other words,

$$T|\mathbf{x} \sim \text{WEIB}\left(\theta = \exp(\beta_0 + \beta_1 x_1 + \cdots + \beta_k x_k), \kappa\right). \tag{6.37}$$

This is the same model that was called *Weibull regression* in Section 6.5.

6.9 Checking Model Assumptions

The usual assumptions behind any of the regression models discussed in this chapter include the following:

1. the response variable (e.g. failure time) have the specified distribution (e.g. lognormal, Weibull, etc.),
2. each data point is independent,
3. the function that specifies (or links) the independent variables or factors to the response is correctly specified.

Distribution profiling and residual analysis are common approaches for checking these assumptions.

To illustrate the process, we focus on a model of lifetimes using the Weibull distribution. The assumptions behind Weibull regression imply the following:

1. the lifetimes have a Weibull distribution,
2. the lifetimes are independent,
3. for any two values of the predictor variables, say, $\mathbf{x}_1$ and $\mathbf{x}_2$, the hazard functions $h(t|\mathbf{x}_1)$ and $h(t|\mathbf{x}_2)$ are proportional,
4. the hazard function is related to the predictor variables through the factor $\exp(-\kappa\mathbf{x}'\boldsymbol{\beta}_1)$.

Notice that we break the correct function specification into two parts: first their relative impact on the predictor variables and second the structure of the linking function (exp). The last two assumptions can be checked using the residuals from the model.

6.9.1 Residual Analysis

Let t_i denote an event time, which could be a failure time or a censoring time and let $y_i = \log t_i$. Thus, if t_i is a censoring time from the Weibull distribution, then y_i will be a censoring time from the SEV distribution. Then $y_1, y_2, \ldots, y_n$ are (possibly censored) event times from the SEV(μ_i, σ_i) distribution. The usual definition of a residual is the difference between the actual value and the predicted value. Standardized residuals are defined as

$$r_i = \frac{y_i - \hat{\mu}}{\hat{\sigma}}. \tag{6.38}$$

Then, assuming that the model is correct, these residuals should resemble (possibly censored) observations from the SEV(0, 1) distribution. Of course, if t_i is censored, then y_i is censored, and so also is the residual r_i.

The Cox–Snell residuals are defined to be

$$r_{CS} = \exp(r_i) = \exp\left(\frac{\log t_i - \hat{\mu}_i}{\hat{\sigma}}\right) = \left(\frac{t_i}{\exp(\hat{\mu}_i)}\right)^{1/\hat{\sigma}_i}. \tag{6.39}$$

Plots of Cox–Snell residuals against the predictor variables or the predicted values can indicate violations of the assumptions. If experiments were performed in a given time-order, then a plot of residuals against time can uncover time effects. We illustrate the model checking using the battery life data from Section 4.4.1, an example with no censoring.

Example 6.13 In Section 4.4.1 we analyzed the battery life under the assumption of a normally distributed lifetime. There was no censoring and the normal distribution seemed to be reasonable for these data. For this example, apply Weibull regression, compute the Cox–Snell residuals, and plot them.

Solution:
The battery lifetimes are plotted in Figure 6.21 against both predictors. We will apply Weibull regression with both variables assumed to be categorical. The data can be loaded using the code from Section 4.4.1.

We can then run the Weibull regression model with the following code:

```
survreg.model = survreg( Surv(LifeTime) ~ MaterialType.f +
                    Temp.f + MaterialType.f:Temp.f )
summary( survreg.model )
```

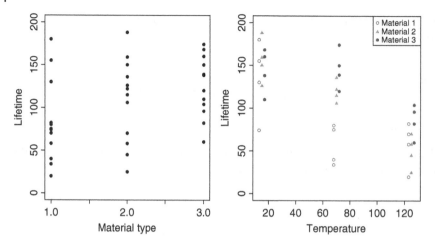

Figure 6.21 Battery life against material type and temperature.

The output from the `summary(survreg.model)` function is

```
Call:
survreg(formula = Surv(LifeTime) ~ MaterialType.f + Temp.f +
    MaterialType.f:Temp.f)
                            Value Std. Error       z        p
(Intercept)                5.0200     0.1004   50.02 < 2e-16
MaterialType.f2            0.0668     0.1410    0.47   0.6356
MaterialType.f3           -0.0085     0.1410   -0.06   0.9519
Temp.f70                  -0.8006     0.1409   -5.68 1.3e-08
Temp.f125                 -0.7923     0.1409   -5.63 1.9e-08
MaterialType.f2:Temp.f70   0.5160     0.1993    2.59   0.0096
MaterialType.f3:Temp.f70   0.8054     0.1992    4.04 5.3e-05
MaterialType.f2:Temp.f125 -0.2409     0.1993   -1.21   0.2268
MaterialType.f3:Temp.f125  0.2906     0.1992    1.46   0.1446
Log(scale)                -1.6136     0.1431  -11.28 < 2e-16

Scale= 0.199

Weibull distribution
Loglik(model)= -164.7   Loglik(intercept only)= -188.7
        Chisq= 48.14 on 8 degrees of freedom, p= 9.3e-08
Number of Newton-Raphson Iterations: 10
n= 36
```

These results suggest that there are significant two-factor interactions between material type and temperature. Two of the four interaction terms have P-values are less than 0.01. Since these higher-order term are in the model, the corresponding lower-order terms should also be included (by the hierarchy principle). Thus, temperature and material type should both be included in a model. Note that both `Temp=70` and `Temp=125` are significantly different from the baseline of `Temp=15`.

To compute the Cox–Snell residuals, we must first extract the estimated coefficients from the `survreg.model` object. We can then define a function to return the predicted value for $\mu = \log\theta$.

```
beta0hat         = survreg.model$coefficients[1]
beta1.2hat       = survreg.model$coefficients[2]
beta1.3hat       = survreg.model$coefficients[3]
beta2.2hat       = survreg.model$coefficients[4]
beta2.3hat       = survreg.model$coefficients[5]
beta1.2_2.2hat   = survreg.model$coefficients[6]
beta1.3_2.2hat   = survreg.model$coefficients[7]
beta1.2_2.3hat   = survreg.model$coefficients[8]
beta1.3_2.3hat   = survreg.model$coefficients[9]

muhat.function = function( material, temp )
{
  mat2  = as.numeric( material == 2 )
  mat3  = as.numeric( material == 3 )
  temp2 = as.numeric( temp == 70 )
  temp3 = as.numeric( temp == 125 )
  z = beta0hat + beta1.2hat*mat2 + beta1.3hat*mat3 +
      beta2.2hat*temp2 + beta2.3hat*temp3 +
      beta1.2_2.2hat*mat2*temp2 +
      beta1.3_2.2hat*mat3*temp2 +
      beta1.2_2.3hat*mat2*temp3 +
      beta1.3_2.3hat*mat3*temp3
  return( z )
}
sigmahat = survreg.model$scale
kappahat = 1/sigmahat
n = length( LifeTime )
muhat = muhat.function( MaterialType.f, Temp.f )
```

We see that the point estimate for the scale parameter of the SEV distribution is $\hat{\sigma} = 0.199$. Thus the estimate for κ, the shape parameter of the Weibull distribution (assumed to be constant across all levels of predictors) is $\hat{\kappa} = 1/0.199 = 5.02$. This is quite a bit larger than 1 (the value for which the Weibull reduces to the exponential distribution) indicating that wearout plays an important role in the life of the batteries. We can test whether σ is equal to 1, using the output from `summary(survreg.model)`. The last line of the table of estimates gives information about `Log(scale)`. The point estimate is -1.6136; note that $\exp(-1.6136) = 0.199$, or equivalently, $\log(0.199) = -1.6136$. The standard error for the estimate of σ is 0.1431, yielding a Wald statistic of $z = -11.28$ and a tiny P-value. Thus, there is evidence that $\log\sigma$ is not 0, or equivalently, there is evidence that σ is not 1.

Finally, given the predicted values for μ, we can compute and plot the Cox–Snell residuals. This is done with the following code.

```
y = log( LifeTime )
r = ( y - muhat ) / sigmahat
rCS = exp( r )     ## Cox--Snell Residuals
```

Figure 6.22 shows three plots of the Cox–Snell residuals. The top and middle graphs show the Cox–Snell residuals plotted against the two predictor variables: material type and temperature. The bottom plot shows a plot of the residuals against the predicted value. These plots show stable patterns of residuals, which is expected if the assumed model is correct.

In terms of determining what variables (and interactions) affect the lifetime of the battery, the normal distribution and the Weibull distribution yield similar results. Different distributional assumptions

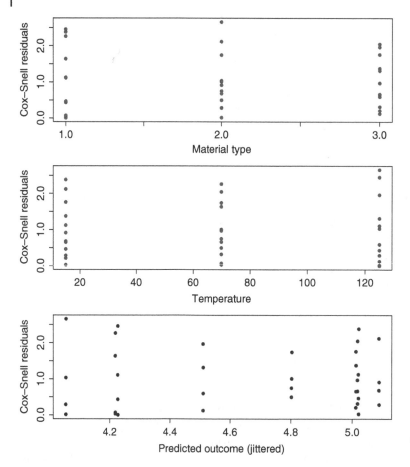

Figure 6.22 Cox–Snell residuals for the battery life data. The top and middle plots show the residuals against the predictor variables, material type and temperature. The bottom plot shows the residuals against the predicted value of the battery life.

will often lead to similar conclusions regarding the center of a distribution, as we have studied here. If interest focuses on the tails of the distribution, the different distributions can yield quite different results. For example, if for warranty reasons, we were interested in the 0.10 quantile of the distribution (i.e. the time value for which 90% of batteries exceed) then the normal and the Weibull distribution may yield vastly different estimates.   ■

Next, we illustrate the Cox–Snell residuals with the transmission data from Example 6.8. Because these residuals can be censored, it is important to plot censored values using a different plot character. Heavy censoring can make these residuals difficult to interpret.

Example 6.14   For the transmission data, plot and interpret the Cox–Snell residuals for the first- and second-order models.

Solution:

Initially, we fit a first-order model with all seven predictor variables. The Cox–Snell residuals can be computed from (6.39). Figure 6.23 shows the plots of the residuals against each of the predictor variables. In this plot, we applied some jitter in the predictor variable to help distinguish points that were overplotted.

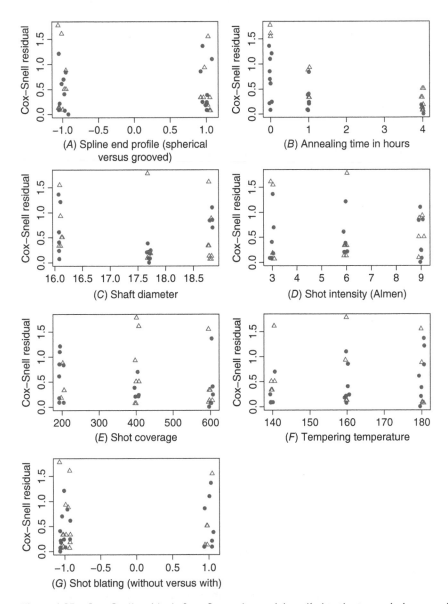

Figure 6.23 Cox–Snell residuals for a first-order model applied to the transmission experiment plotted against each predictor variable. Triangles indicate censored times.

Most of the predictors were numeric with three levels; only factors (A) and (G) were categorical. At first glance, the top right graph suggests that residuals are smaller for the highest level of annealing time. However, a closer examination uncovers that at $B = 4$ there are 10 censored values. (There were only 17 censored values in the entire data set.) The first two levels, $B = 0$ and $B = 1$ had 3 and 4 censored values, respectively. This heavy censoring at $B = 4$ indicates that there might be a positive relationship between B and the survival time. The third graph in Figure 6.23 indicates that there might be a nonlinear effect due to factor C, the shaft diameter. Among the 12 residuals at the level $B = 17.7$, 11 were below 0.5 (although two of these were censored), and just one, a censored residual, was larger than 0.5. Residuals at $C = 16.1$ and $C = 18.8$ are generally larger, suggesting a concave up quadratic effect may exist between C and the lifetime.

Often, the variability of the responses (lifetimes in our case) increases as the predicted value increases. Here we use the predicted value of μ, which is the location parameter for the logarithm of lifetimes. This departure from the assumptions can be checked by plotting the Cox–Snell residual against the predicted value. For the first-order model, we plot the residuals against the predicted value in Figure 6.24. A first

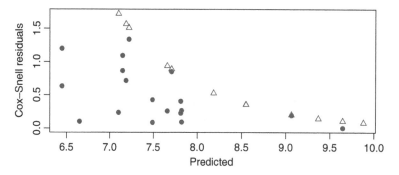

Figure 6.24 Cox–Snell residuals for a first-order model applied to the transmission experiment plotted against the predicted time. Triangles indicate censored times.

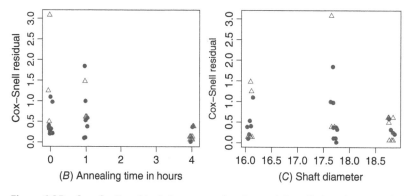

Figure 6.25 Cox–Snell residuals for a second-order model applied to the transmission experiment plotted against each predictor variable. Triangles indicate censored times.

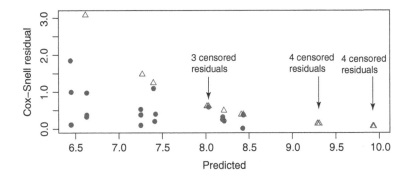

Figure 6.26 Cox–Snell residuals for a second-order model applied to the transmission experiment plotted against the predicted time. Triangles indicate censored times.

glance indicates that the residuals are getting smaller as the predicted value increases, but again, a closer examination of the censored values is needed. For predicted values of μ beyond 8.0, there are only two observed failure times and 12 censored values. Thus, it is probably not the case that the residuals decrease as the predicted value increases.

Now consider the second-order model fit to factors B and C. Figure 6.25 shows the Cox–Snell residuals against both predictor variables. Once again, we must keep in mind that 10 of the 17 censored residuals occur at $B = 4$, and knowing this we conclude that the residuals are stable across the levels of factor B. Similarly, 8 of the 17 residuals occur at the highest level of factor C while 6 of the 17 occur at the lowest level. Again, the residuals seem to be stable across the levels of factor C.

Figure 6.26 shows the plots of the Cox–Snell residuals against the predicted lifetime. As was the case for the first-order model, most of the censored values occur for the largest predicted times. ∎

6.9.2 Distribution Selection

One of the key modeling decisions and assumptions that needs to be checked is what is the appropriate distribution for the model (e.g. Lognormal, Exponential, Weibull, Gamma, etc.). The selection of an appropriate distribution for modeling failure time data is frequently informed by probability plotting. For failure times, probability plots have the probability of failure, $F(t)$, on the y-axis and failure time, t on the x-axis. Probability plots are useful for both assessing the adequacy of a particular distribution to fit the data and comparing the data to the results of a particular model fit. Historically, probability plots have also been used to develop graphical estimates of distribution parameters by approximating a best fit line through the data. However, those methods have been replaced by parametric and nonparametric estimates widely available in software.

Example 6.15 Table 6.1 shows 28 failures times in minutes generated from a lognormal distribution with $\mu = 2$ and $\sigma = 0.5$. The histogram shown in Figure 6.27 generated from the Analyze, Distribution function in JMP shows the maximum likelihood fit to the data for the lognormal distribution and the

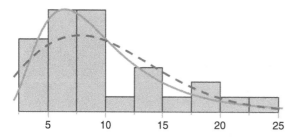

Figure 6.27 Histogram of the 28 failure times compared with lognormal fit (solid line) and Weibull fit (dashed line).

Weibull distribution. While it looks like the lognormal might be a better fit, it is hard to make a comparison based on histograms alone. Provide a probability plot justification for analyzing the data using the lognormal distribution.

Solution:

Figure 6.28 compares the probability plots for the lognormal fit (a) and the Weibull fit (b) to the data. The shaded region is the confidence region for the best fit distribution based on maximum likelihood estimates of the distribution parameters. Notice that the y-axis provides the probability quantiles and the scale is selected such that the data should fall on a straight line centered on expected value if the distributional choice is appropriate (e.g. lognormal panel (a) or Weibull panel (b)). The scales on the two plots were optimized to show trends (curvature) in the structure of data. However, that results in different scales so it is important to read axes closely when comparing values. Clearly, all of the points on the right panel fall close the lognormal fit to the data. Additionally, there are no features in the data suggesting that the lognormal would be a poor fit (e.g. curvature). On the Weibull plot (panel (b)) on the other hand there is slight curvature in the points that suggest the Weibull distribution is failing to explain some of the variability in the data. Additionally, the points fall further away from the line comparatively.

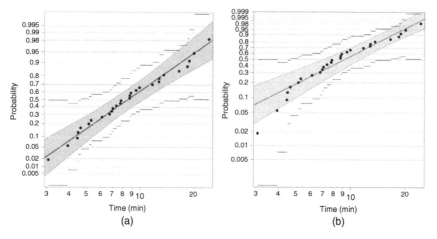

Figure 6.28 (a) Lognormal probability plot, shown on lognormal quantile scale. (b) Weibull probability plot, shown on log scale.

Problems

6.1 Nelson (1982) provides failure times of diesel generator fans in hours. Table D.7 in Appendix D provides the full dataset. Using all of the data (censored and failure data) calculate the expected number of hours when 75% of the fans are expected to fail assuming the failures follow a normal distribution. Next calculate the expected number of hours when 75% of the fans are expected to fail assuming the failures follow a lognormal distribution. How does your conclusion change?

6.2 Using the diesel generator fan data from Table D.7, compare the normal, lognormal, gamma, exponential, and Weibull fits to the failure fan data. What distribution is the best fit to the data. Provide plots to justify your answer.

6.3 Nelson (1990) describes an experiment for 26 cylindrical nickel-based superalloy that are stress tested under various levels of Pseudostress (ksi). The full dataset is given in Table D.6. Notice there are 4 censored observations. First, remove the censored observations from the data set and analyze the data using a log transformation on the number of cycles until failure. Recall this is done by first taking the log of the cycles to failure, then using standard linear regression. Provide the regression model to include the coefficient estimates on log-scale. Based on this analysis what is the expected lifespan for cylinder place under a pseudo stress of 100 ksi? What about 160 ksi?

6.4 Repeat the analysis in Problem 6.3, but use the lognormal distribution via the log-location scale parameterize to fit the model directly on cycles to failure. How can you show this model is equivalent to the model in Problem 6.1? Provide the expected lifespan for a cylinder place under a pseudo stress of 100 and 160 ksi. Provide 95% confidence intervals for the expected lifespans.

6.5 Repeat the analysis in Problem 6.4, but consider the full dataset (including censored observations), use the lognormal distribution via the log-location scale parameterize to fit the model directly on cycles to failure. You will need to include an indicator variable for censoring. Provide the expected lifespan for a cylinder place under a pseudo stress of 100 and 160 ksi. Provide 95% confidence intervals for the expected lifespans. Describe the impact of including the censored data on the model and expected lifespans.

6.6 Zelen (1959) discusses a factorial experiment designed to determine the effect of voltage and temperature on the lifespan of a glass capacitor. Zelen describes the experimental setup as "n components are simultaneously placed on test." Zelen uses two different temperature settings (170 and 180 °C) and four different voltages (200, 250, 300, and 350 V) in the experiment for a total of eight treatment combinations. Eight capacitors are tested on each test stand, and Zelen uses Type II censoring after the first four failures. The full dataset is given in Table D.2. Assume the failures follow a lognormal distribution. Fit a lognormal regression model using either a transformation or the log-location scale approach directly. What are your conclusions of the effect of temperature and voltage on capacitor lifespan? Is there a significant interaction effect between temperature and voltage?

6.7 Repeat the analysis in Problem 6.1 with the Zelen data, but this time assume the failure times follow a Weibull distribution. How do your estimates change? Do your conclusions about the effect of temperature and voltage change? If so how?

6.8 For the Zelen data, which assumption is the better assumption, failure times are lognormal distributed, or failure times are Weibull distributed? Include plots to illustrate your conclusion.

6.9 Rerun the analysis shown in Example 6.5 on the ball bearing experiment. This time assume the ball bearing failure times follow an exponential distribution. How do your results change? Compare the Weibull and exponential distribution probability plots. Which distribution is a better fit to the data? What is the impact of assuming the scale parameter of the Weibull is fixed at $\kappa = 1$?

6.10 This example is taken from a paper by Bisgaard and Fuller (1994–1995). The experiment involves a 16 run two-level fractional factorial design in nine factors. The response variable is the count of the number of defects in an automobile grille. The purpose of this experiment is to screen out the insignificant factors. The generators for this 2^{9-5} Resolution III design are $E = BD$, $F = BCD$, $G = AC$, $H = ACD$, and $J = AB$. Every one of the nine main effects is aliased with two of the

Table 6.15 Data for Problem 6.10.

Obs	A	B	C	D	E	F	G	H	J	$\hat{c}$	$\sqrt{\hat{c}}$	FT Mod.
1	−1	−1	−1	−1	1	−1	1	−1	1	56	7.48	7.52
2	1	−1	−1	−1	1	−1	−1	1	−1	17	4.12	4.18
3	−1	1	−1	−1	−1	1	1	−1	−1	2	1.41	1.57
4	1	1	−1	−1	−1	1	−1	1	1	4	2.00	2.12
5	−1	−1	1	−1	1	1	−1	1	1	3	1.73	1.87
6	1	−1	1	−1	1	1	1	−1	−1	4	2.00	2.12
7	−1	1	1	−1	−1	−1	−1	1	−1	50	7.07	7.12
8	1	1	1	−1	−1	−1	1	−1	1	2	1.41	1.57
9	−1	−1	−1	1	−1	1	1	1	1	1	1.00	1.21
10	1	−1	−1	1	−1	1	−1	−1	−1	0	0.00	0.50
11	−1	1	−1	1	1	−1	1	1	−1	3	1.73	1.87
12	1	1	−1	1	1	−1	−1	−1	1	12	3.46	3.54
13	−1	−1	1	1	−1	−1	−1	−1	1	3	1.73	1.87
14	1	−1	1	1	−1	−1	1	1	−1	4	2.00	2.12
15	−1	1	1	1	1	1	−1	−1	−1	0	0.00	0.50
16	1	1	1	1	1	1	1	1	1	0	0.00	0.50

two-factor interactions. The experimental responses in this case are counts of defects per grille, which we assume to be a Poisson distributed variable. The design matrix, the count of defects, which they labeled $\hat{c}$, are shown in Table 6.15. Bisgaard and Fuller analyzed these data by taking the square root of the counts and also by using the Freeman and Tukey (FT) modified transformation $(\sqrt{\hat{c}} + \sqrt{\hat{c} + 1})/2$.

(a) Repeat the Bisgaard and Fuller analysis. Assume that two-factor interactions are negligible. Which modeling approach do you prefer?

(b) Analyze the count data using a GLM with a Poisson distribution and the log link. Do you prefer this model to either of the ones found in part (a)?

6.11 Analyze the grille defects data using the Poisson distribution and a square root link. Compare this model with the ones you found in Problem 6.10.

7

Semi-parametric Regression Models

7.1 The Proportional Hazards Model

Another approach to modeling the relationship of the predictor variables $\mathbf{x}$ to the lifetime T is through the hazard function. Recall that the hazard function is

$$h(t) = \lim_{\Delta t \to 0} \frac{P(t < T \le t + \Delta t \mid T > t)}{\Delta t}.$$

This can be interpreted as the probability of failure in the next small interval of time divided by the length of the interval. (Technically, it is the limit of this ratio.) The larger the value of the hazard, the greater the chance of an imminent failure. In the proportional hazards model we assume that the predictor variables exert a multiplicative effect on the hazard. If we let $h_0(t)$ denote the hazard at the baseline condition $\mathbf{x}_0$, that is, $h_0(t) = h(t|\mathbf{x}_0)$, the proportional hazards model assumes

$$h(t \mid \mathbf{x}) = P(\mathbf{x})h_0(t). \tag{7.1}$$

Notice how the proportional hazards model partitions the effects due to the predictor variables and time:

$$h(t \mid \mathbf{x}) = \underbrace{P(\mathbf{x})}_{\text{a function of } \mathbf{x} \text{ only}} \times \underbrace{h_0(t)}_{\text{a function of } t \text{ only}}.$$

The function $P(\mathbf{x})$ must satisfy the conditions

- $P(\mathbf{x}) > 0$ for every condition $\mathbf{x}$, and
- $P(\mathbf{x}_0) = 1$.

Note, these are the same assumptions as for the accelerating factor in parametric Weibull regression.

The baseline hazard and the $P(\mathbf{x})$ function may both contain unknown parameters. Here we make the assumption that

$$P(\mathbf{x}) = \exp\left(-\beta_1 x_1 - \beta_2 x_2 - \cdots - \beta_p x_p\right) = \exp\left(-\mathbf{x}'\boldsymbol{\beta}\right).$$

Some authors, such as Lawless (2003), write this as $\exp\left(\mathbf{x}'\boldsymbol{\beta}\right)$; the reason for our choice is that the result is similar to the SAFT model described in Chapter 6. (This will be seen later in this section.) The Cox proportional hazards (CPH) model, to be discussed in Section 7.2, makes no parametric assumption about the form of the baseline hazard $h_0(t)$, although similar assumptions are made for the function $P(\mathbf{x})$. For these

Design of Experiments for Reliability Achievement, First Edition.
Steven E. Rigdon, Rong Pan, Douglas C. Montgomery, and Laura J. Freeman.
© 2022 John Wiley & Sons, Inc. Published 2022 by John Wiley & Sons, Inc.
Companion website: www.wiley.com/go/rigdon/designexperiments

reasons, the Cox model is often described as **semiparametric**. Note that a positive slope coefficient β_i leads to a lower hazard function as the predictor x_i increases; this in turn leads to longer lifetimes. Note also the absence of a constant term in the exponent of the formula for $P(\mathbf{x})$. The constant is considered to be part of the baseline hazard, leaving the sole role of $P(\mathbf{x})$ to increase or decrease the hazard function (with a multiplicative effect) that depends on the value of $\mathbf{x}$.

Whatever functional forms are assumed for $P(\mathbf{x})$ and $h_0(t)$, we can relate the survival function to the baseline survival function as follows:

$$
\begin{aligned}
S(t|\mathbf{x}) &= \exp\left(-\int_0^t h(u|\mathbf{x})\ du\right) \\
&= \exp\left(-P(\mathbf{x})\int_0^t h_0(u)\ du\right) \\
&= \left[\exp\left(-\int_0^t h_0(u)\ du\right)\right]^{P(\mathbf{x})} \\
&= [S_0(t)]^{P(\mathbf{x})},
\end{aligned}
\tag{7.2}
$$

where $S_0(t)$ is the survival function at baseline. Thus,

$$
\log S(t|\mathbf{x}) = P(\mathbf{x})\log S_0(t).
\tag{7.3}
$$

If $P(\mathbf{x}) = 1$ for some particular $\mathbf{x}$, then $S(t|\mathbf{x}) = S_0(t)$ and we get identical distributions for the value of the predictor $\mathbf{x}$ and the baseline. Otherwise (7.2) and (7.3) imply that the survival functions (and therefore the cumulative distribution functions (CDFs)) cannot cross. Thus one of $S(t|\mathbf{x})$ and $S_0(t)$ is uniformly above the other.

If we make the additional assumption that the baseline distribution is WEIB(θ, κ), then

$$
h_0(t) = \frac{\kappa}{\theta}\left(\frac{t}{\theta}\right)^{\kappa-1}, \qquad t > 0
$$

so

$$
h(t|\mathbf{x}) = e^{-\mathbf{x}'\beta}\frac{\kappa}{\theta}\left(\frac{t}{\theta}\right)^{\kappa-1}, \qquad t > 0.
$$

The survival function is therefore

$$
\begin{aligned}
S(t|\mathbf{x}) &= \exp\left(-\int_0^t e^{-\mathbf{x}'\beta}\frac{\kappa}{\theta}\left(\frac{u}{\theta}\right)^{\kappa-1} du\right) \\
&= \exp\left(-e^{-\mathbf{x}'\beta}\left(\frac{t}{\theta}\right)^{\kappa}\right) \\
&= \exp\left(-\left(\frac{t}{\theta\exp(\mathbf{x}'\beta/\kappa)}\right)^{\kappa}\right).
\end{aligned}
$$

If we set $\beta_0 = \kappa\log\theta$, we can then write

$$
S(t|\mathbf{x}) = \exp\left(-\left(\frac{t}{\exp\left(\frac{\beta_0}{\kappa} + \frac{\beta_1}{\kappa}x_1 + \cdots + \frac{\beta_p}{\kappa}x_p\right)}\right)^{\kappa}\right)
$$

which shows that

$$T|\mathbf{x} \sim \text{WEIB}\left(\theta = \exp\left(\frac{\beta_0}{\kappa} + \frac{\beta_1}{\kappa}x_1 + \cdots + \frac{\beta_p}{\kappa}x_p\right), \kappa\right). \tag{7.4}$$

Compare this result with the Weibull regression formula obtained in (6.37). The parameter κ plays the same role, but the slope parameters differ by a factor of $1/\kappa$.

The other commonly used parameterization of the Weibull distribution has survival function

$$S(t) = \exp(-\lambda t^\kappa)$$

and hazard function

$$h(t) = \lambda \kappa t^{\kappa - 1}.$$

With this parameterization, the proportional hazards model becomes

$$h(t|\mathbf{x}) = e^{-\mathbf{x}'\beta} \lambda \kappa t^{\kappa - 1}$$

and the survivor function is

$$S(t|\mathbf{x}) = \exp\left(-e^{-\mathbf{x}'\beta} t^\kappa\right)$$
$$= \exp\left(-t^\kappa \exp\left(-\beta_0 - \beta_1 x_1 - \beta_2 x_2 - \cdots - \beta_p x_p\right)\right)$$

where $\beta_0 = -\log \lambda$. Thus,

$$T|\mathbf{x} \sim \text{WEIB}\left(\lambda = \exp\left(-\beta_0 - \beta_1 x_1 - \beta_2 x_2 - \cdots - \beta_p x_p\right), \kappa\right). \tag{7.5}$$

Compare the seemingly different results in (6.37), (7.4), and (7.5). These results are not contradictory, but rather they arise from different formulations of the model. It is important for a user to know that different formulations exist, because various software packages use different model formulations.

7.2 The Cox Proportional Hazards Model

The Weibull regression model from Section 6.5 assumes that the lifetime of a unit with predictor $\mathbf{x}_i$ has distribution

$$T_i \sim \text{WEIB}\left(\theta = e^{\beta'\mathbf{x}_i}, \kappa\right)$$

We saw previously that Weibull regression has the proportional hazards property; this means that the ratio of the hazard functions for units that have vectors of predictor variables $\mathbf{x}_i$ and $\mathbf{x}_j$ is independent of time t. In other words the hazard for $\mathbf{x}_i$ is a multiple of the hazard for $\mathbf{x}_j$. The ratio of hazards is

$$\frac{h(t|\mathbf{x}_i)}{h(t|\mathbf{x}_j)} = \frac{\frac{\kappa}{\exp(x_i'\beta)}\left(\frac{t}{\exp(x_i'\beta)}\right)^{\kappa-1}}{\frac{\kappa}{\exp(x_j'\beta)}\left(\frac{t}{\exp(x_j'\beta)}\right)^{\kappa-1}} = \frac{\kappa \exp(-\kappa x_i'\beta)t^{\kappa-1}}{\kappa \exp(-\kappa x_j'\beta)t^{\kappa-1}} = \exp\left(-\kappa x_i'\beta + \kappa x_j'\beta\right), \qquad t > 0,$$

which is independent of t. Let the vector of predictor variables be $\mathbf{x}_i = [1, x_{i1}, x_{i2}, \ldots, x_{ip}]^T$. We can consider the hazard when all predictor variables are equal to zero, that is, $\mathbf{x}_0 = [1, 0, 0, \ldots, 0]^T$, to be the baseline hazard function, which is equal to

$$h_0(t) = \frac{\kappa}{e^{\beta_0}} \left(\frac{t}{e^{\beta_0}} \right)^{\kappa-1}.$$

The ratio of an arbitrary hazard function to the baseline hazard is therefore

$$\frac{h(t|x_{i1}, x_{i2}, \ldots, x_{ip})}{h_0(t)} = \exp\left(-\kappa(\beta_1 x_1 + \beta_2 x_2 + \cdots + \beta_p x_p)\right). \tag{7.6}$$

Problem 7.14 asks you to derive this.

We can then write

$$
\begin{aligned}
h(t|x_{i1}, x_{i2}, \ldots, x_{ip}) &= h_0(t) \exp\left(-\kappa(\beta_1 x_1 + \beta_2 x_2 + \cdots + \beta_p x_p)\right) \\
&= h_0(t) \exp\left(\beta_1^* x_1 + \beta_2^* x_2 + \cdots + \beta_p^* x_p\right)
\end{aligned}
$$

where $\beta_j^* = -\kappa\beta_j$. This relationship illustrates how the predictor variables serve to scale the baseline hazard function $h_0(t)$.

The CPH model assumes that the hazard function as a function of the time t and the predictor variable $\mathbf{x}$ is of the form

$$h(t|\mathbf{x}_i) = h_0(t) \exp\left(\beta_1 x_1 + \beta_2 x_2 + \cdots + \beta_p x_p\right) = h_0(t) e^{\boldsymbol{\beta}' \mathbf{x}_i}, \qquad t > 0. \tag{7.7}$$

Thus, just as is the case for Weibull regression, the effect of the predictor variables is to scale (i.e. multiply) the baseline hazard function by a factor that depends on the predictor variables. As a result, the hazard functions (as a function of time t) for the various treatments can never intersect.

Inference for the CPH is usually based on a partial likelihood, a likelihood that depends on the ordering of the failure times and not on the actual times of the failures. To illustrate the partial likelihood, suppose that eight units with predictor vectors $\mathbf{x}_1, \mathbf{x}_2, \ldots, \mathbf{x}_8$ are placed on test and the failure times (indicated by an "X") and censoring times (indicated by a circle) are as shown in Figure 7.1. Let's consider, for example, unit 1, which failed at time t_1. To begin the development of the partial likelihood, we ask the question:

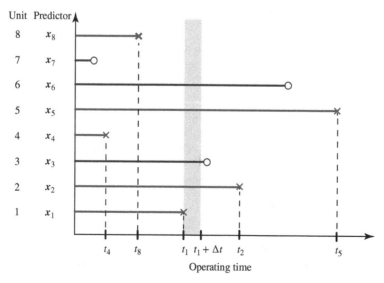

Figure 7.1 Illustration of likelihood contribution for Cox proportional hazards model.

"What is the probability that it is unit 1 that failed in the interval $[t_1, t_1 + \Delta t)$ given that one of the units still operating at time t_1 failed in this interval?" Note that units 1, 2, 3, 5, and 6 were still operating at time t_1; the others have already either failed or been censored (it doesn't matter which). The probability of failure in a small interval, conditioned on survival to the beginning of the interval, is approximately equal to the hazard function at that time times the length of that interval. Let $R(t)$ denote the **risk set at time** t, that is, the set of those units still operating at time t. For this example,

$$R(t_1) = \{1, 2, 3, 5, 6\}.$$

The desired probability is thus

$$P\left(\text{unit 1 fails in } [t_1, t_1 + \Delta t) \mid \text{one of the 5 units in } R(t_1) \text{ fails in } [t_1, t_1 + \Delta t)\right)$$

$$= \frac{P\left(\text{unit 1 fails in } [t_1, t_1 + \Delta t)\right)}{P\left(\text{one of the 5 units in } R(t_1) \text{ fails in } [t_1, t_1 + \Delta t)\right)}$$

$$= \frac{e^{\beta' x_1} h_0(t_1) \Delta t}{\left[e^{\beta' x_1} h_0(t_1) \Delta t + e^{\beta' x_2} h_0(t_1) \Delta t + e^{\beta' x_3} h_0(t_1) \Delta t + e^{\beta' x_5} h_0(t_1) \Delta t + e^{\beta' x_6} h_0(t_1) \Delta t\right]}$$

$$= \frac{e^{\beta' x_1}}{e^{\beta' x_1} + e^{\beta' x_2} + e^{\beta' x_3} + e^{\beta' x_5} + e^{\beta' x_6}}$$

since $h_0(t_1)$ and Δt factor out of both the numerator and denominator. This can be thought of as the probability that unit 1 fails at time t_1 given that there was a failure at time t_1. Note that other than accounting for which units are operating at time t_1, the expression earlier is independent of t_1. Table 7.1 shows which units were still operating just before each failure time: t_1, t_2, t_4, t_5 and t_8.

The partial likelihood is then the product of the probabilities of failure at each observed failure time conditioned on there being a failure (among those units still operating) at each observed failure time t_i. For our simple eight-unit example, this is

$$L(\beta \mid t_1, t_2, \dots, t_n, \mathbf{x}_1, \mathbf{x}_2, \dots, \mathbf{x}_n)$$

$$= \frac{e^{\beta' x_1}}{e^{\beta' x_1} + e^{\beta' x_2} + e^{\beta' x_3} + e^{\beta' x_5} + e^{\beta' x_6}} \times \frac{e^{\beta' x_2}}{e^{\beta' x_1} + e^{\beta' x_5} + e^{\beta' x_6}}$$

$$\times \frac{e^{\beta' x_4}}{e^{\beta' x_1} + e^{\beta' x_2} + e^{\beta' x_3} + e^{\beta' x_4} + e^{\beta' x_5} + e^{\beta' x_6} + e^{\beta' x_8}}$$

$$\times \frac{e^{\beta' x_5}}{e^{\beta' x_5}} \times \frac{e^{\beta' x_8}}{e^{\beta' x_1} + e^{\beta' x_2} + e^{\beta' x_3} + e^{\beta' x_5} + e^{\beta' x_6} + e^{\beta' x_8}}.$$

Table 7.1 Units still operating just before each observed failure.

Failed unit	Units still operating
1	1,2,3,5,6
2	2,5,6
4	1,2,3,4,5,6,8
5	5
8	1,2,3,5,6,8

The general formula for the partial likelihood is

$$L(\beta|\mathbf{t},\mathbf{x}) = \prod_{\substack{\text{all failure} \\ \text{times } i}} \frac{e^{\beta'\mathbf{x}_i}}{\sum_{j \in R(t_i)} e^{\beta'\mathbf{x}_j}}.$$

where $R(t)$ is the risk set at time t. The partial maximum likelihood estimate (PMLE) of β can then be approximated using numerical methods applied to the partial likelihood. Such methods are performed automatically in R and JMP.

The CPH model is nonparametric in the sense that no parametric form for the lifetimes (e.g. exponential or Weibull) is assumed. The method of maximum likelihood is invariant to any monotonic transformation of t. For example, taking the square root of every failure or censoring time has no effect on the point estimates of the β parameters.

For the case where there is only one predictor and it is an indicator variable, the likelihood function depends on only a single parameter. Let x_i be 1 if unit i is in group A and 0 if it is not in group A. Suppose for illustration that four units are assigned to one treatment, which we label as group A, and another four units are assigned to the other treatment; we'll say that these are in group B. Data are as shown in Table 7.2. In this case, $e^{\beta x_i} = e^\beta$ for the four units in group A and $e^{\beta x_i} = e^0 = 1$ for the four units in Group B. In this case, the partial likelihood function simplifies to

$$L(\beta|t_1, t_2, \ldots, t_n, \mathbf{x}_1, \mathbf{x}_2, \ldots, \mathbf{x}_n) = \frac{1}{3 + 2e^\beta} \frac{1}{1 + 2e^\beta} \frac{1}{4 + 3e^\beta} \frac{1}{1} \frac{e^\beta}{3 + 3e^\beta}.$$

Figure 7.2 shows the graph of $\log L(\beta|t_1, t_2, \ldots, t_n, \mathbf{x}_1, \mathbf{x}_2, \ldots, \mathbf{x}_n)$ versus β. The maximum of the log likelihood function, and therefore the maximum of the likelihood function itself, occurs at $\hat\beta = -1.1423$. This agrees with the estimate of β from R's output to the coxph function. The code

```
t = c(7,10,8,2,16,14,1,4)
delta = c(1,1,0,1,1,0,0,1)
x = c(0,0,0,0,1,1,1,1)

library( survival)
coxph( Surv(t,delta) ~ x)
```

Table 7.2 Data set where units 1–4 were assigned to treatment group A, and 5–8 were assigned to treatment group B.

Unit i	Predictor x_i	Time t_i	Status δ_i
1	0	7	1
2	0	10	1
3	0	8	0
4	0	2	1
5	1	16	1
6	1	14	0
7	1	1	0
8	1	4	1

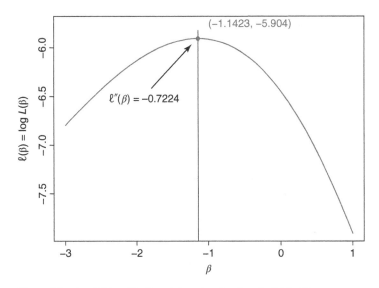

Figure 7.2 Partial log-likelihood function for data in Table 7.2.

produced the output below

```
Call:
coxph( Surv(t,delta) ~ x)

  n= 8, number of events= 5

     coef   exp(coef) se(coef)     z  Pr(>|z|)
x  -1.1423   0.3191    1.1772   -0.97   0.332

   exp(coef) exp(-coef) lower.95 upper.95
x    0.3191     3.134    0.03176   3.206

Concordance= 0.618  (se = 0.149)

Likelihood ratio test= 1.08  on 1 df,   p=0.3
Wald test           = 0.94  on 1 df,   p=0.3
Score (logrank) test = 1.03  on 1 df,   p=0.3
```

7.3 Inference for the Cox Proportional Hazards Model

In Section 7.2 we described the CPH model, the construction of the partial likelihood, and the maximum likelihood estimators for the parameters in the semi-parametric regression model. For the CPH model, we must rely on asymptotic normal distributions for the estimates of the model parameters $\beta_1, \beta_2, \ldots, \beta_p$. The observed information matrix is

$$\hat{I}(\hat{\beta}) = \left[-\frac{\partial^2 \ell}{\partial \beta_i \, \partial \beta_j} \right]_{\hat{\beta}}$$

where $\ell(\beta)$ is the log of the partial likelihood function. The asymptotic covariance matrix for the estimator $\hat{\beta}$ is related to the observed information matrix by

$$\hat{V}(\hat{\beta}) = \left[\hat{I}(\hat{\beta})\right]^{-1}$$

The approximate variance for the estimator $\hat{\beta}_j$ is just the (j,j) entry in this matrix. Thus, for large sample sizes

$$\frac{\hat{\beta}_j - \beta_j}{\sqrt{(j,j) \text{ entry in } \hat{V}(\hat{\beta})}} \overset{\text{approx}}{\sim} N(0,1).$$

This result can be used to find an approximate 95% confidence interval for β_j or test $H_0 : \beta_j = \beta_{j0}$; often we are interested in testing whether β_j is equal to 0 because $\beta_j = 0$ indicates that predictor j has no effect on the lifetime.

The approximate confidence interval for β_j is

$$\left(\hat{\beta}_j - z_{\alpha/2} \frac{1}{\sqrt{(j,j) \text{ entry in } \hat{V}(\hat{\beta})}}, \hat{\beta}_j + z_{\alpha/2} \frac{1}{\sqrt{(j,j) \text{ entry in } \hat{V}(\hat{\beta})}} \right).$$

The Wald test uses this normal approximation to test $H_0 : \beta_j = 0$ using the statistic

$$z_{\text{Wald}} = \frac{\hat{\beta}_j - 0}{\sqrt{(j,j) \text{ entry in } \hat{V}(\hat{\beta})}} \overset{\text{approx}}{\sim} N(0,1).$$

The null hypothesis $H_0 : \beta_j = 0$ is rejected if $z < -z_{\alpha/2}$ or $z > z_{\alpha/2}$. R gives the result of Wald's test in the row that gives the point estimate and estimated standard error. The column has heading z and the *P*-value for a two-sided test is given in the last column with heading `Pr(>|z|)`.

When there is more than one predictor variable, it is possible to test the omnibus hypothesis that none of the predictors affect the lifetime; in other words, all of the β_j coefficients are zero. That is, we are testing the hypotheses

$$H_0 : \beta_1 = \beta_2 = \cdots = \beta_p = 0 \text{ versus } H_1 : \text{ at least one } \beta_j \text{ is not zero.}$$

R gives three test statistics for testing this null hypothesis. One is analogous to Wald's test and uses the result that

$$\left(\hat{\beta} - 0\right)' \hat{I}(\hat{\beta}) \left(\hat{\beta} - 0\right) \overset{\text{approx}}{\sim} \chi^2(p). \tag{7.8}$$

We would reject the null hypothesis if this statistic exceeds $\chi^2_\alpha(p)$. For the special case where there is just a single predictor, this statistic is just the square of the z from Wald's test given earlier.

A second test is the likelihood ratio test statistic

$$\lambda = \frac{\max_{\beta \in H_0} L(\beta)}{\max_\beta L(\beta)} \tag{7.9}$$

where the numerator is the partial likelihood evaluated under the null hypothesis, and the maximum in the denominator is taken over the entire space. Since the numerator can be thought of as taking the maximum over the null hypothesis (which is a single point for this null hypothesis) and the denominator is the maximum taken over a larger set (the entire parameter space in this case), the likelihood ratio must be less than 1. For large sample sizes

$$-2 \log \lambda \overset{\text{approx}}{\sim} \chi^2(p). \tag{7.10}$$

In general, the degree of freedom is equal to the difference in the dimension of the parameter space in the denominator and the numerator. Here, the difference is $p - 0 = p$. We reject for values that exceed $\chi_\alpha^2(p)$.

A third test is the score test, which is based on the score function

$$S(\beta) = \ell'(\beta).$$

We compute the test statistic

$$S(0)'\hat{I}(0)^{-1}S(0)$$

and reject for values that exceed $\chi_\alpha^2(p)$. When there is just one predictor variable and it is an indicator variable, then the score test is equivalent to the log-rank test, a nonparametric test of whether the survival curves for two groups are the same. See Moore (2016).

Example 7.1 For the data in Table 7.2, compute the Wald statistic, the likelihood ratio statistic, and the score statistic to test whether the parameter β is equal to zero.

Solution:
The point estimate for the (one and only) parameter was found to be $\hat{\beta} = -1.1423$. If we evaluate the second derivative of the partial likelihood function, we find that

$$\ell''(\hat{\beta}) = -0.7224.$$

The estimated standard error of $\hat{\beta}$ is therefore

$$\text{se}(\hat{\beta}) = \frac{1}{\sqrt{\hat{I}(\hat{\beta})}} = \sqrt{-\frac{1}{-0.7224}} = 1.177.$$

This agrees with the output from R shown previously.

In the case of a single predictor, there is just one parameter so the omnibus test that parameters are equal to zero is equivalent to the hypothesis $H_0 : \beta = 0$. Results from the Wald test for the individual parameter and for the omnibus test should agree, and we see that the Wald statistic from (7.8) is

$$(\hat{\beta} - 0)'\hat{I}(\hat{\beta})(\hat{\beta} - 0) = z_{\text{Wald}}^2 = (-0.97)^2 = 0.94$$

which agrees with R's output.

The score statistic was found to be

$$S(0) = \ell'(0) = -0.995$$

and the second derivative of the log partial likelihood function was found to be

$$\ell''(0) = -0.957.$$

The information at $\beta = 0$ is then $\hat{I}(0) = 0.957$. The score statistic is then

$$S(0)\hat{I}(0)^{-1}S(0) = (-0.995)(1/0.957)(-0.995) = 1.034.$$

This also agrees with the output from R's `coxph` function.

The likelihood ratio test statistic is

$$-2\log\lambda = -2\left(\ell(0) - \ell(\hat{\beta})\right) = -2(-6.446 - (-5.904)) = 1.084.$$

Once again, this agrees with the output from R.

All three of the omnibus test statistics are compared with the $\chi^2(1)$ distribution, so in this case all three reach the same conclusion since they all fall short of the $\chi^2_{0.05}(1) = 3.84$. There is no evidence that the treatment affects survival. ∎

Note that the three omnibus tests, the Wald test, the score test, and the likelihood ratio test, are based on different approximations and will give different test statistics. Usually, the differences will be small and the conclusions consistent. There are times, though, when one might lead to a significant result while the others do not.

Next, let's apply the CPH function to a larger and more realistic data set.

Example 7.2 Data from an epoxy electrical insulation unit were tested at three accelerated voltage conditions as shown in Table 7.3. Failure times are in minutes, and there are six censored times. Apply the CPH model to see whether voltage has an effect on the lifetime. Also, apply the Weibull regression model and compare the results.

Solution:
Let's first plot the data to get some idea of the problem being addressed. The scatter plot of Voltage on the horizontal axis and lifetime on the vertical axis is shown in Figure 7.3, which suggests that, as expected, the higher voltage corresponds to a shorter lifetime.

The CPH model assumes that the hazard as a function of t and Voltage is

$$h(t|\text{Voltage}) = h_0(t)e^{\beta_1 \text{Voltage}}.$$

In R, the commands

```
CoxRegEpoxy = coxph( Surv(T,C) ~ Voltage)
summary( CoxRegEpoxy)
betahatCoxPH = CoxRegEpoxy$coefficients
```

will execute the CPH function coxph on the survival object Surv(T,C) and use Voltage as the regressor. The third command will extract the parameter estimate for β_1. The output from the summary command is

```
Call:
coxph(formula = Surv(T, C) ~ Voltage)

n= 60, number of events= 54

            coef exp(coef) se(coef)     z Pr(>|z|)
Voltage 0.25109   1.28543  0.07669 3.274  0.00106 **
---
Signif. codes:  0 '***' 0.001 '**' 0.01 '*' 0.05 '.' 0.1 ' ' 1

          exp(coef) exp(-coef) lower.95 upper.95
Voltage       1.285      0.778    1.106    1.494
```

Table 7.3 Lifetimes of epoxy in minutes as a function of voltage.

Voltage	T	C	Voltage	T	C	Voltage	T	C
52.5	4690	1	55	258	1	57.5	510	1
52.5	740	1	55	114	1	57.5	1000	0
52.5	1010	1	55	312	1	57.5	252	1
52.5	1190	1	55	772	1	57.5	408	1
52.5	2450	1	55	498	1	57.5	528	1
52.5	1390	1	55	162	1	57.5	690	1
52.5	350	1	55	444	1	57.5	900	0
52.5	6095	1	55	1464	1	57.5	714	1
52.5	3000	1	55	132	1	57.5	348	1
52.5	1458	1	55	1740	0	57.5	546	1
52.5	6200	0	55	1266	1	57.5	174	1
52.5	550	1	55	300	1	57.5	696	1
52.5	1690	1	55	2440	0	57.5	294	1
52.5	745	1	55	520	1	57.5	234	1
52.5	1225	1	55	1240	1	57.5	288	1
52.5	1480	1	55	2600	0	57.5	444	1
52.5	245	1	55	222	1	57.5	390	1
52.5	600	1	55	144	1	57.5	168	1
52.5	246	1	55	745	1	57.5	558	1
52.5	1805	1	55	395	1	57.5	288	1

```
Concordance= 0.63   (se = 0.033)
Likelihood ratio test= 11.1   on 1 df,    p=9e-04
Wald test              = 10.72   on 1 df,    p=0.001
Score (logrank) test = 11.38   on 1 df,    p=7e-04
```

The point estimate for the parameter β_1 is $\hat{\beta}_1 = 1.28543$, with a standard error of 0.07669. Thus an approximate confidence interval for β_1 is

$$(1.2854 - 1.96 \times 0.0767, 1.2854 + 1.96 \times 0.0767) = (0.1008, 0.4014).$$

which excludes the value 0, indicating that 0 is not a plausible value for β_1. Thus, there is evidence that Voltage affects the in a positive way; that is, when Voltage is larger, the hazard is larger. When the hazard is larger, the lifetime is smaller. Thus, higher Voltage is associated with shorter lifetimes. Note that exponentiating the lower and upper confidence limits yields the confidence interval (1.106, 1.494) for β_1, which agrees with the result in R's output.

We can use R's survfit function to compute the estimated survival function for particular values of Voltage. For example, to estimate the survival function when Voltage is equal to 52.5, we would define

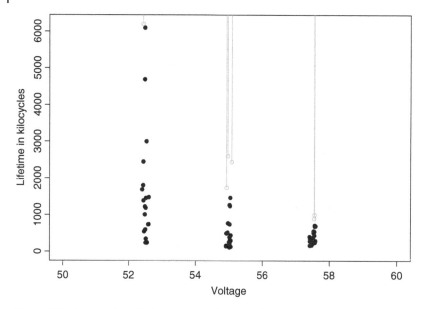

Figure 7.3 Scatter plot of Voltage versus Lifetime. Gray lines indicate possible lifetimes for the six units that were censored.

a new data frame (which must have the same variable names as the data frame we used in the `coxph` function call):

```
VoltageNew = 52.5
newData = data.frame( VoltageNew, 1, 1)
colnames(newData) = c("Voltage","T","C")
CoxRegEpoxy = survfit( CoxRegEpoxy, newdata=newData,
                       se.fit=TRUE, conf.int=.95)
```

The resulting plots, shown in Figure 7.4, clearly indicate that items last longer for shorter values of `Voltage`.

In JMP the function to be used is "Fit Proportional Hazards" in the function group of "Analyze" → "Reliability and Survival." After properly specifying the Cox model in the model dialog window, we obtain the following results of parameter estimation and risk ratio estimation (see Figure 7.5) and they match the results from R.

Let's next apply the Weibull regression model and compare the results. Weibull regression can be performed in R with the following commands:

```
WeibullRegEpoxy = survreg( Surv(T,C) ~ Voltage)
summary( WeibullRegEpoxy)
beta0hat = WeibullRegEpoxy$coefficients[1]
beta1hat = WeibullRegEpoxy$coefficients[2]
kappahat = 1/WeibullRegEpoxy$scale
```

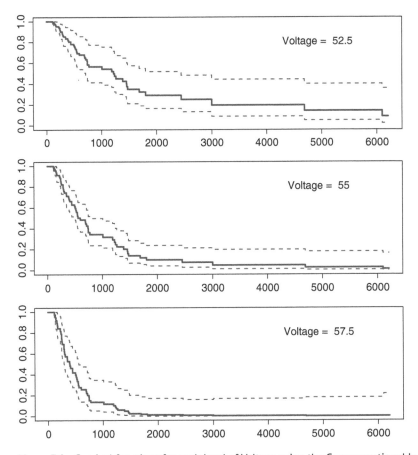

Figure 7.4 Survival functions for each level of Voltage using the Cox proportional hazards model.

Figure 7.5 The JMP PH model parameter estimation outputs. Source: SAS Institute Inc.

▾ **Parameter Estimates**

Term	Estimate	Std Error	Lower 95%	Upper 95%
Voltage	0.25062734	0.076721	0.1023227	0.4045956

▾ **Risk Ratios**

▾ **Unit Risk Ratios**

Per unit change in regressor

Term	Risk Ratio	Lower 95%	Upper 95%	Reciprocal
Voltage	1.284831	1.107741	1.498696	0.7783124

The first line runs the Weibull regression with the survival object `Surv(T,C)` being the outcome and Voltage being the predictor. The second line produces a summary of the output, and the next three lines extract the parameter estimates. The output is

```
Call:
survreg(formula = Surv(T, C) ~ Voltage)
             Value Std. Error     z        p
```

```
(Intercept)  22.0312     3.0456   7.23 4.7e-13
Voltage      -0.2746     0.0553  -4.96 6.9e-07
Log(scale)   -0.1679     0.1069  -1.57    0.12

Scale= 0.845

Weibull distribution
Loglik(model)= -425.8    Loglik(intercept only)= -434.6
        Chisq= 17.75 on 1 degrees of freedom, p= 2.5e-05
Number of Newton-Raphson Iterations: 5
n= 60
```

The estimated coefficients are $\hat{\beta}_0 = 22.0312$ and $\hat{\beta}_1 = -0.2746$. Thus, the estimated characteristic life is

$$\hat{\theta} = \exp\left(\hat{\beta}_0 + \hat{\beta}_1 \text{Voltage}\right)$$
$$= \exp\left(22.0312 - 0.2746 \text{ Voltage}\right).$$

Thus, larger Voltage leads to smaller characteristic life times. This is the same conclusion as for the CPH model, except signs for the estimated β_1 coefficients are different. The reason for this apparent contradiction involves the parameterization that we have chosen for the Weibull distribution. With our parameterization, the Weibull distribution has hazard

$$h^{(W)}(t) = \frac{\kappa}{\theta}\left(\frac{t}{\theta}\right)^{\kappa-1}$$
$$= \frac{\kappa}{\left(\exp\left(\beta_0^{(W)} + \beta_1^{(W)}x\right)\right)^{\kappa}} t^{\kappa}$$
$$= \frac{\kappa}{\exp\left(\kappa\beta_0^{(W)}\right)} t^{\kappa} \times \frac{1}{\exp\left(\kappa\beta_1^{(W)}x\right)}$$
$$= h_0^{(W)} \times \exp\left(-\kappa\beta_1^{(W)}x\right) \tag{7.11}$$

where $h_0^{(W)}(t)$ is the baseline hazard for the Weibull regression, that is, the hazard function assuming the predictor variable is equal to 0. The hazard function for the CPH model is

$$h^{(C)}(t) = h_0^{(C)}(t) \exp\left(\beta_1^{(C)}x\right). \tag{7.12}$$

Thus, in both cases the hazard is equal to a baseline hazard (i.e. a hazard for when the predictor variable is 0) times an exponential function: $\exp\left(-\kappa\beta_1^{(W)}x\right)$ in the case of Weibull regression, and $\exp\left(\beta_1^{(C)}x\right)$ for the case of the CPH model. While the baseline hazards are estimated in different ways, we would expect them to be similar, especially if the sample size is large and the Weibull model is a good fit. This means that we would expect $\exp(-\kappa\beta_1^{(W)}x)$ and $\exp(\beta_1^{(C)}x)$ to be similar, which implies

$$-\kappa\beta^{(W)}x \approx \beta_1^{(C)}x.$$

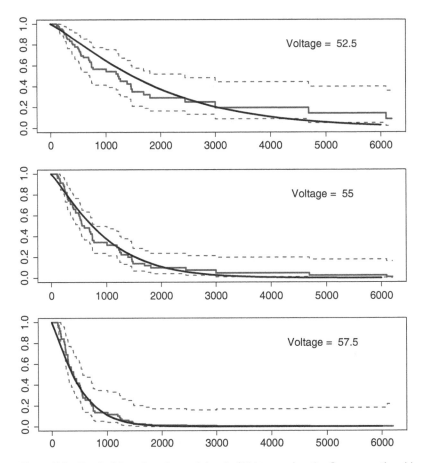

Figure 7.6 Survival functions for each level of Voltage using the Cox proportional hazards model and the Weibull regression model. The smooth curve is the result from Weibull regression.

We can compare the fit of the Weibull regression with that of the CPH model by plotting the estimated survival curves from the Weibull regression on the graphs containing the survival estimates from the Cox model. This has been done in Figure 7.6. The Weibull survival curves are the smooth curves in this figure. The estimated survival function for Voltage equal to 52.5 seems to overestimate the survival probability for small values of t and underestimate the survival probability for large value of t. The same phenomenon occurs when Voltage equals 55, but to a lesser extent. The fit for Voltage equal to 57.5 seems fine. Lawless (2003) suggested that the shape parameter κ could also be modeled as a function of Voltage according to a form such as $\kappa = \gamma \text{Voltage}^{\delta}$, which would address the lack of fit that we have observed here. ∎

The CPH model is often called **semi-parametric** because a functional form is assumed for the dependence of the hazard on the predictor variables, yet no distributional function is assumed for the lifetimes. It can be an effective technique when there are a relatively large number of observations and few different treatments. For this reason, the Cox model is often applied in the medical or health domain where there is often a single factor with just a few levels. In industrial settings there are often several factors, each at several levels, leading to a large number of treatments. In these cases, the parametric approach of assuming a particular distribution for lifetimes may be a more efficient approach.

7.4 Checking Assumptions for the Cox PH Model

The Weibull regression model, discussed in the Chapter 6, has a hazard function that can be written as

$$h(t|\mathbf{x}) = \frac{\kappa}{\theta}\left(\frac{t}{\theta}\right)^{\kappa-1} = \kappa t^{\kappa-1} \exp(\beta_0 + \mathbf{x}'\boldsymbol{\beta})^{-\kappa} = \exp(-\kappa\mathbf{x}'\boldsymbol{\beta}_1) \times \kappa \exp(-\beta_0\kappa)t^{\kappa-1}$$

where $\boldsymbol{\beta}_1$ is the vector containing all of the slope parameters (i.e. all of the β parameters except the intercept β_0. Thus, the Weibull regression model makes three key assumptions:

1. for any two values of the predictor variables, say, $\mathbf{x}_1$ and $\mathbf{x}_2$, the hazard functions $h(t|\mathbf{x}_1)$ and $h(t|\mathbf{x}_2)$ are proportional,
2. the hazard function is related to the predictor variables through the factor $\exp(-\kappa\mathbf{x}'\boldsymbol{\beta}_1)$
3. the lifetimes have a Weibull distribution.

The assumed hazard function for the CPH model is

$$h(t|\mathbf{x}) = \exp(\mathbf{x}'\boldsymbol{\beta}) \times h_0(t).$$

Note that β plays different roles within the Weibull regression and CPH models. The context should determine whether β indicates the parameter in Weibull regression or the parameter in the CPH model. Thus, the CPH assumes

1. for any two values of the predictor variables, say, $\mathbf{x}_1$ and $\mathbf{x}_2$, the hazard functions $h(t|\mathbf{x}_1)$ and $h(t|\mathbf{x}_2)$ are proportional, and
2. the hazard function is related to the predictor variables through the factor $\exp(\mathbf{x}'\boldsymbol{\beta})$.

The main difference is that Weibull regression assumes a Weibull distribution for lifetimes, whereas the CPH model does not.

Checking the assumptions for the CPH model therefore involves two checks: (i) the proportionality of the hazard functions and (ii) the adequacy of the assumed relationship between the predictors and the hazard function through the factor $\exp(\mathbf{x}'\boldsymbol{\beta})$. A graphical goodness of fit test can be obtained by plotting the log of the cumulative hazard function versus the log of survival time. Figure 7.7 shows this plot for the voltage data of Example 7.2. In such a plot, the three groups (exhibited by the three different plot characters) should follow parallel curves if the proportional hazards assumption is correct. In this case, the `Voltage = 55` group's curve does not seem to be parallel to the other two. This calls into question the assumption of proportional hazards across all three groups.

Residual analyses can also be done for the Cox proportional model. See Moore (2016) for further details.

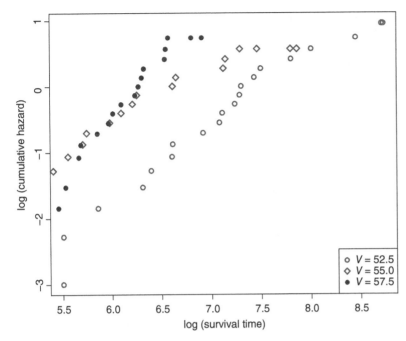

Figure 7.7 Log survival time versus log cumulative hazard function for the epoxy data. The nonparallelness of these lines suggests that the proportional hazards assumption does not hold.

Problems

7.1 Plot the hazard functions for the distributions EXP(20), EXP(30), and EXP(40). Are the latter two hazard functions proportional to the first?

7.2 Plot the hazard functions for the distributions WEIB(1.5, 20), WEIB(1.5.30), and WEIB(1.5, 40). Are the latter two hazard functions proportional to the first?

7.3 Plot the hazard functions for the distributions GAM(2,20), GAM(2,30), and GAM(2,40). Are the latter two hazard functions proportional to the first?

7.4 Plot the hazard functions for the distributions LOGN(2,2), GAM(2,3), and GAM(2,4). Are the latter two hazard functions proportional to the first?

7.5 Plot the hazard functions for the distributions WEIB(1.0, 20), WEIB(1.5.25), and WEIB(2.0, 30). Are the latter two hazard functions proportional to the first?

7.6 Apply the Cox proportional hazards model to the snubber data set described in Example 3.3 and presented in Table 3.2. Let $x_i = 0$ if the snubber was built from the old design and $x_i = 1$ if it was built from the new design. Plot the estimated hazard functions for each design, analogous to Figure 7.4.

7.7 Fit Weibull distributions separately to the old and new designs of the snubber and graph the estimated survival functions. Compare with the results of Problem 7.6.

7.8 Apply Weibull regression to the snubber data with the predictor variable x as defined in Problem 7.7. Plot the estimated survival functions and compare with the results of Problem 7.7.

7.9 Consider the data in Table 7.4. Apply Weibull regression using the temperature in degrees C, plus
two indicator variables for the printed circuit board (PCB) Type.

Table 7.4 Lifetimes in cycles solder joints on a printed circuit board (PCB).

PCB	Temp C	Cycles	PCB	Temp C	Cycles	PCB	Temp C	Cycles
1	20	218	2	20	685	3	20	791
1	20	265	2	20	899	3	20	1140
1	20	279	2	20	1020	3	20	1169
1	20	282	2	20	1082	3	20	1217
1	20	336	2	20	1207	3	20	1267
1	20	469	2	20	1396	3	20	1409
1	20	496	2	20	1411	3	20	1447
1	20	507	2	20	1417	3	20	1476
1	20	685	2	20	1470	3	20	1488
1	20	685	2	20	1999	3	20	1545
1	60	185	2	60	593	3	60	704
1	60	242	2	60	722	3	60	827
1	60	254	2	60	859	3	60	925
1	60	280	2	60	863	3	60	930
1	60	305	2	60	956	3	60	984
1	60	353	2	60	1017	3	60	984
1	60	381	2	60	1038	3	60	1006
1	60	504	2	60	1107	3	60	1166
1	60	556	2	60	1264	3	60	1258
1	60	697	2	60	1362	3	60	1362
1	100	7	2	100	188	3	100	98
1	100	46	2	100	248	3	100	154
1	100	52	2	100	266	3	100	193
1	100	82	2	100	269	3	100	230
1	100	90	2	100	291	3	100	239
1	100	100	2	100	345	3	100	270
1	100	101	2	100	352	3	100	295
1	100	105	2	100	381	3	100	332
1	100	112	2	100	385	3	100	491
1	100	151	2	100	445	3	100	532

Predictor variables are PCB type (1 = copper-nickel-tin, 2 = copper-nickel-gold, and 3 =
copper-tin-lead.

7.10 Apply the Cox proportional hazards model to the data in Table 7.4. Create plots of the estimated survival function for each temperature/PCB type combination.

7.11 For the toaster snubber data given in Example 3.3, give a plot of the estimated survival plot for both the old and new designs assuming separate Weibull models for the two designs. Is the assumption of proportional hazards reasonable for this data set?

7.12 The data in Table 6.9 give the lifetimes of rolling ball bearings for a 2^3 unreplicated factorial design. Apply the Cox proportional hazards model.

7.13 For the data in Table 7.5, define four indicator variables and run the Cox proportional hazards model with the indicator variables as the predictors. These data were first given by Brown and Potts (1977) and discussed in McCool (2012). Construct the plots of the estimated survival function for each type of steel.

Table 7.5 Rolling contact fatigue life times in millions of cycles for steel specimens made from five types of steel.

		Type of steel		
A	B	C	D	E
3.03	3.19	3.46	5.88	6.43
5.53	4.26	5.22	6.74	9.97
5.60	4.47	5.69	6.90	10.39
9.30	4.53	6.54	6.98	13.55
9.92	4.67	9.16	7.21	14.45
12.51	4.69	9.40	8.14	14.72
12.95	5.78	10.19	8.59	16.81
15.21	6.79	10.71	9.80	18.39
16.04	9.37	12.58	12.28	20.84
16.84	12.75	13.41	25.46	21.51

7.14 Derive the relationship in Eq. (7.6).

7.15 Extend the comparison of the Cox model and Weibull regression given in Eq. (7.11) and (7.12) to the case of p predictor variables.

Part IV

Experimental Design for Reliability Studies

8

Design of Single-Testing-Condition Reliability Experiments

In Chapters 8 and 9 we discuss how to use experimental design techniques to improve product reliability. First of all, it is of prominent importance to product developers and manufacturers to be able to accurately estimate their product's reliability before it is delivered to customers. This often requires a life test or accelerated life test to obtain enough failure data so as to infer the product's lifetime distribution. Second, a product development process involves many design and use variables to be considered, such as product functionality, manufacturability, use conditions, etc., for their impacts on product failures. Therefore, we explore the methodology of statistical experimental design to study efficiently the effects of these variables and to choose better product and process designs for reliability improvement.

To a reliability engineer, these types of experimental design activities are called reliability test planning. Reliability tests are often aimed at making reliability prediction, so the response variable is typically a product's lifetime. A test, if it is conducted under a single testing condition to obtain product failure time data, is called a life test. Note that a life test does not mean that the test is conducted under the product's normal operating conditions. Since modern engineering products are often durable enough for years of use, testing them under their normal-operation conditions is usually impractical. Accelerated life testing (ALT) is a method that subjects test units to an elevated operational stress condition, typically by applying higher environmental stresses, such as high temperature, humidity, or voltage, on test units in order to force them to fail faster. As long as we know the acceleration factor, we can extrapolate the results from ALT to the normal-use condition to predict product reliability. This chapter deals with both product life tests and accelerated life tests under a single testing condition. To plan such tests, one needs to determine test duration, sample size (i.e., the number of test units), censoring scheme, and expected number of failures, as this type of test may take a long time with a limited number of test units, while still running a risk of having very few or even zero failure observations. Therefore, one needs to specify these test planning parameters with a great care so as to make the test plan statistically efficient. An efficient test plan is expected to provide adequate lifetime information with a shorter testing duration and fewer test units. The lifetime distributions and the statistical inference methods for estimating distribution parameters previously studied in Chapters 2 and 3 will be extensively used in this chapter.

Design of Experiments for Reliability Achievement, First Edition.
Steven E. Rigdon, Rong Pan, Douglas C. Montgomery, and Laura J. Freeman.
© 2022 John Wiley & Sons, Inc. Published 2022 by John Wiley & Sons, Inc.
Companion website: www.wiley.com/go/rigdon/designexperiments

8.1 Life Testing

Through life testing, engineers obtain failure time data so as to infer the lifetime distribution of a product. Note that many life tests are executed at a higher-than-normal-use stress level in order to accelerate the product's failure process. We do not introduce acceleration factors in this section; this topic will be discussed in Section 8.2. Instead, we assume that the test to be carried out will involve only one stress level, regardless of whether it is the normal use stress level or an accelerated stress level, and the test planning parameters of interest are the number of test units, their testing duration, the allowable number of failures, etc.

A common purpose of conducting life tests is to quantify a lifetime distribution. Some examples are

- the failure rate of a product;
- the mean lifetime of a product;
- a lifetime percentile (oftentimes it is a low percentile such as the 10th percentile, or 0.10 quantile, denoted $t_{0.10}$), which is the time corresponding to a specific failure percentage.

With these distribution parameters, we can answer the questions such as

- What is the expected lifetime of a new light-bulb?
- How many hours of flight would we expect until one percent of the fan blades fail?
- How long should a product's warranty period be such that only 10% of failures occur before the warranty expires?

To plan a life test we will determine the required number of test units or the sufficient duration of the test, or maybe other test settings, in order to be able to answer the aforementioned questions at a certain level of statistical significance. Clearly, the answers to these questions depend on the objective of the test. For example, estimating the median lifetime is harder than estimating the 10th percentile, because more failures are required.

As lifetime censoring is unavoidable in life tests, we need to know what type of censoring scheme is to be implemented in the test before planning other aspects of the test. The most commonly used censoring scheme is Type I censoring or time censoring. Under the Type I censoring scheme a life test will be terminated at a prespecified time, so the lifetimes of survived test units are right censored. Another censoring scheme is Type II censoring or count censoring. By the Type II censoring scheme a life test is terminated at the time when a prespecified number of failures have been observed. Therefore, with the Type I censoring scheme the testing time is fixed and the number of failures is a random variable, while with the Type II scheme the number of failures is fixed and the testing time is random. Clearly, by Type I censoring we can easily control the time spent on a test, but if this testing time is not long enough, there could be very few or even no failure observations. The opposite applies to Type II censoring: although we may be able to obtain more failure information by a Type II censored test plan, the testing time could be too long to be practical. Therefore, more complicated censoring plans have been proposed. For example, in a hybrid plan that combines both Type I and Type II schemes, we terminate the test when either there are enough failures observed or a long enough testing time has been spent, depending on which event occurs first.

Another scenario that can complicate test planning occurs when some test units are removed from the test before the test is scheduled to be completed. The reason of these removals could be to save some expensive test units, or to examine live units for a better understanding of failure physics, or to free up testing facilities, etc. This scenario, if pre-planned, creates a progressive censoring scheme. In this chapter we will not consider these complicated censoring schemes. There is a vast literature having detailed discussions on various hybrid lifetime censoring models, their inferential results, and applications; and interested readers are referred to Cohen (1963), Ng et al. (2009), and Balakrishnan and Kundu (2013), etc.

8.1.1 Life Test Planning with Exponential Distribution

Although the exponential distribution has many restrictive assumptions such as constant failure rate and the memory-less property as discussed in Chapter 2, it is the simplest lifetime distribution with only one distribution parameter, and it is often adequate for modeling failure times of electronic components. More importantly, with the exponential distribution, it is possible to derive closed-form analytical solutions for planning life test parameters such as sample size and test duration. Therefore, we will start with test planning for the exponential lifetime distribution, and then broaden the discussion to other lifetime distributions.

8.1.1.1 Type II Censoring

As the statistical properties of Type II censored tests are easier to derive than those of Type I censored tests, we will first present the statistically optimal test plans under Type II censoring. Assume that the lifetime distribution of a product is $EXP(\theta)$ and there are n units to be tested. The test will be stopped at the time of the rth failure. Let $T_1, T_2, \ldots, T_n$ denote the unordered failure times, which are assumed to be i.i.d. $EXP(\theta)$, and let $T_{(1)}, T_{(2)}, \ldots, T_{(r)}$ be the failure times ordered from smallest to largest. Note that, when $r = n$, this life test terminates when *all* test units have failed. As explained in Chapter 2, the mean of the $EXP(\theta)$ distribution is θ, which is also called the mean-time-to-failure, $MTTF$, and the reciprocal of the mean, $\lambda = 1/\theta$, is called the failure rate.

In Chapter 3, we have shown that the maximum likelihood estimators of mean and failure rate of an exponential distribution are given by, respectively,

$$\hat{\theta} = \frac{TTT}{r}$$

and

$$\hat{\lambda} = \frac{r}{TTT},$$

where

$$TTT = \sum_{i=1}^{r} T_{(i)} + (n - r)T_{(r)},$$

is the total time on test. Note that TTT is a random variable, because the $T_{(i)}$'s are order statistics of a random sample from the exponential distribution. It can be shown that TTT follows a gamma distribution (see p. 152 of (Lawless, 2003)):

$$TTT \sim GAM(\theta, r). \tag{8.1}$$

Thus, the MTTF estimator $\hat{\theta}$ has a sampling distribution as, $\hat{\theta} \sim GAM(\theta/r, r)$. Since r is an integer, it can be further shown that $2r\hat{\theta}/\theta$ has a chi-square distribution with $2r$ degrees of freedom, i.e.

$$\frac{2r\hat{\theta}}{\theta} \sim \chi^2(2r). \tag{8.2}$$

Similarly, the sampling distribution of $\hat{\lambda}$ can be obtained as

$$\frac{2r\lambda}{\hat{\lambda}} \sim \chi^2(2r). \tag{8.3}$$

To obtain the two-sided confidence interval of θ or λ at the $100(1-\alpha)\%$ confidence level, based on Eqs. (8.2) and (8.3), we take, respectively,

$$P\left(\chi^2_{2r,1-\alpha/2} < \frac{2r\hat{\theta}}{\theta} < \chi^2_{2r,\alpha/2}\right) = 1 - \alpha,$$

and

$$P\left(\chi^2_{2r,1-\alpha/2} < \frac{2r\lambda}{\hat{\lambda}} < \chi^2_{2r,\alpha/2}\right) = 1 - \alpha,$$

where $\chi^2_{2r,\alpha/2}$ is the right percentile point of the chi-square distribution. Thus, the exact confidence interval for $MTTF$ at the $100(1-\alpha)\%$ confidence level is given by

$$\left(\frac{2TTT}{\chi^2_{2r,\alpha/2}}, \frac{2TTT}{\chi^2_{2r,1-\alpha/2}}\right). \tag{8.4}$$

Similarly, the confidence interval of failure rate is given by

$$\left(\frac{\chi^2_{2r,1-\alpha/2}}{2TTT}, \frac{\chi^2_{2r,\alpha/2}}{2TTT}\right). \tag{8.5}$$

Reliability engineers often want to know at least how many failures are required from a given number of test units such that the width of the confidence interval of failure rate at the $100(1-\alpha)\%$ statistical confidence level is less than a threshold value; or, conversely, how many test units are needed for a specific number of failures. Therefore, it requires

$$\frac{\chi^2_{2r,\alpha/2} - \chi^2_{2r,1-\alpha/2}}{2TTT} < w,$$

where w is the width of the confidence interval.

The total testing time in the aforementioned formula can be replaced by its expected value at the stage of test planning, i.e. $r \times E[T] + (n - r) \times E[T_{(r)}]$. From here, the minimum sample size or the minimum number of failures can be found. It is obvious that for a given number of test units, more failures are needed (thus a longer testing period is needed) to achieve a higher estimation precision.

8.1.1.2 Type I Censoring

The Type I censoring scheme is more commonly used in life testing than the Type II censoring because Type II censoring cannot guarantee a fixed testing period while Type I can. But, unfortunately, the exact solution of the confidence interval of θ or λ for Type I censoring is not available; instead, it is often approximated by altering the Type II censoring result with a more conservative bound. For example, the confidence interval of the failure rate is given by Tobias and Trindade (2011) as

$$\left(\frac{\chi^2_{2r,1-\alpha/2}}{2TTT}, \frac{\chi^2_{2(r+1),\alpha/2}}{2TTT}\right). \tag{8.6}$$

Note that the degrees of freedom for the chi-square distribution in the upper bound have been adjusted as if there were $r + 1$, not r, failures to be expected; thus this bound in Eq. (8.6) is more conservative than the one in Eq. (8.5). This conservative bound is often used in test planning.

For example, given a targeted failure rate value, a reliability engineer may want to use life tests to show whether a product's failure rate is less than the target at a certain confidence level. Let the test duration be t and the allowable number of failures be r; then, the number of test units n is to be determined. To meet the failure rate requirement, the upper bound of the confidence interval of failure rate should be less than or equal to the targeted value. A one-sided bound is applied, which is

$$\hat{\lambda}_U = \frac{\chi^2_{2(r+1),\alpha}}{2TTT}. \tag{8.7}$$

The total testing time can then be estimated by $(n - r)t + rt/2$, because, for the units that last longer than the test duration, their testing times are t, and for the units that fail during the test, the exponential distribution theory says that their ordered failure times follow a uniform distribution over the range of the testing period. Therefore, we have the following formula:

$$\frac{\chi^2_{2(r+1),\alpha}}{2(n-r)t+rt} \leq \lambda_{target}. \tag{8.8}$$

The left-hand side of Eq. (8.8) involves three planning parameters – sample size n, number of failures r, and test duration t. Using this formula, we can determine the minimally required value of any one of these planning parameters given the values of the other two.

Example 8.1 An electronic device is designed to have a failure rate of 500 FITs (i.e. 0.0005 failures per thousand hours, or 5 failures per 10 million hours). If the test duration is set to be two thousand hours (which is less than three months) and only one failure is allowed through the test, then what is the sample size requirement for this test so that the engineer may claim that the product's failure rate is less than 500 FITs at the 95% confidence level?

Solution:
In this example, the targeted failure rate λ_{target} is $0.0005/Khr$, test duration t is $2\ Khr$, and the number of failures r is 1. The chi-square right percentile $\chi^2_{4,0.05}$ is found to be 9.49. By Eq. (8.8), we would need 4746 units to be put on this test. ∎

Example 8.2 If, for the previous example, there are only 2000 test units available, how long should the test be extended to, in order to demonstrate the same level of product reliability?

Solution:
In this case, sample size n is 2000. By Eq. (8.8), we would need 4746 hours (which is about 6.5 months). ∎

8.1.1.3 Large Sample Approximation
Let us come back to the large sample approximation technique discussed in Chapter 3. Using the large sample lognormal approximation (see Eq. (3.22) in Section 3.3.4), we have the upper bound of one-sided interval estimation of failure rate to be

$$\hat{\lambda}_U = \frac{r}{TTT} \exp(z_\alpha/\sqrt{r}).$$

Using the expected total testing time, we obtain the following formula to meet the failure rate target:

$$\frac{r}{(n-r)t + rt/2} \exp(z_\alpha/\sqrt{r}) \leq \lambda_{target}. \tag{8.9}$$

When the sample size is very large and the testing duration is relatively short, the total testing time can be approximated by nt. Then (8.9) becomes

$$\frac{r}{nt} \exp(z_\alpha/\sqrt{r}) \leq \lambda_{target}.$$

Example 8.3   Using the previous example and the large sample formula from (8.9), calculate the required number of test units that can demonstrate the same reliability level of 500 FITs with one failure at the 95% confidence level.

Solution:
By Eq. (8.9), we would need to solve the equation, $1.6 + \log 2 - \log(n - 1/2) = \log 0.001$. The solution for n is approximately 9900, which is much larger than the solution provided by the chi-square distribution based formula.   ∎

8.1.1.4   Planning Tests to Demonstrate a Lifetime Percentile

The general procedure to derive the sample size requirement for a lifetime percentile demonstration test is the same as what was discussed for a distribution parameter. That is, we first find the sampling distribution of the percentile statistic of interest, and then find the sample size or number of failures requirement based on the confidence bound of this statistics. For example, to show the pth lifetime percentile is larger than a targeted value at the $100(1 - \alpha)\%$ confidence level, it is sufficient to show that the lower bound of this lifetime percentile estimation is larger than the targeted value. From the exponential distribution function, it is easy to derive the lifetime percentile as

$$t_p = \frac{-\log(1-p)}{\lambda}.$$

Therefore, the lower bound of the $100 \times p$-th lifetime percentile estimate is associated with the upper bound of failure rate estimate, i.e.

$$\hat{t}_{p,L} = \frac{-\log(1-p)}{\hat{\lambda}_U}. \tag{8.10}$$

Given the one-sided failure rate upper bound as Eq. (8.7), we obtain

$$-2TTT \log(1 - p) = t_{p,target} \chi^2_{2(r+1),\alpha}. \tag{8.11}$$

If there are many test units and the test duration is relatively short, the total testing time can be approximated by $n \times t$, then the sample size and testing time requirement becomes

$$n \times t = \frac{t_{p,target} \chi^2_{2(r+1),\alpha}}{-2\log(1-p)}. \tag{8.12}$$

Example 8.4   We use an example of electrical appliance life test to show the relationship between the required number of test units (sample size), the number of failures, and the testing duration of a life test. This dataset is presented in Appendix D. The computation is done by using the reliability module of JMP. Many other statistical software packages have similar functions.

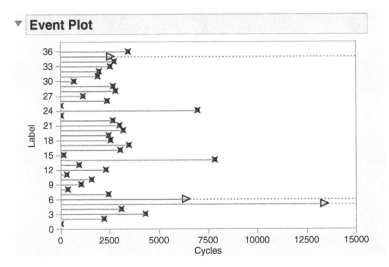

Figure 8.1 The event plot of 36 electrical appliance lifetimes. Source: SAS Institute Inc.

The dataset consists of 36 electrical appliance lifetimes (in cycles), which were originally reported in Nelson (1970). Out of these 36 lifetime observations, there were three right censored observations (survival times). Although the original data indicate that there are more than one failure mode occurred on these failed units, in the following analysis we ignore these different failure modes and only investigate the lifetime distribution that fits all failure (and survival) times. Figure 8.1 plots these failure/survival data. We will first find a suitable lifetime distribution for this dataset, and then properly design life test plans based on the fitted distribution.

Solution:
We fit both the exponential distribution and the Weibull distribution to this dataset. Based on their AIC and BIC values (see Figure 8.2), it is found that the exponential distribution provides a better fit, with an MTTF of 3007. Although the Weibull distribution has one more parameter than exponential distribution, its shape parameter is estimated to be 0.95, which is not statistically different from 1. Therefore, the exponential distribution, with fewer parameters, should be chosen.

Now, given this distribution model, EXP(3007), we would like to design a life test to show that, at the 95% confidence level, this product can work at least to 1000 cycles with a failure probability of 0.1. (That is, at least 90% of all products can survive up to 1000 cycles.)

To plan this reliability demonstration test, we may fix two of the three planning parameters – maximum number of failures allowed (e.g. setting it at 1) and testing duration (e.g. setting it at 3000 cycles), and then find the minimum number of test units required. It is found that this life test requires at least 16 test units (see Figure 8.3). In other words, we can put 16 electrical appliances on test and test each of them for 3000 cycles, if only one appliance or no appliances failed by the end of the test, then we can claim that the product meets its reliability target; otherwise, it fails to meet this target.

On the other hand, if we want to be able to precisely estimate MTTF such that the absolute width of two-sided interval is less or equal to 1000 cycles and we again plan the life test duration to be 3000 cycles, then the required sample size is 17. See Figure 8.4. ∎

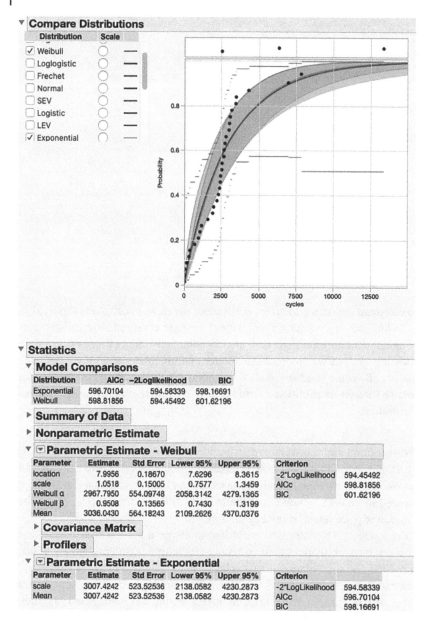

Figure 8.2 Comparing the Weibull and exponential distributions fitted the electrical appliance lifetime data. Source: SAS Institute Inc.

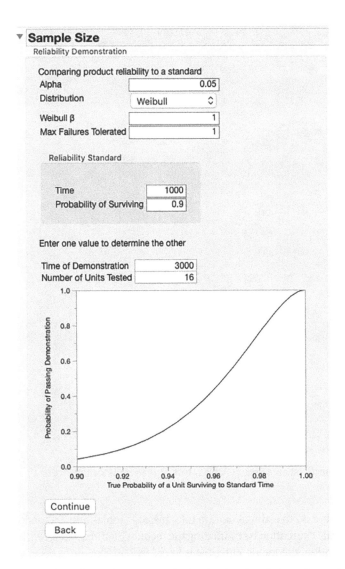

Figure 8.3 Determine the sample size for the electrical appliance demonstration test with a lifetime percentile target. Source: SAS Institute Inc.

8.1.1.5 Zero Failures

When a reliability test produces no failures, does it mean that the failure rate of the tested product is zero? If so, it implies that the product will never fail, which is obviously not true. Imagine that we compare two sets of test results, both of them generate zero failures, but one test has 10 test units in the testing chamber for 10 hours and the other one has 100 test units for 100 hours. Will the failure rate estimated from these two tests be the same? Intuitively, we know that we should assign a smaller failure rate to the second test because it has more test units and longer testing duration.

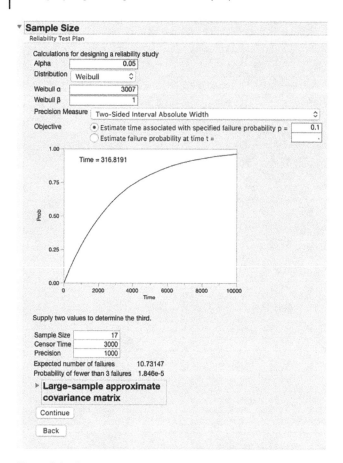

Figure 8.4 Determine the sample size for the electrical appliance reliability test with an MTTF confidence interval requirement. Source: SAS Institute Inc.

The MLE method cannot be applied to this case to estimate failure rate. Instead, Tobias and Trindade (2011) suggest an approximation of failure rate estimation by estimating the median lifetime. Again, with the exponential lifetime distribution assumption, the median lifetime is found as

$$F(\tilde{t}) = 1 - e^{-\lambda \tilde{t}} = 0.5,$$

where $\tilde{t}$ is the median failure time. So,

$$\lambda = \frac{\log 2}{\tilde{t}}.$$

If there are no failures from a test with a total testing time of TTT, it is reasonable to estimate the median failure time by TTT; that is, there is an equal chance of failure before or after TTT, but we happen to observe a survival event. Therefore, the point estimate of the failure rate is given by

$$\hat{\lambda} = \frac{\log 2}{TTT}. \tag{8.13}$$

This formula can be derived from another argument based on the upper bound of the failure rate. As presented in Eq. (8.7), the upper bound of the failure rate for a test of n units without failure is given by

$$\hat{\lambda}_U = \frac{\chi^2_{2,1-\alpha}}{2TTT} = -\frac{\log \alpha}{TTT}.$$

When $\alpha = 0.5$, this is the 50% chance of zero failure. The chance of having no failures with this failure rate value is like flipping a fair coin to get a head.

Therefore, for a demonstration test of duration t with no failures, the minimum number of test units can be derived by equating Eq. (8.13) to the targeted failure rate value; then we have

$$n = \frac{\log 2}{\lambda_{target} t}. \tag{8.14}$$

The zero-failure problem is often met on highly reliable products, which creates difficulties for lifetime parameter estimation and for life test planning. An alternative way to handle this difficulty is via the Bayesian inference approach. Interested readers are referred to Coolen et al. (2005) and Coolen and Coolen-Schrijner (2006) for more information.

Example 8.5 Using Example 8.1, calculate the required number of test units that can demonstrate the same reliability level of 500 FITs with zero failures at the 95% confidence level.

Solution:
By Eq. (8.14), let $\lambda_{target} = 0.0005$ and $t = 2$, we have $n = 301$. Therefore, the required number of test units when no failures are allowed is smaller than the required number when one failure is allowed. ∎

8.1.2 Life Test Planning for Other Lifetime Distributions

Planning life tests with lifetime distributions other than exponential distribution is much harder because, due to censoring, the closed-form solution of confidence interval of the distribution parameter of interest is often elusive. Under certain conditions, we may be able to utilize a data transformation to facilitate data analysis. For example, when the shape parameter of Weibull distribution is known, a power function transformation of time would lead the transformed time becoming exponentially distributed. Then, planning life tests can be carried out based on the formulas presented in Section 8.1.1. As another example, if the lognormal distribution is assumed, making a log transformation of time would produce a normal distribution. Then, to plan a test such that the median lifetime estimation can reach a certain precision, we can work with the mean estimator of the normal distribution. In principle, we can utilize the large sample lognormal approximation of the confidence interval of a distribution parameter (see Chapter 3) to derive the formula for the sample size requirement or testing duration requirement of any life test. However, one should be aware that this approximation is not accurate when the sample size is small or even modest.

Example 8.6 In Example 6.1 a lognormal lifetime distribution was used to fit the data of 28 failure times. The derivation of the distributions of parameter estimates and confidence intervals are provided in Chapter 6. Here, we assume that the estimated lognormal distribution model is indeed the product's lifetime distribution; i.e. $T \sim LogN(\mu, \sigma)$ where $\mu = 2.149$ and $\sigma = 0.529$. Hence, the median life of the product is $t_{0.5} = \exp(\mu) = 8.575$. With this distribution model, we will discuss the life test plan that can reach a required precision for median life estimation.

Solution:

From Example 6.1 we can see that the variance of maximum likelihood estimator $\hat{\mu}$ is given by σ^2/n. Then, based on the delta method, we derive the variance of the estimator for the median to be

$$V(\hat{t}_{0.50}) = e^{2\mu}\sigma^2/n.$$

Or, the standard error is given by $SE(\hat{t}_{0.50}) = 4.536/\sqrt{n}$. Therefore, in order to be able to estimate the median lifetime to be within $\pm 10\%$ of its nominal value, we would need the margin of error to be $2SE$ (0.8575), which in turn requires the sample size to be as $n = 112$. ∎

Note that the derivation above utilizes the standard error of median life estimator, which is proportional to the inverse of square root of sample size. Thus, to make the estimation more precise we have to increase the number of test units by a quadratic speed. Also, this derivation assumes that all test units will be tested to failure, which may not be possible within a limited testing period. To include the consideration of failure time censoring, no closed-form solution to the sample size calculation is available and a software for this purpose is needed.

In JMP, its life test planning tools can be found in the DOE, Sample Size and Power cluster. There, one can see two relevant buttons: Reliability Test Plan and Reliability Demonstration. Clicking the Reliability Test Plan button will activate a calculator, where users can calculate the sample size or test duration, targeting a failure percentile for Weibull, lognormal, loglogistic, and many other distributions. The calculation is based on the large-sample approximation of variance and covariance of parameter estimators of these parameters.

Figure 8.5 shows JMP's sample size calculator. In this example, we select the Objective to be to estimate time associate with failure probability 0.5, i.e. the median lifetime. Then, we set the precision measure to Two-Sided Interval Absolute Width and Precision to be 1.715, which corresponds to the $\pm 10\%$ of nominal median life. When Censor Time is as large as 100, almost all test units are expected to fail during the testing period, then the calculated sample size given by the software is 108, which is close to our previous calculation. Note that if we reduce Censor Time to 10, then about 83 test units are expected to fail and the required sample size increases to 135. Now, continue reducing Censor Time to 1, the required sample size increases dramatically to 67,744,942. Similarly, setting Censor Time back to 100, but changing Precision to 0.8575, which corresponds to the $\pm 5\%$ of nominal median life, the required sample size increases to 431, which is roughly four times the previous sample size requirement for reaching $\pm 10\%$ precision. This is because, as derived earlier, the standard error of median life estimator is inversely proportional to the square root of sample size.

The Reliability Demonstration calculator in JMP gives similar functionality as the Reliability Test Plan calculator, except that, instead of specifying confidence interval or precision requirement on the distribution parameter of interest, Reliability Demonstration asks for the number of failures allowed. A screen shot of this calculator is shown in Figure 8.6.

8.1.3 Operating Characteristic Curves

As discussed in Section 8.1.2, reliability demonstration tests are used to demonstrate that a product has met its lifetime requirement at a given use condition. More formally, reliability demonstration can be formulated as a hypothesis test by specifying a reliability target: "do the data provide enough evidence to reject the null hypothesis that reliability is smaller than the target?" For example,

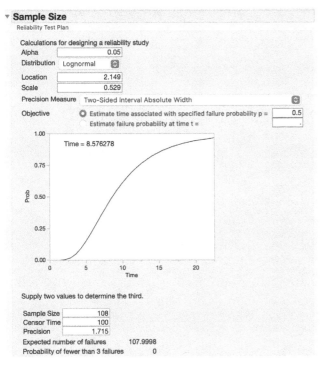

▼ Sample Size
Reliability Test Plan

Calculations for designing a reliability study
Alpha   0.05
Distribution   Lognormal
Location   2.149
Scale   0.529
Precision Measure   Two-Sided Interval Absolute Width
Objective   ⦿ Estimate time associated with specified failure probability p =   0.5
   ○ Estimate failure probability at time t =

Time = 8.576278

Supply two values to determine the third.

Sample Size   108
Censor Time   100
Precision   1.715
Expected number of failures   107.9998
Probability of fewer than 3 failures   0

Figure 8.5 Determine the sample size for the lognormal distribution specified in Example 6.1 by JMP. Source: SAS Institute Inc.

Meeker and Escobar (1998a) consider a product that is advertised to have a mean lifetime of 100 days. A warranty company considering insuring this product wants to know whether this claim can be trusted, so they design a life test. The hypotheses under consideration are:

- Null hypothesis: $MTTF \leq 100$
- Alternative hypothesis: $MTTF > 100$.

This reliability target could be a lifetime distribution percentile too. For example, a company wants to demonstrate that 99% of its products can last longer than five years, or less than 1% of its products may fail within five years. This is about the 0.01 quantile (or equivalently, the 1st percentile) of the product's lifetime distribution, denoted by $t_{0.01}$, which is defined as $F(t_{0.01}) = Pr(T \leq t_1) = 0.01$.

In general, the p-th quantile of a distribution is given by

$$F(t_p) = P(T \leq t_p) = p. \tag{8.15}$$

The median of a distribution is its 0.50 quantile (or 50th percentile), i.e. $t_{0.50}$.

In a reliability study, lower life percentiles such as $t_{0.01}$, $t_{0.05}$, or $t_{0.10}$ are often of interest. In terms of a hypothesis test, to demonstrate the $100 \times p$-th lifetime percentile is longer than five years, the hypotheses under consideration are:

- Null hypothesis: $t_p \leq 5$
- Alternative hypothesis: $t_p > 5$.

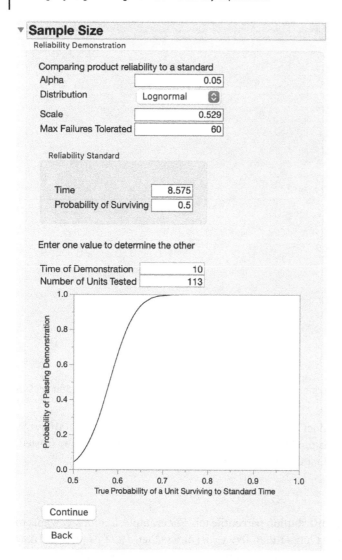

Figure 8.6 Determine the sample size for the lognormal distribution by JMP Reliability Demonstration calculator. Source: SAS Institute Inc.

Note that the null hypothesis is typically a statement that we would like to reject, because rejecting a null hypothesis is a strong decision and there must be tremendous evidence in the data to support such a decision.

Operating characteristic (OC) curves have historically been used to test hypotheses of this structure in quality control applications to determine the defect rate in a batch of products, but they have a useful extension to reliability tests.

For pass/fail requirements, the acceptance sampling plans described in Montgomery (2020) provide a useful methodology for determining the number of tests that are necessary to guarantee a threshold level of performance. Central to the development of these curves is the balancing of *consumer risk* and *producer risk*. Consumer risk is defined as the probability that a bad system (below the threshold reliability) will be accepted, whereas the producer risk is the probability that a good system (above the threshold reliability) will be rejected. These risks are related to the type-I (alpha) and type-II (beta) errors of a hypothesis test procedure. The goal of an OC curve is to illustrate these risks and to determine whether a system performs consistently enough that both the consumer and producer are protected.

In a life test, we may specify the acceptance of the reliability level (i.e. rejecting the null hypothesis) of the product under investigation by stating the number of failures allowed in a test for a given sample size and testing duration. If the number of failures is no more than the allowed value, then we declare that the product has the required reliability. The OC curve plots the probability of acceptance versus the assumed reliability value or a parameter value, such as MTTF, for a given test plan. This probability of acceptance can be derived from the cumulative binomial distribution as follows:

$$P(acceptance) = P(x \le c) = \sum_{x=0}^{c} \binom{n}{x} (1 - R)^x R^{n-x},$$

where $\binom{n}{x} = n!/(x!(n - x)!)$, c is the number of failures allowed, n is the sample size, and R is the reliability value calculated from the assumed lifetime distribution (i.e. assuming the null hypothesis is true). Then, test plans, as defined by (n, c), can be compared by their OC curves. At the same reliability value, the higher the probability of acceptance a test plan has, the higher the consumer risk it presents, because we are more likely to accept the product when in fact its reliability is lower than the acceptable minimum.

Example 8.7   Consider a product that is designed to have a *MTTF* of 100 hours. Then, the purpose of life testing is to confirm one of the following hypotheses:

- Null hypothesis: $MTTF \le 100$
- Alternative hypothesis: $MTTF > 100$

What is the OC curve of this test?

Solution:
Assuming an exponential lifetime distribution, the probability of acceptance can then be computed for a given (n, c) pair. Figure 8.7 presents two cases – (100, 10) and (300, 30). Note that we keep a constant ratio of allowable number of failures to sample size, i.e. $c/n = 0.1$. Comparing these two OC curves, one can see that the (300, 30) test plan has a lower probability of acceptance at a lower reliability value, such as 0.37, which corresponds to the mean failure time. Therefore, this test plan presents a lower consumer risk. In general, the steeper an OC curve is, the greater is its discriminatory power. The ideal reliability OC curve would look like a step function, but in practice, this can never be obtained. Also, note that both test plans control their α value at 0.05. This is equivalent to keeping the producer risk at the same level. ∎

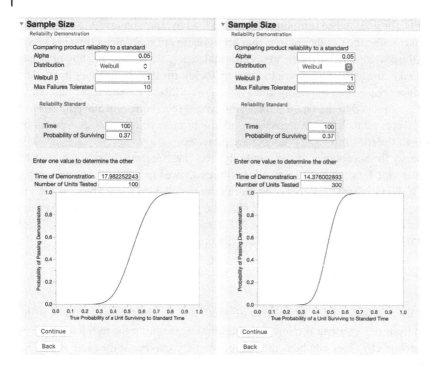

Figure 8.7 The probability of acceptance of two life testing plans: (100, 10) and (300, 30). Source: SAS Institute Inc.

It is easy to see that changing sample size or test duration will affect the conclusion that we may draw from a reliability test. In general, with a larger sample size or a longer test duration, we expect to see more failures, thus a more accurate estimation of a lifetime parameter, so a higher discriminatory power of a hypothesis test could be obtained.

8.2 Accelerated Life Testing

Because it is often impossible to observe a sufficient number of failures under a product's normal use condition within a limited testing period, most life tests are in fact ALTs. In ALTs, elevated stress conditions are applied on test units to accelerate their failure processes, thus shortening lifetimes. These elevated stress conditions can be achieved by increasing test units' usage loading or by heightening their environmental stress level. To be able to utilize ALT data to infer product reliability at its normal-use stress level, we need to define an acceleration factor, which is simply a factor that expands the lifetime under the ALT testing stress condition to the lifetime under the normal-use stress condition.

If life acceleration is achieved by increasing a product's usage loading, it is easy to calculate the acceleration factor, because it should be proportional to the loading frequency. For example, a washing machine is typically used in a household once per day and it takes 30 minutes per run; however, to test its durability, the manufacturer may run test machines continuously for seven hours and then rest one hour.

This eight-hour testing cycle is equivalent to 14 days (two weeks) of household use, which means the acceleration factor is 42.

This section, however, will mainly focus on the environmental stress variables and how they may accelerate the failure process of a product. These environmental stress variables include temperature, relative humidity, pressure, vibration, voltage, current, etc. Based on failure physics, it is known that these stresses can affect material properties, and consequentially some product functions, thereby affecting the product's lifetime.

Note that the ALTs to be discussed in this section assume that the failure mechanism at the test stress level is the same as at the normal-use stress level. Only by this assumption can we extrapolate the failure data collected from ALTs to the normal-use stress level and quantify the lifetime distribution at the normal-use stress level. One needs to be aware that some highly accelerated life tests (HALTs) implemented at a product's early design stage aim to stress-out the product to reveal design faults and any potential, even unlikely, failure modes. The failure information collected from these tests are not used for product lifetime distribution inference; instead, they are used, often in conjunction with failure modes and effects analysis (FMEA), for identifying design faults and weak components. Although HALT is a very important tool for product reliability improvement, it is not discussed in this section. Readers interested in the HALT technique are referred to Hobbs (2005).

8.2.1 Acceleration Factor

Under a more severe stress condition, we expect a product to fail earlier than under its normal-use stress condition. We introduced the scale accelerated failure time (SAFT) model in Section 6.8. This model assumes that the amount of time for a unit to reach a predetermined level of deterioration is a multiple of the baseline stress. The reciprocal of this multiple is called the acceleration factor; that is,

$$AF(s_A) = \frac{T(s_U)}{T(s_A)},$$

where $T(s)$ is equal to the amount of time it takes to reach a fixed level of deterioration for the stress level s. The acceleration factor $AF(s_A)$ is a function of the stress level, but is otherwise constant across time. For example, if $AF(s_A) = 2$ then time passes at twice the rate, so a unit tested at s_U for 100 hours will be in the same condition as this unit tested at s_A for 50 hours. Thus, the survival function (or reliability function) is as

$$S(100|s_U) = S(50|s_A),$$

and in general

$$S(t|s_U) = S\left(\frac{t}{AF(x_A)}|s_A\right).$$

If we assume a WEIB(θ_U, κ_U) distribution for lifetimes, then

$$e^{-(t/\theta_A)^{\kappa_A}} = e^{-(AF(s_A) \, t/\theta_U)^{\kappa_U}}.$$

If we also assume $\kappa_A = \kappa_U$, i.e. stress variables do not affect the shape parameter of the Weibull distribution, then this leads to

$$AF(s_A) = \left(\frac{\theta_U}{\theta_A}\right).$$

Again, with the assumption that stress variables do not affect the shape parameter of Weibull distribution, it can be show that when the SAFT model is applied on a Weibull distribution, the characteristic life of a Weibull distribution is a function of stress variables, in the form of

$$\log \theta = \beta_0 + \beta_1 s. \tag{8.16}$$

For the lognormal distribution, we assume that only the median lifetime parameter is affected by stress variables in the loglinear function form of Eq. (8.16), while the shape parameter σ is a constant. Note that these distributions belong to the log-location-scale distribution family and they can be transformed to a location-scale distribution after a natural logarithm transformation of random variable (see Chapter 2); therefore, the SAFT model introduced in Chapter 6 will lead to a scale-up of the scale parameter and it is quantified by a log-linear function such as Eq. (8.16).

8.2.2 Physical Acceleration Models

In this section we give a brief description of some commonly used physical acceleration models. These models are built upon the physics-of-failure approach, which studies the failure mechanisms of materials under various types of stresses. *Why* a unit fails and *how* it fails are investigated through scientific experiments on materials and the theoretical development of physical-chemical processes of failures. This approach is particularly important to the electronic industry. Some physical acceleration models are well documented by JEDEC Solid State Technology (see (JEDEC, 2001, 2003a, b, 2004)).

8.2.2.1 Arrhenius Model

The Arrhenius acceleration model concerns the thermal stress and its effects on material properties. The Arrhenius theory was introduced in 1887 by the Swedish scientist, Svante Arrhenius, who studied ionization in water. It essentially gives the quantitative basis of the relationship between the activation energy and the rate at which a chemical reaction proceeds. In the context of accelerated failure time (AFT) model, the scale parameter of the lifetime distribution, no matter whether it is the characteristic life of Weibull distribution or the median life of lognormal distribution, is an exponential function of temperature, given by (here, in the form of median life, $t_{0.50}$)

$$t_{0.50} = A \exp\left(\frac{E_a}{kT}\right), \tag{8.17}$$

where A is a scale factor determined by experiments, E_a is called the activation energy which is unique to each failure mechanism, k is the Boltzmann's constant, which is given as $k = 8.617 \times 10^{-5} \text{eV/K}$ or $1.380 \times 10^{-16} \text{ergs/K}$, or equivalently, $1/k = 11{,}605 K/\text{eV}$, and T is temperature in degrees Kelvin. Note that degrees Kelvin is equal to degrees Celsius plus 273.15. Taking the logarithm on both sides of Eq. (8.17) leads to

$$\log t_{0.50} = \log A + \frac{E_a}{k}\frac{1}{T},$$

indicating that $\log t_{0.50}$ is linearly related to the inverse of temperature, $1/T$.

Denote T_U to be the temperature at a product's normal-use condition and T_A be the temperature at an accelerated test condition ($T_A > T_U$); then the acceleration factor is

$$AF = \frac{t_{0.50,U}}{t_{0.50,A}} = \exp\left[E_a\left(\frac{1}{kT_U} - \frac{1}{kT_A}\right)\right].$$

If we define $-1/(kT)$ as the natural stress for temperature, denoted by s, then we have

$$\log AF = E_a(s_A - s_U).$$

Example 8.8 A material's normal-use temperature is 25°C and the laboratory testing temperature is 120°C. Assume its activation energy is 0.3, what is the acceleration factor?

Solution:
The absolute temperatures of normal use and laboratory testing condition are $25 + 273.15 = 298.15$ and $120 + 273.15 = 393.15$, respectively. By the Arrhenius model,

$$
\begin{aligned}
AF &= \exp\left[E_a(1/(kT_U) - 1/(kT_A))\right] \\
&= \exp\left[0.3 \times (1/(8.62 \times 10^{-5} \times 298.15) - 1/(8.62 \times 10^{-5} \times 393.15))\right] \\
&= \exp(2.82) \\
&= 16.8.
\end{aligned}
$$

Therefore, the 50th percentile of failure times at use temperature is expected to be 16.8 times longer than the 50th percentile observed under an accelerated live test at 120°C. ∎

Example 8.9 Suppose a material's MTTF is 10,000 hours at its 20°C normal use temperature condition, but shortens to 500 hours at the 100°C testing temperature condition. What is its activation energy value?

Solution:
The AF is $10,000/500 = 200$. So,

$$
\begin{aligned}
E_a &= \log(AF)/[1/(kT_U) - 1/(kT_A)] \\
&= \log(200)/[1/(8.62 \times 10^{-5} \times 293.15) - 1/(8.62 \times 10^{-5} \times 373.15)] \\
&= 0.624
\end{aligned}
$$

∎

8.2.2.2 Eyring Model

The Eyring model was named after Henry Eyring, a Mexican-born American theoretical chemist (Eyring, 1935). This model applies to either a thermal stress variable or a thermal stress along with another stress variable. The effect of temperature includes the exponential term as in the Arrhenius model and an inverse power term to explain the remaining effect than cannot be covered by the Arrhenius model. This model assumes

$$t_{0.50} = AT^{-\gamma} \exp\left(\frac{E_a}{kT}\right), \tag{8.18}$$

where the exponent γ is an extra parameter to be estimated by experiment.

Adding another stress variable, such as voltage, humidity, etc., the two-stress-variable Eyring model can be written as

$$t_{0.50} = AT^{-\gamma} \exp\left(\frac{E_a}{kT}\right) V^{-\beta}. \tag{8.19}$$

Note that this formula does not model the interaction between temperature and the other stress variable, so their effects on lifetime are assumed to be independent. Another variant of the Eyring model does include the interaction term of two stress variables. It is given by

$$t_{0.50} = AT^{-\gamma} \exp\left(\frac{E_a}{kT}\right) \exp\left(\left(\beta + \frac{C}{T}\right)s\right), \tag{8.20}$$

where $s = -\log V$ and C is the coefficient of the interaction term.

8.2.2.3 Peck Model

The Peck model is named after D. Stewart Peck, an American engineer and consultant, who did a comprehensive study of the impact of temperature and humidity on epoxy packages (Peck, 1986). This model can be expressed as

$$t_{0.50} = A(RH)^{-\beta} \exp\left(\frac{E_a}{kT}\right), \tag{8.21}$$

where RH is relative humidity. For the epoxy packaging, β is often taken as 2.7 and E_a is 0.79eV.

The Peck model is similar to the Eyring model in the sense that both of them are the combination of the Arrhenius function of thermal stress and one or two power functions of other stresses. It is easy to show that, by taking the natural logarithm on both sides of Eq. (8.21), the log-lifetime parameter is a linear function of two or more natural stress variables.

8.2.2.4 Inverse Power Model

From both the Eyring model and the Peck model, we see that for a stress variable other than temperature, an inverse power function is commonly used to connect the lifetime parameter and stress variable. In fact, the inverse power model is often found in applications such as electrical insulation and dielectrics (voltage endurance), ball and roller bearings, incandescent lamp filaments, flash lamps, etc., where the stress variable is voltage, current, or mechanical loading. The basic form of this model is given by

$$t_{0.50} = A \frac{1}{V^{\beta}}, \tag{8.22}$$

where A is a scale parameter, V is the stress level, and β is the power parameter that is associated with the failure mechanism and is determined by experiments.

8.2.2.5 Coffin–Manson Model

The Coffin–Manson acceleration model was proposed to model the fatigue life of metals due to thermal cycling or thermal shocks (Coffin, 1954; Manson, 1953). It is often used on solder joint thermal cycling fatigues and its basic form is given by

$$t_{0.50} = \frac{A}{(\Delta T)^{\beta}}, \tag{8.23}$$

where ΔT is the range of temperature cycle (temperature in degrees Kelvin), and $t_{0.50}$ represents the median number of cycles-to-failure. One can see that it is basically an inverse power model applied to the temperature range stress variable.

A modified model adds the consideration of the cycle frequency and the Arrhenius effect of maximum temperature. This model assumes

$$t_{0.50} = A f^{\gamma} \frac{1}{(\Delta T)^{\beta}} \exp \left(\frac{E_a}{k T_{max}} \right),$$ (8.24)

where f is the frequency of temperature cycle and T_{max} is the maximum temperature.

There are many other failure acceleration models that work for specific materials or failure processes, such as the Epaarachchi–Clausen model for composite materials and Miner's rule for fatigue damages. Interested readers are referred to the relevant publications for more information of these models (e.g. Epaarachchi and Clausen, 2003). Overall, these physical models describe the failure mechanism and how fast a product may deteriorate under a given environmental stress condition. In ALTs, we elevate these environmental stresses to a higher-than-normal level, so as to make units fail faster. Then, the failure data obtained under the accelerated test condition can be used to infer the product lifetime distribution under a normal-use environmental stress condition.

8.2.3 Relationship Between Physical Acceleration Models and Statistical Models

In some cases there is a direct connection between the physical acceleration model and the statistical model used to model the lifetime as a function of the acceleration factor. Two models for which this is possible are the inverse power model and the generalized Eyring model. Other models can still be fit with a statistical model, but to get a mapping between coefficients one will need to use a non-linear model.

The equivalence between the inverse power model and the generalized Eyring model to the log-location scale model and therefore the SAFT model for the Weibull or lognormal distribution is a key characteristic that makes it well suited to model lifetime data. To illustrate the equivalence, recall from Chapter 6 that $t_p = \exp \left[\mu + \Phi^{-1}(p)\sigma \right]$ for any log-location scale model, where μ is the linear predictor expressed as $\mu = \beta_0 + \beta_1 x_1 + \cdots + \beta_k x_k$. If one can re-organize any of the aforementioned life-stress relationships to match this model form it is possible to fit the physics-based relationship with the statistical model and estimate the coefficients of the relationship.

As previously defined, let the acceleration factor be

$$AF = \frac{t_{0.50,\text{use}}}{t_{0.50,\text{accel}}},$$

where $t_{0.50,\text{use}}$ is the 50th percentile at use conditions and $t_{0.50,\text{accel}}$ is the 50th percentile under accelerated conditions. Plugging in the expression for $t_{0.50}$ for the log-location scale family gives:

$$AF = \frac{\exp \left[\mu_{\text{use}} + \Phi^{-1}(0.5)\sigma \right]}{\exp \left[\mu_{\text{accel}} + \Phi^{-1}(0.5)\sigma \right]},$$

$$= \frac{\exp \left[\mu_{\text{use}} \right]}{\exp \left[\mu_{\text{accel}} \right]}$$

Notice that this form assumes the acceleration factor only impacts the linear predictor and not the scale parameter and when rearranging the terms including the scale parameter cancel for this constant scale parameter assumption.

Now let's assume a single variable voltage acceleration where $\mu = \beta_0 + \beta_1 x$ and $x = log(voltage)$. Then we can plug in a rearrange as follows:

$$
\begin{aligned}
AF &= \frac{\exp\left[\beta_0 + \beta_1 x_{\text{use}}\right]}{\exp\left[\beta_0 + \beta_1 x_{\text{accel}}\right]} \\
&= \frac{\exp\left[\beta_1 \log(Volt_{\text{use}})\right]}{\exp\left[\beta_1 \log\left(Volt_{\text{accel}}\right)\right]} \\
&= \left(\frac{Volt_{\text{use}}}{Volt_{\text{accel}}}\right)^{\beta_1}.
\end{aligned}
$$

It is easy to show this is the same AF for the inverse power relationship, where β_1 is the power parameter that is associated with the failure mechanism. It was previously mentioned that this parameter could be determined by experiments; this connection to the log-location scale distribution provides a mechanism for estimation.

As another example of accelerating relationship for which the log-location scale equivalence holds, consider a generalized Eyring relationship given as follows:

$$
t_{0.50} = A \exp\left(\frac{E_a}{kT}\right) \exp\left(\left(\beta + \frac{C}{T}\right)s\right),
$$

where $s = -\log(Volt)$ and C is the coefficient of the interaction term. The effect of the accelerating factors on the time to failure can be modeled using a log-location scale distribution where $\mu = \beta_0 + \beta_1 x_1 + \beta_2 x_2 + \beta_3 x_1 x_2$. The variable of the log-location scale model maps back to the generalized Eyring relationship by letting:

$$
\beta_1 = E_a,
$$
$$
\beta_2 = \beta,
$$
$$
\beta_3 = C \times k,
$$
$$
x_1 = \frac{1{,}1605}{Temp(K)},
$$
$$
x_2 = \log(Volt)
$$

where k is Boltzmann's constant, equal to $1/11{,}605$, and temperature is measured in degrees Kelvin.

The mapping between the statistical log-location scale model and the physics-based failure model is convenient for obtaining estimates of the accelerating relationship and planning experiments based on such coefficients.

8.2.4 Planning Single-Stress-Level ALTs

To plan a reliability test under a single testing condition, regardless of whether it is a life testing or accelerated life testing, the planning parameters of concern are the same – sample size, test duration, and allowable number of failures. The planning techniques discussed in Sections 8.1.1 and 8.1.2 are still applicable to single-stress-level ALTs; only the distribution parameter needs to be modified by AF. Here, we assume that the AF is known. It is often the case in reliability test planning that some lifetime distribution parameters are presumed to be known before planning reliability experiments, because reliability models are generally nonlinear and those optimal test plans are model parameter dependent.

Example 8.10 Following Example 8.8, the acceleration factor at the laboratory testing temperature is 16.8. Suppose it is required that the MTTF of the material at its normal-use temperature is required to be higher than 1000 hours at the 80% confidence level. Derive the formula for the minimum sample size, test duration and allowed number of failures to demonstrate the material reliability at its laboratory testing condition.

Solution:
The acceleration factor for this problem is given. Showing the MTTF at the normal use stress level is equivalent to showing the corresponding MTTF at the accelerated stress level. The MTTF at the laboratory testing condition is $1000/16.8 = 60$ hours. Then, with the exponential lifetime assumption, we use Eq. (8.8) to calculate the minimum sample size, test duration and allowable number of failures, i.e.

$$\frac{\chi^2_{2(r+1),0.2}}{2(n-r)t + rt} \leq \frac{1}{60}.$$

∎

Consider a case where the AF value is unknown or some parameters of the physical acceleration models need to be determined by experiments. We need at least two testing conditions to be designed. The planning parameters in this case include the stress levels, number of samples at each level or the proportion of total sample size at each level, testing duration, censoring scheme, etc. To tackle this problem, we can utilize the lifetime regression and design of experiments techniques presented in the Chapters 4–7. This topic will be discussed in Chapter 9.

Consider another case where the lifetime distribution at the normal-use stress condition is known, and we want to determine the size of the acceleration factor of a certain accelerated life testing with a desired statistical precision. Then, again, it requires a reliability demonstration test at the accelerated testing condition.

Example 8.11 Let us continue the previous example. Assume the activation energy of this material is 0.3 and its lifetime at 25°C follows the exponential distribution with the MTTF of 1000 hours. In order to study the temperature-related lifetime acceleration factor, we would like to run an experiment at a higher temperature. What is the lowest temperature we may choose such that with 100 test units and 20 hours, we can demonstrate the AF at the 80% confidence level?

Solution:
At the accelerated stress test, the MTTF equals $1000/AF$ hours. To demonstrate this MTTF with 80% confidence, by Eq. (8.8) we have

$$\frac{\chi^2_{2(r+1),0.2}}{2(n-r)t + rt} \leq \frac{AF}{1000}.$$

Given that $n = 100$ and $t = 20$, if we let the allowable number of failures to be 1, then

$$AF \geq 1000 \times \frac{\chi_{4,0.2}^2}{3980}.$$

So, the lowest AF is calculated to be 1.5.

Now, we can determine the testing temperature by finding the solution of the following equation:

$$AF = \exp\left[E_a(1/(kT_U) - 1/(kT_A))\right]$$
$$= \exp\left[0.3 \times (1/(8.62 \times 10^{-5} \times 298.15) - 1/(8.62 \times 10^{-5} \times T_A))\right].$$

Plugging in the value of AF, we obtain that the testing temperature can be as low as 36°C. ∎

Problems

8.1 Consider the lifetime data of 10 pieces of equipment presented in Example 2.1.
    (a) Fit an exponential distribution for this dataset.
    (b) Construct the two-sided interval estimates of failure rate at 90% confidence level.
    (c) If a life test is required to demonstrate the failure rate to be at most 0.025 at 90% confidence level, while one failure is allowed during the test, derive the formula for finding the minimal requirements on sample size and test duration.
    (d) If, instead, two failures are allowed during the life test, revise the formula for minimum sample size and test duration.
    (e) If the failure rate demonstration is to be shown at the 80% confidence level and one failure is allowed, revise the formula for minimum sample size and test duration.
    (f) Discuss the relationship between sample size requirement, test duration, number of allowable failures and the level of confidence of a life test.

8.2 Consider the electrical appliance data included in Appendix D (also see Example 8.4). For now, we ignore the failure code except code 0. The code 0 indicates a right censoring (i.e. a survival observation). So, we now have two datasets: the first dataset includes all 36 observations with 3 right censored observation and the second dataset includes only 33 failure observations.
    (a) Assume the exponential lifetime distributions for both datasets. Estimate their MTTFs and provide the interval estimates.
    (b) Discuss the impact of censoring on the point and interval estimates of MTTF.
    (c) Using the lifetime distribution models fitted for the first and second datasets, derive the life test plans for these two models to demonstrate the MTTF to be at least 3500 with 80% confidence.
    (d) If no failures are allowed in the life test, should the required sample size or test duration be increased or decreased? Why?

8.3 Consider the first dataset of the previous problem (with right censored data) and the lifetime distribution model fitted this dataset.
    (a) Estimate the 10% lifetime percentile, or L_{10}, and its 80% confidence interval.
    (b) Plan a life test so as to demonstrate that 90% of all products have their lifetimes longer than 2500 at the 80% confidence level.

8.4 Consider the deep groove ball bearing data presented in Table 3.3 and the Weibull distribution fitted to this dataset.

(a) Plan a life test so as to demonstrate that the characteristic life of Weibull lifetime is longer than 80 at the 80% confidence level.

(b) Plan a life test so as to demonstrate that the 10th percentile, $t_{0.10}$, is longer than 40 at the 80% confidence level.

8.5 Consider the deep groove ball bearing data presented in Table 3.3 and the Lognormal distribution fitted to this dataset.

(a) Plan a life test so as to demonstrate that the median life of this product is longer than 65 at the 80% confidence level.

(b) Plan a life test so as to demonstrate that the 10th percentile, $t_{0.10}$, is longer than 40 at the 80% confidence level.

8.6 Comparing the two curves in Figure 8.7, the curve on the right is steeper than the curve on the left. What does this mean and why does this happen?

8.7 With a Weibull lifetime distribution model, we assume the shape parameter is a constant and only the characteristic life is affect by the stress variable. Present the Weibull-Arrhenius model for thermal stress. How do we define the acceleration factor for this model?

8.8 A product's normal-use temperature is $30°C$ and the accelerated life testing temperature is $100°C$. Let the activation energy be 0.5. What is the acceleration factor value? If a Weibull lifetime distribution is assumed, how do we incorporate this acceleration factor into the lifetime distribution model?

8.9 Assume a Weibull lifetime distribution and only its characteristic life is affected by environmental stress variables. There are two stress variables – temperature and humidity – and the physical lifetime acceleration model follows the Peck model. Define the natural stress variables for temperature and humidity, and show the log-linear relationship between the distribution's parameter and these natural stress variables.

8.10 Show that both the exponential distribution and the Weibull distribution with constant shape parameter have the proportional hazard property discussed in Chapter 7. For a general PH model (the baseline hazard function is unspecified), how do we define the acceleration factor for an accelerated life test? What is the relationship between the covariates in a PH model and the environmental stress variables?

9

Design of Multi-Factor and Multi-Level Reliability Experiments

In the final chapter of this book, we return to the original question asked at the beginning of the book, which is, how do we use experimental design techniques to improve a product's reliability? Given a product's performance model, the prediction of its performance corresponding to its specific design and process parameters can be made, then it is possible to optimize reliability by selecting an optimal set of these parameters. This model-based prediction and optimization methodology is based on an accurate performance model, which in turn is estimated from an experimental study. Like many other quality improvement experiments, in reliability experiments the experimental design can help experimenters try many factor-level combinations in an efficient way, while their effects on responses can be adequately studied. But unlike most quality improvement experiments, reliability experiments focus on the long term impact of these factors by modeling either failure time or failure count as the response of interest.

In the first section of this chapter we discuss some unique features of reliability experiments and their implications on experimental designs. Section 9.2 presents some commonly used lifetime regression models. This section provides a brief summary of the models introduced in Chapters 6 and 7. After the presentation of lifetime regression models, we come to the optimal design and robust design theories and we will use several examples to demonstrate the characteristics of design for reliability due to the unique features of reliability experiments. Most of these designs can be obtained from JMP. We also use the `ALTopt` package of R, which can be downloaded from the R CRAN repository, to generate optimal ALT designs (see Seo and Pan, 2015). Some derivation details of expected Fisher information matrices are shown in Appendix C.

The materials to be presented in this chapter are closely related to the topic of optimal test planning for ALTs in the reliability test literature. The research on optimal ALT planning can be traced back to the mid-to-late 1970s when Nelson and his collaborators first introduced the optimal experimental design (OED) theory into planning ALTs (Nelson and Kielpinski, 1976; Nelson and Meeker, 1978). Later, various types of lifetime distribution models and censoring schemes have been considered (e.g. Yum and Choi (1989), Seo and Yum (1991), Islam and Ahmad (1994), Ng et al. (2004), and Dahmen et al. (2012)). Escobar and Meeker (1995), Sitter and Torsney (1995), and Park and Yum (1996) presented the optimal test plans with two stress factors. For a summary of the ALT literature up to 2015, readers are referred to Nelson (2005, 2015). In Limon et al. (2017), over 180 publications of accelerated testing (including accelerated life testing and accelerated degradation testing) are examined and categorized. More recently, this topic has been further explored from the OED for the generalized linear model (GLM) point of view (see Monroe et al., 2011; Yang and Pan, 2013; Pan et al., 2015; Seo and Pan, 2018,

Design of Experiments for Reliability Achievement, First Edition.
Steven E. Rigdon, Rong Pan, Douglas C. Montgomery, and Laura J. Freeman.
© 2022 John Wiley & Sons, Inc. Published 2022 by John Wiley & Sons, Inc.
Companion website: www.wiley.com/go/rigdon/designexperiments

2020). This approach is based on the proportional hazard (PH) regression model with censored lifetime data (Aitkin and Clayton, 1980; Finkelstein, 1986).

In traditional ALT planning, the main concerns are with the determination of sample size, the testing duration, the setting of environmental stress variables, etc. In this chapter, however, we will consider not only environmental stress variables, but also product design variables, manufacturing process variables, and the interactions among these variables. In general, these experiments have multiple experimental factors and each factor can be experimented at multiple levels. They are typically carried out under the accelerated life testing condition with the purpose of identifying the significant factors and possible interactions between these factors so as to better design the product to improve its reliability.

9.1 Implications of Design for Reliability

Failure times and failure counts are the typical response variables being studied by reliability engineers to predict product reliability. These variables often cannot be modeled by a normal distribution. Failure time follows a lifetime distribution, such as the exponential, Weibull, gamma, or lognormal distribution. Failure count is a discrete variable and often follows a binomial or negative binomial distribution. Note that, in a regular experimental design, the normal distribution is usually assumed for modeling experimental outcomes. As discussed in Chapter 6, some data transformation techniques may be able to transform the non-normal response data to be approximately normally distributed. However, it is often desirable for experimenters to work directly with a native distribution (Myers et al., 2012). One reason is that the native distribution model will be more interpretable, as the effects of individual factors and their interactions on the distribution parameter of interest can be directly explained, which is of great value to engineers.

Data censoring is another feature of reliability experiments, arguably the most unique feature. As we have seen in Chapter 8, to plan a life test or accelerated life test, the planner must take the sample size and test duration into consideration. It is expected that not all test units will fail during testing the period. Censored observations (right censoring, left censoring, and interval censoring) give us only partial information about a product's lifetime. This deficiency causes some common DOE methods, such as factorial designs with equally allocated number of test units, to be less efficient than they are for a normally distributed response variable without censoring.

Thirdly, the experimental factors in reliability experiments are often of diverse types. For example, there may be some product design and manufacturing process factors that are categorical, but also some environmental stress variables that are continuous. As explained in Chapter 8, the purpose of conducting a reliability experiment under an elevated environmental stress condition is to accelerate the product failure process so that we can obtain enough failure data within a reasonable time frame. On the other hand, for improving product reliability, we are interested in optimizing the product's design and process factors and may also look into the interactions of these factors with environmental stress variables.

Lastly, it must be acknowledged that, as most of practical reliability experiments are conducted under accelerated testing conditions, response extrapolation must be conducted in order to infer the lifetime at the normal-use condition; therefore, the stress condition of interest is often placed outside of the experimental design region. The extra prediction uncertainty induced by data extrapolation should be properly taken into account during the experimental design process.

Most existing design for reliability work has employed, at best, standard regression approaches with the occasional use of Taguchi robust parameter design techniques. The rest of this chapter will address the adjustments that need to be made to these techniques (and the similarities to some recent work in design for non-normal data). In particular, we introduce environmental stress variables and their effects on product lifetime. Extending our previous discussion on ALTs, we will construct experimental designs for when the acceleration factor is unknown and the relationship between a lifetime parameter and stress variables needs to be quantified. Then, we will expand the scope of our model to include other variables such as product design variables and manufacturing process variables, and discuss the interactions between these variables with stresses and how to achieve a robust product design with high reliability. Note that Nelson (1984) covers a wide range of topics on the statistical methods for design and analysis of ALTs. This chapter does not intend to explain ALTs; instead, we want to explain how to use reliability tests (most of which are ALTs) to improve product reliability.

9.2 Statistical Acceleration Models

In Chapter 8 we described several physical acceleration models; however, we have to admit that these models' utilities are limited because (i) oftentimes they cannot fully explain a complex failure mechanism and (ii) they may not be able to fit the actual failure data from experiments well. Statistical acceleration models, in the form of a parametric regression function, can often provide adequate fit to data, but they may not be able to give a meaningful interpretation of the failure mechanism to scientists and engineers. Therefore, we need to combine these two types of modeling techniques to take advantage of their strengths. As mentioned previously, the log-linear relationship between lifetime and a natural stress variable is very common (although it may not be suitable for heterogeneous materials). In this section, we revisit the lifetime regression models introduced in Chapters 6 and 7. The OEDs to be explained in the remainder of this chapter will be built upon these regression models.

9.2.1 Lifetime Regression Model

First, a lifetime distribution is defined as $D(\theta, \delta)$, where parameter θ is the scale parameter of this distribution model and δ is an auxiliary parameter that is assumed to be a constant across all factor levels. Note that for the Weibull distribution, θ can represent either the characteristic life or intrinsic failure rate; for the lognormal distribution, it could be the median lifetime. Also, if this distribution belongs to the log-location-scale family (both Weibull and lognormal distribution are), a natural logarithm transformation of the lifetime variable will result in a new variable with location-scale distribution, in which the location parameter is as $\mu = \log \theta$ and the scale parameter as $\sigma = 1/\delta$. Therefore, similar to regular regression models, at the next step we can model the natural logarithm of lifetime scale parameter as a function of natural stress variables (see the definition of natural stress variable in Chapter 8). The simplest function is the linear function

$$\log \theta = \beta_0 + \mathbf{s}^T \boldsymbol{\beta}, \tag{9.1}$$

where $\boldsymbol{\beta}$ and $\mathbf{s}$ are vectors of regression coefficients and natural stress variables, respectively.

Note that if the distribution model is the normal distribution and we let $\log \theta$ be the median of the normal distribution, the regression model earlier is exactly a multiple linear regression model. Similar to linear regression, the regression function, Eq. (9.1), can be expanded to include interaction terms and higher-order polynomial terms, such as

$$\log \theta = \beta_0 + \beta_1 s_1 + \beta_2 s_2 + \beta_{12} s_1 s_2,$$

or

$$\log \theta = \beta_0 + \beta_1 s + \beta_2 s^2.$$

Now, with this lifetime regression model, to estimate regression coefficients, β, as well as the auxiliary parameter δ, the maximum likelihood method can be applied. The incomplete failure time data caused by, for example, right censoring or interval censoring can be handled by the probabilities of such events as explained in Chapter 3. For OEDs, our task is to choose the settings of s such that the expected experimental outcomes contain the maximum amount of information relevant to the objective of the experiment. Although the detailed discussion of OEDs will be given in Section 9.3.1, here we give a brief explanation to show how it is related to the expected Fisher information matrix of an experimental design.

A common experimental objective is to quantify the log-linear function such as Eq. (9.1). An efficient experiment is expected to be able to estimate the regression coefficients precisely; therefore, it requires maximizing the determinant of the expected Fisher information matrix of coefficient estimators, which is the inverse of the variance-covariance matrix of these estimators. The entries of Fisher information matrix are obtained by taking the second derivative of the log-likelihood function of the regression model. The details of deriving expected Fisher information matrices are given in Appendix C.

Example 9.1   In this example, we use a dataset from Zelen (1959) to demonstrate a two-factor reliability experiment and to show that an orthogonal design is indeed the optimal design in this case.

The data were obtained from a glass capacitor reliability test with two stress variables – temperature and voltage. The temperature variable has two levels (170°C and 180°C) and the voltage variable has four levels (200, 250, 300, and 350 volts); thus, a full factorial design consists of eight different testing conditions. There are eight specimens being tested at each testing condition and these tests are Type II censored at the time of fourth failure. The full experimental data are presented in Appendix D, Table D.2.

Solution:

Using this dataset, we can fit a lognormal lifetime regression model and the result is

$$T_i \sim \text{LOGN}(t_{0.50,i}, 0.516)$$

where

$$\log t_{0.50,i} = \mu_i = 3.379 + 0.497 s_{1i} - 1.728 s_{2i}.$$

Here s_1 and s_2 are the Arrhenius transformation of temperature (i.e. $s_1 = 11,605/($Temperature in Celsius$+273.15))$ and the natural logarithm transformation of voltage (i.e. $s_2 = \log(\text{Voltage})$), respectively, and $i = 1, 2, ..., 8$ is the index for the testing condition. Note that, given the temperature and voltage ranges, we can standardize these regressors to $[-1, 1]$. This gives

$$\mu_i = 6.612 - 0.144x_{1i} - 0.484x_{2i},$$

where x_1 and x_2 are standardized regressors with the range of $[-1, 1]$ by converting s to

$$x = \frac{s - (\max(s) + \min(s))/2}{\text{range}(s)/2}.$$

This is a standard multiple linear regression model such as

$$Y_i = \log T_i \sim N(\mu_i, \sigma^2)$$

and

$$\mu_i = \beta_0 + \beta_1 x_{1i} + \beta_2 x_{2i}.$$

Extending the result in Appendix C, we can show that for a reliability test that tests all specimens to failure, its expected Fisher information is given by

$$\mathbf{I}(\mathbf{X}) = \frac{n}{\sigma^2} \begin{bmatrix} 1 & \sum_{i=1}^{k}\sum_{j=1}^{n_i} x_{1ij} & \sum_{i=1}^{k}\sum_{j=1}^{n_i} x_{2ij} & 0 \\ \sum_{i=1}^{k}\sum_{j=1}^{n_i} x_{1ij} & \sum_{i=1}^{k}\sum_{j=1}^{n_i} x_{1ij}^2 & \sum_{i=1}^{k}\sum_{j=1}^{n_i} x_{1ij}x_{2ij} & 0 \\ \sum_{i=1}^{k}\sum_{j=1}^{n_i} x_{2ij} & \sum_{i=1}^{k}\sum_{j=1}^{n_i} x_{1ij}x_{2ij} & \sum_{i=1}^{k}\sum_{j=1}^{n_i} x_{2ij}^2 & 0 \\ 0 & 0 & 0 & 2 \end{bmatrix}.$$

Here, k is the number of testing conditions, n_i is the number of specimens planned at each testing condition and $n = \sum_i n_i$, while x_{1ij} and x_{2ij} are the settings of two stress variables for each specimen assigned to each testing condition. It can be shown that the determinant of this Fisher information matrix will be maximized when we choose four testing conditions located at the four corners of the design space, with an equal number of specimens assigned to each testing condition. That is, the optimal design is an orthogonal design with the four design points:

$$\begin{array}{cc} x_1 & x_2 \end{array}$$

$$\begin{bmatrix} 1 & 1 \\ 1 & -1 \\ -1 & 1 \\ -1 & -1 \end{bmatrix}.$$

■

We call the experimental design of the aforementioned example a D-optimal design, because this design maximizes the *determinant* of expected Fisher information matrix. Indeed, one can see that this design is a factorial design and only two levels $(-1, +1)$ of each stress variable are needed. Therefore, the previous design of using four levels for the voltage variable is unnecessary for this linear model. Also notice that, to maximize the determinant of expected Fisher information matrix, we do not need the regression coefficients' values. This indicates that the *D*-optimal design is model parameter-independent for the case of all specimens being tested to failure. However, if censoring is expected from the reliability test (e.g. right censoring is expected when the total testing duration has to be limited due to time and/or budget constraints), the OED is often model parameter-dependent.

Similarly, if the lifetime regression model is a Weibull regression model, from Appendix C one can see that maximizing the determinant of the Fisher information matrix does not involve model parameter values for the test-to-failure case. A factorial design will become a *D*-optimal design.

From this example one can see that if the lifetime distribution is the lognormal or Weibull distribution, both of which belong to a log-location-scale family, and a reliability experiment tests specimens to failure, then with a natural logarithm lifetime transformation, the lifetime regression model becomes a linear regression model on a regular location-scale distribution. Its expected information matrix does not depend on the coefficient values in the regression model. This model parameter independence property is a well-known property for normal-linear models. It is the reason that factorial experimental designs can often achieve many desirable optimal properties.

However, when the lifetime distribution does not belong to the location-scale family and/or specimens cannot be tested to failure due to the limited testing period or test budget, the expected information matrix would indeed depend on the regression coefficient values. This model parameter dependency problem for OED causes the resulting optimal design to be in fact *locally* optimal. Now, a contradiction appears: before we conduct experiments and collect and analyze data, how do we know these coefficient values? Therefore, at the experimental design stage, these coefficient values are at most our best guess, maybe derived from past engineering experience, existing field failure data, or a pilot study. We called them the planning values. Model parameter dependency is commonly met in optimal designs for nonlinear models, such as lifetime regression models and GLMs. To consider the uncertainties in these planning values and their impact on design optimality, robust experimental designs have been developed in literature (e.g. Monroe et al., 2010; Nasir and Pan, 2017). We will give an introduction to robust experimental designs in Section 9.4.

Example 9.2 To illustrate the model parameter-dependent property of an ALT test plan with censoring consideration, we may look into a simple exponential regression model such as

$$T_i \sim EXP(\theta_i)$$

and

$$\log \theta_i = \beta_0 + \beta_1 s.$$

Here θ is the MTTF of the exponential distribution. Let $\beta_0 = -50$ and $\beta_1 = 1.6$, while the environmental stress is related to temperature, and is given by

$$s = \frac{11,605}{\text{Temperature in Celsus} + 273.15}.$$

Then the mean-time-to-failure of this exponential distribution at the normal-use temperature of 25°C is 214,782. Assuming the time unit is hours, this number is equivalent to 24.5 years. Obviously, it is impossible to conduct reliability tests under the normal use temperature. If we conduct ALTs at 50°C and 60°C, then the MTTFs are 1736 (72 days) and 309 (13 days), respectively.

Now, suppose there are a total of 100 specimens to be tested and these ALTs are planned for a test duration of 720 hours (30 days). We will expect some Type I censoring at both the lower and higher stress levels. Using JMP's Accelerated Life Test Plan function, the *D*-optimal test plan is found to be

Temperature	Number of specimens	Expected failures
50	38	12.9
60	62	55.9

Note that this is not a balanced design.

Suppose we are able to conduct these tests for 10,000 hours; then nearly all specimens will fail. The *D*-optimal design is found to be

Temperature	Number of specimens	Expected failures
50	50	49.8
60	50	50

This is a balanced design.

Suppose the model parameter β_0 is slightly changed to $\beta_0 = -51$. (Notice that it is actually a substantial change to failure time model, because the MTTF is the exponentiation of linear function of natural stress.) The *D*-optimal design becomes

Temperature	Number of specimens	Expected failures
50	43	29.1
60	57	56.9

Figure 9.1 shows JMP's ALT planning user interfaces for these three cases. They also include the expected parameter variances from these optimal designs, where one can see that these variances are much reduced from the initial balanced design. ∎

9.2.2 Proportional Hazards Model

As discussed in Chapter 7, the Cox PH model is a semiparametric model in which the baseline hazard function may take a nonparametric form but the effect of covariates on the hazard function is parameterized with an exponential function. Applying this model to accelerated life tests, we assume that a higher environmental stress level will accelerate the product's failure rate, in proportion to its baseline failure rate function. It is different from a lifetime regression model in the sense that the lifetime distribution is not fully defined and the model only quantifies the effect of stress variable on hazard function.

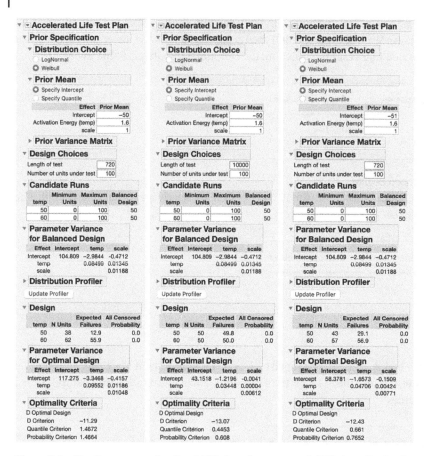

Figure 9.1 The three cases of optimal ALT plans for exponential lifetime distributions with different planning parameter values. Source: SAS Institute Inc.

Or, we may regard the acceleration as hastening the product's failure rate, in contrast to shortening its lifetime directly. However, if the baseline hazard function has indeed been specified, the resulting PH model will become a fully parameterized model. For the purpose of finding an OED, we typically need a fully parameterized model.

Let us examine an exponential lifetime regression model such as $T \sim \text{EXP}^*(\lambda(\mathbf{s}))$. Here EXP* is the alternative parameterization of the exponential distribution (see Chapter 2, Eq. (2.8)). The parameter is λ, and the PDF is

$$f(t|\lambda) = \lambda \exp(-\lambda t), \qquad t > 0.$$

The parameter λ is related to θ by $\lambda = 1/\theta$. By applying the log-linear function on failure rate, we have

$$\lambda(\mathbf{s}) = e^{\beta_0 + \mathbf{s}^T \beta} = \lambda_0 e^{\mathbf{s}^T \beta},$$

where $\lambda_0 = e^{\beta_0}$ and it is the baseline failure rate. Clearly, this lifetime regression model is equivalent to a PH model with a constant baseline hazard function.

Like the exponential distribution, the Weibull distribution also has an alternative parameterization. When the Weibull is parameterized in terms of the hazard λ and the shape parameter ϕ, the PDF is

$$f(t|\lambda, \phi) = \lambda \phi t^{\phi-1} \exp(-\lambda t^{\phi}), \qquad t > 0$$

and the survival function is

$$S(t|\lambda) = \exp(-\lambda t^{\phi}), \qquad t > 0.$$

We will denote the Weibull parameterized this way as $\text{WEIB}^*(\lambda, \phi)$ (see Chapter 2, Eq. (2.24)). The parameters are related by

$$\phi = \kappa$$
$$\lambda = \theta^{-\kappa}.$$

Since $\phi = \kappa$, we will parameterize the WEIB^* distribution as $\text{WEIB}^*(\lambda, \kappa)$.

The aforementioned derivation shows that, with a properly defined baseline hazard function, an AFT model may become a PH model. This conclusion can be applied to the Weibull distribution with a constant shape parameter. The WEIB^* hazard function at an accelerated stress level can be written in the form

$$h(t, \mathbf{s}) = h_0(t)e^{\mathbf{s}^T \beta^*} = \lambda_0 \kappa t^{\kappa-1} e^{\mathbf{s}^T \beta^*}.$$

Here, the baseline hazard function is specified as $h_0(t) = \lambda_0 \kappa t^{\kappa-1}$ and λ_0 is the intrinsic failure rate at the baseline stress level. If we reparameterize this baseline intrinsic failure rate as $\lambda_0 = e^{\beta_0}$, then the reliability function becomes

$$S(t, \mathbf{s}) = [S_0(t)]^{\exp(\mathbf{s}^T \beta^*)} = e^{(\beta_0^* + \mathbf{s}^T \beta^*)t^{\kappa}}.$$

This corresponds to the intrinsic failure rate being a log-linear function of natural stress variables; i.e.

$$\log \lambda(\mathbf{s}) = \beta_0^* + \mathbf{s}^T \beta^*,$$

or the characteristic life being a log-linear function of natural stress variable,

$$\log \theta(\mathbf{s}) = \beta_0^* + \mathbf{s}^T \beta,$$

where

$$\beta = -\beta^*/\kappa,$$

or equivalently,

$$\beta^* = -\kappa\beta$$

Therefore, the PH model of Weibull distribution is equivalent to its AFT model. One may also want to review Section 6.5 for the properties of Weibull regression.

Example 9.3 In Example 6.5 the rolling ball bearing failure times were analyzed by fitting a Weibull regression model. This estimated model was found to be

$$T_i \overset{\text{estimated}}{\sim} \text{WEIB}(\theta_i, 3.295)$$

where

$$\log \theta_i = 3.6080 + 0.4454x_{1i} + 0.4412x_{2i} + 0.0995x_{3i}.$$

Reformulate it to be a PH model.

Solution:

The parameters are related by

$$\hat{\phi} = \hat{\kappa} = 3.295$$
$$\hat{\beta}_0^* = -\hat{\kappa}\hat{\beta}_0 = -3.295 \times 3.607975 = -11.8884$$
$$\hat{\beta}_1^* = -\hat{\kappa}\hat{\beta}_1 = -3.295 \times 0.445395 = -1.467591$$
$$\hat{\beta}_2^* = -\hat{\kappa}\hat{\beta}_2 = -3.295 \times 0.445395 = -1.453854$$
$$\hat{\beta}_3^* = -\hat{\kappa}\hat{\beta}_3 = -3.295 \times 0.445395 = -0.327762$$

The estimate of the intrinsic failure rate function is therefore

$$\hat{\lambda}(\boldsymbol{x}) = e^{-11.8884 - 1.467591x_1 - 1.453854x_2 - 0.327762x_3}$$

The hazard function is therefore

$$\hat{h}(t|\boldsymbol{x}) = \hat{\lambda}(\boldsymbol{x}) \, t^{\hat{\phi}-1}.$$

The baseline hazard function, which occurs when $\boldsymbol{x} = \boldsymbol{0}$, is therefore

$$\hat{h}_0(t) = \exp(-11.8884) \, t^{3.295-1} = 0.00000687 \, t^{2.295}. \qquad \blacksquare$$

9.2.3 Generalized Linear Model

Besides failure times, another type of response variable from an ALT is a failure count, which is usually obtained when periodic inspections are conducted, so failure times are interval censored. Then, the binomial distribution arises to be a proper distribution choice for this response variable. Assume that during an interval (t_{i-1}, t_i) there are k_i failures out of n_i test units; we model the failure count distribution as

$$K_i \sim \text{BIN}(n_i, p_i)$$

with

$$p_i = P(t_{i-1} \le T < t_i | T > t_{i-1}) = \frac{F(t_i) - F(t_{i-1})}{S(t_{i-1})}.$$

Note that $F(t_{i-1})$ and $F(t_i)$ are the lifetime distribution function evaluated at the beginning and the end of time interval, so they are affected by stress variables.

For example, for the exponential failure time distribution, we obtain

$$1 - p_i = e^{-\lambda(\mathbf{s})(t_i - t_{i-1})} = e^{-\lambda_0 \Delta t_i e^{\mathbf{s}^T \beta}},$$

where λ_0 is the baseline hazard and Δt_i is the length of the time interval. Here, we have also applied the log-linear function to model the relationship between failure rate and natural stress variables. It is clear now that a complimentary log–log link function can be established; i.e.

$$\log[-\log(1 - p_i)] = \log(\lambda_0 \Delta t_i) + \mathbf{s}^T \boldsymbol{\beta}.$$

To follow the GLM formulation as defined in Chapter 5, we have (i) a binomial response distribution, (ii) a complimentary log-log link function between mean response and predictors, and (iii) a linear predictor function.

Similarly, this GLM formulation can be established for Weibull lifetime distribution if we assume the shape parameter is a constant. For other lifetime distributions, their link functions may be more complicated. In fact, this conversion is possible for any distribution that possesses the PH property.

Example 9.4 The solder joint failure time data are presented in Appendix D, Table D.3. Here, we use the PCB type 1 data only, and we create a case of interval censoring by assuming that the test units are inspected every 100 time units. Note that there are a total of 30 test units and 10 of them are tested at each of the temperature stress levels of 100°C, 60°C, and 20°C. Then, the interval times and failure counts are summarized in Table 9.1. Use R's icenReg and glm packages to fit a lifetime regression model with interval censoring and a GLM model, respectively, and compare their outputs.

Table 9.1 The interval censored solder joint failure time data.

Start	End	x	Count
0	100	100	6
100	200	100	4
100	200	60	1
200	300	60	3
300	400	60	3
500	600	60	2
600	700	60	1
200	300	20	4
300	400	20	1
400	500	20	2
500	600	20	1
600	700	20	2

Solution:

The following code applies the ic_par function in the icenReg package:

```
Y <- with(PCBtype1_interval_data, Surv(start, end,
+ event=rep(3, nrow(PCBtype1_interval_data)), type="interval"))
s <- with(PCBtype1_interval_data, 11605/(x+273.15))
w <- with(PCBtype1_interval_data, count)
(m1 <- ic_par(Y~s, data=PCBtype1_interval_data,
+ model="ph", dist="exponential", weights=w))
```

The output from this code is as follows:

```
Model: Cox PH
Dependency structure assumed: Independence
```

```
Baseline: exponential
Call: ic_par(formula = Y ~ s, data = PCBtype1_interval_data, model = "ph",
    dist = "exponential", weights = w)

          Estimate Exp(Est) Std.Error z-value
log_scale   5.6980 298.1000   0.19420  29.350 0.000000
s          -0.1764   0.8383   0.06295  -2.802 0.005075

final llk =  -57.74892
Iterations =  8
```

Here, `PCBtype1_interval_data` is the data frame used for storing the interval-censored data in the previous table. We can see that the coefficient of the natural stress variable is estimated to be −0.1764. By our Arrhenius transformation function, $s = 11,605/(x + 273.15)$, a higher temperature value becomes a lower natural stress value; thus, with a negative coefficient, the failure rate will increase by a factor of $e^{0.1764}$ for one unit decrease of natural stress value.

Now, to convert this analysis to a GLM formulation, we need a data table such as Table 9.2. The "failure" and "survival" columns record the number of test units that failed or survived to the end of testing in each time interval, respectively. Then, we use the R package `glm` with the binomial distribution and complementary log–log link function to fit the data.

Table 9.2 The data format of solder joint failure times for fitting a GLM model.

Start	End	x	Failure	Survival
0	100	100	6	4
100	200	100	4	0
0	100	60	0	10
100	200	60	1	9
200	300	60	3	6
300	400	60	3	3
400	500	60	0	3
500	600	60	2	1
600	700	60	1	0
0	100	20	0	10
100	200	20	0	10
200	300	20	4	6
300	400	20	1	5
400	500	20	2	3
500	600	20	1	2
600	700	20	2	0

```
s <- with(PCBtype1_count_data, 11605/(x+273.15))
(m2<-glm( cbind(failure, survival) ~ s,
          data = PCBtype1_count_data,
+ family = binomial(link=cloglog)))

Call:  glm(formula = cbind(failure, survival) ~ s,
+ family = binomial(link = cloglog), data = PCBtype1_count_data)

Coefficients:
(Intercept)            s
    5.2918       -0.1764

Degrees of Freedom: 15 Total (i.e. Null); 14 Residual
Null Deviance:          50.61
Residual Deviance:      42.53     AIC: 66.07
```

One can see that the estimated coefficient value for natural stress is the same as the one from the lifetime regression model earlier, as expected. ∎

Although the GLM approach provides an alternative data analysis method for censored failure times, it is not commonly used because this approach requires the PH property and the iterative least squares estimation method employed by the GLM solution often produces a larger standard error for the coefficient estimate. However, this formulation can be very useful for planning ALTs, particularly for the cases with multiple stress factors at multiple stress levels and when data censoring is unavoidable. In such case, we need to choose carefully the combinations of factor and level and properly allocate test units to each testing condition.

9.2.4 Converting PH Model with Right Censoring to GLM

Failure time censoring is a salient feature of ALT. In Section 9.2.3 we show that interval censored ALTs produce failure counts that can be modeled by binomial distributions. This is due to the change of data view from continuous failure times to discrete failure events. In fact, with the PH regression model, ALTs with right censoring can be modeled with GLMs too. Right censoring is common in reliability testing. Considering a reliability test that will be terminated at a fixed time point, it is almost certain that there will be some unfailed test units at the end of the test. Now, instead of modeling failure times, we may model failure events, which is binary – failure or survival. Interestingly, under the PH assumption, this will result in a Poisson distribution.

Aitkin and Clayton (1980) first showed that the PH regression model with censored survival data can be re-written in a GLM form and this model is applicable to multiple distributions, including the exponential, Weibull, and extreme value distributions. Barbosa et al. (1996) analyzed ALT results using a piecewise exponential distribution with the GLM approach. Utilizing this approach, Monroe et al. (2010, 2011) derived the optimal ALT planning process for a problem with three stress variables and restricted design space. Their methodology allows a simultaneous planning of stress-level selection (i.e. selecting experimental conditions) and test unit allocation. Seo and Pan (2018) further extended this approach to a generalized linear mixed model (GLMM) to include the consideration of random effects in ALT planning. The R package `ALTopt` uses the GLM approach to generate optimal test plans with multiple stress factors and Type I and interval censoring schemes.

Assume that a Type I censored ALT experiment is conducted, so there are some test units that have survived to the end of the reliability test; i.e. their failure times are censored at the test termination time. The likelihood function for the ALT data can be constructed. The contributions of these censored units to the total likelihood function is their reliability functions, while for those observed failure times, the contributions are their PDFs. The total likelihood function is given by

$$L = \prod_{i=1}^{n}(S_i(y_i))^{1-c_i}(f(y_i))^{c_i} = \prod_{i=1}^{n}(h(y_i))^{c_i}S(t_i), \tag{9.2}$$

where subscript i represents the index of test unit and c_i is an indicator variable with $c_i = 1$ if the observed time y_i is a failure time, or 0 if it is a survival time.

Based on the Cox PH model, Eq. (7.7), the log-likelihood function can be derived as

$$\log L = \sum_{i=1}^{n}[c_i(\log h_0(y_i) + x_i^T\beta) + e^{x_i^T\beta}\log S_0(y_i)]. \tag{9.3}$$

Let

$$\mu_i = e^{x_i^T\beta}(-\log S_0(y_i));$$

then

$$\log L = \sum_{i=1}^{n}[c_i\log\mu_i - \mu_i + c_i(\log h_0(y_i) - \log(-\log S_0(y_i)))]. \tag{9.4}$$

Note that the first two terms of the right-hand-side of Eq. (9.4) consist of a Poisson distribution likelihood function, while the last term is a constant, or invariant to any stress variable x. Therefore, by considering the first two terms only, the indicator variable c_i can be treated as following a Poisson distribution with mean μ_i. By maximizing this Poisson likelihood function, the estimation of stress variable effects, β's, can be obtained.

Recasting the whole formula into a GLM fashion (Nelder and Wedderburn 1972), it becomes

- The response variable, c_i, are independently Poisson distributed as POIS(μ_i);
- The linear predictor is given by $\eta_i = x_i^T\beta$;
- The link function is a canonical link function given by $\log\mu_i = \eta_i + a_i$, where $a_i = \log[-\log S_0(y_i)]$ is an offset term.

To substantiate this formulation further, suppose failure times follow a Weibull distribution with the reliability function of Eq. (2.23); then the offset term becomes

$$a_i = \log\lambda_0 + \gamma\log y_i. \tag{9.5}$$

Note that this offset term will not affect the estimation of stress variable effects if the partial likelihood function maximization method is used. However, it does affect the full likelihood function and is needed for estimating the Weibull shape parameter. In addition, to design ALT experiments, if the concern is about how to estimate the stress effects efficiently or how to predict the characteristic life or MTTF at the normal-use condition more precisely, this offset term becomes a nuisance term. In Section 9.3, the design issues of ALT will be discussed based on this GLM formulation.

Example 9.5 In the glass capacitor reliability testing example (Zelen 1959) previously discussed in Example 9.1, the data consist of two stress variables – temperature and voltage – and 64 test units, while eight units are allocated to each testing condition and Type II censoring scheme is applied with failure times are censored at the occurrence of fourth failure.

It can be shown that a power transformation of $T^{2.814}$ can transform the failure/survival times to be suitable for exponential distribution models. Then, we apply both exponential AFT model and Poisson GLM model to find the estimates of stress coefficients.

```
Y<-with(glass_capacitor_offset, Surv(Y, Failure))
s1<-with(glass_capacitor_offset, 11605/(Temp+273.15))
s2<-with(glass_capacitor_offset, log(Voltage))
(m1<-survreg(Y~s1+s2, dist = "exponential"))

Call: survreg(formula = Y ~ s1 + s2, dist = "exponential")

Coefficients:
(Intercept)            s1            s2
   5.408863      1.507285     -4.567681


Y<-with(glass_capacitor_offset, Failure)
o<-with(glass_capacitor_offset, offset)
(m2<-glm(Y~s1+s2, family=poisson(link="log"), offset = o))

Call: glm(formula = Y ~ s1 + s2,
          family = poisson(link = "log"), offset = o)

Coefficients:
(Intercept)            s1            s2
     -5.409        -1.507         4.568
```

Again, the estimation results are identical, except their signs. This is because the exponential regression model models the relationship between mean-time-to-failure and stress variables, while the GLM model models the relationship between failure rate and stress variables. ∎

9.3 Planning ALTs with Multiple Stress Factors at Multiple Stress Levels

Planning ALT experiments is important, because a typical ALT experiment is very expensive. It uses specialized testing chambers and takes a long testing time. If the experiment is not well designed, the results from an ALT could be disappointing, or even useless, because of failure time censoring or other unexpected events (e.g. machine breakdown). The general guidelines for designing experiments as listed in Montgomery (2020), pp. 14-21, can be used to advise any type of scientific experiment, but design and analysis of reliability experiments, such as ALTs, bring in more challenges as we described in Section 9.1 regarding the unique features of reliability experiments.

Designing ALT plans is an active research topic in the reliability engineering field. Nelson (2005, 2015) summarized hundreds of published papers that are related to accelerated testing. They classify these papers to the areas of statistical models, data analysis, test planning, etc. In this section, we will present the principles of optimal ALT planning and the software tools for implementing these principles. In a 1998 article entitled "Pitfalls of Accelerated Testing," Meeker and Escobar (1998b) put forward several issues that need to be carefully attended when planning ALTs. First of all, the experimenter needs to understand what type of failure mode might be triggered by the accelerated stress factor and whether or not this failure mode might occur on the product in its normal use environment. If an ALT experiment induces units to fail with some totally different failure modes from what might happen at the normal use stress level, then the ALT data cannot be used for inferring the failure behavior at the use condition. Secondly, we need to be aware of the effect of interaction between stress variables and failure modes. A good test plan should be able to either avoid unwanted interaction or explore and study the interaction effect effectively. In addition, the authors reminded us of the danger of uncertainties in parameter estimation and model extrapolation. Therefore, a well-designed ALT should be able to reduce these uncertainties (increase statistical efficiency) by properly choosing sample size, testing duration and testing conditions, and be able to tolerate some deviations of the assumed model from the true model (robust to model assumptions).

In a follow-up article, "More Pitfalls of Accelerated Tests," Meeker et al. (2013) further emphasized the following misunderstandings and misuses of ALTs in industrial practice:

1. Use of equal unit allocation at all levels of the accelerating variable in an ALT;
2. Choosing the wrong accelerating variable or having a test plan that will not provide useful information;
3. Testing at too severe stresses (requiring extreme extrapolation);
4. Using unnecessarily complicated testing and data analysis;
5. Testing at high levels of accelerating variables that cause new failure modes;
6. Focusing only on the obvious failure mode(s);
7. Improper use of system level ALTs;
8. Inadequate monitoring of failures during an ALT;
9. Not inspecting survivors at the end of an accelerated test (or not fully using the information gained from such inspections);
10. Ignoring the impact of idle time;
11. ALTs that generate failures that will not arise in field use;
12. Attempting to estimate a life distribution from a HALT;
13. Using an inadequate acceleration model;
14. Failure to recognize a non-parametric accelerating variable relationship;
15. Use of inaccurate information about an activation energy (for an Arrhenius model);
16. Ignoring or inadequately treating interaction;
17. Not comparing failure modes in the field with failure modes in the laboratory;
18. Failure to utilize available degradation data.

In particular, items 1–7 are issues related to ALT planning. Based on these concerns, we provide following suggestions to practitioners who want to use ALTs to either understand or predict failures.

Understand the purpose of an ALT experiment: Each ALT experiment may carry a different purpose. One experiment may aim at establishing the relationship between failures and stress variables, so the primary concern is to estimate the regression coefficients in the life-stress function. Another one may be targeting the reliability prediction, with, for example, the prediction of mean failure time at the use condition being desired. Different test plans should be designed to best serve these purposes.

Anticipate the impact of data censoring: Censoring is unavoidable even in accelerated tests. The general effect of censoring is that we will have only partial failure information from censored test units. At the lower test stress level more censoring is expected, so it is wise to allocate more test units at the lower stress level than at the higher stress level. In particular, when the purpose of the experiment is to predict reliability at the use stress level, the lower test stress level is closer to the use stress level, thus adding more test units at the lower stress level will help increase the accuracy and precision of reliability prediction.

Assess the robustness of a test plan to assumptions: To derive an optimal test plan, we typically would need to know the form of the regression function of a failure parameter to stress variables (e.g., with or without interaction of stress variables) and the values of regression coefficients, which are also called planning values. This planning knowledge could be gathered from existing failure data or engineering experience, but in general, there are uncertainties associated with it. A good test plan should be robust to these uncertainties, in the sense that even if the assumed model is a bit different from the true model, the statistical efficiency of the selected plan should not be greatly affected.

Be aware of the danger of model extrapolation: Extrapolation often has to be performed when one needs to predict reliability at the use stress condition, but extrapolation is dangerous. To be able to finish a test on time, we may not be able to avoid extrapolation completely, but it is highly desirable to set the lower test stress level to be as close to the use stress level as possible. In addition, failure modes at high stress levels could be found to be different from the failure modes that occur under the use condition. To make a valid inference on the reliability at the use condition, we want to have the test stress levels not to be too far away from the use condition.

Plan for the uncertainty of use condition: Often a product's use condition is not a single well-defined stress level, but a range of stress levels. For example, a computer chip will endure different temperatures at different computing loads. Considering the uncertainty in use conditions, we may either plan an ALT for the average case scenario or for the worst case scenario. Understanding the product's use condition can help us fine-tune the ALT experiment to target specific product market and/or for specific product use.

Incorporate practical engineering/management constraints into the plan: For any experiment, engineering and management constraints always have an impact on experimental design. For examples, a heat chamber may be able to hold only 20 test units at a time, or the heat distribution in the chamber is not uniform. It is necessary to consider these constraints in the planning stage, so we would not be surprised by the problems at the time of executing these tests or analyzing test data.

9.3.1 Optimal Test Plans

Optimal reliability test plans are developed with the purpose of collecting product failure data more efficiently from a reliability experiment. For ALTs, experimenters may design the stress factors and their levels as in a typical factorial design. But, unlike most traditional DOEs, reliability experiments also need to consider other design parameters such as the stressing limit, testing duration, and test unit allocation at each stress level. In addition, due to non-normal response distribution and response censoring, the statistical models to be used are not Gaussian-linear models, thus the factorial designs and fractional factorial designs introduced in Chapter 4 may not be optimal in terms of statistical sampling efficiency. In this section, we discuss several optimality criteria that are often used for planning ALTs and use examples to demonstrate and compare optimal ALT plans with ordinary experimental designs.

Two commonly used optimality criteria in optimal ALT planning are *D*-optimality and *I*-optimality. *D*-optimality is achieved by minimizing the determinant of the variance-covariance matrix of the estimators of all or some of the model parameters. This is equivalent to maximizing the determinant of the expected Fisher information matrix of these estimators from a design. Therefore, this criterion focuses on parameter estimation; i.e. to find an experimental design such that the experimenter can expect that the model parameters can be more precisely estimated by this design than any other design. By contrast, the *I*-optimality focuses on the quality of prediction of response. It minimizes the average response prediction variance over the whole experimental design region. In ALTs, the experimental design variables are the environmental stress variables and the normal use stress region is typically outside of the experimental stress region. Thus, to emphasize that a model-based extrapolated response prediction would be made, some researchers call the minimization of response prediction variance at the use stress level as *U*-optimal or U_c-optimal (Monroe et al., 2010).

In Chapter 8, we have discussed the optimal test plan for life testing, where sample size and test duration are the planning parameters. In this section, we will derive the optimal test plan for multiple-stress accelerated life testing. Here, we assume that the total number of test units and the test duration are fixed, but a test planner needs to select the testing conditions, i.e. the combinations of stress variables and their level settings, and the allocation of test units to these testing conditions. The balanced factorial design, as introduced in Chapter 4, becomes sub-optimal in this scenario because of non-normal response distributions and the anticipated failure time censoring. To find the optimal test plan, we will utilize the GLM formulation for censored failure times.

Weibull regression is used to illustrate the *D*- and *I*-optimality criteria for optimal ALT planning. Assuming that the stress variables may only affect the characteristic life parameter of the Weibull distribution while the shape parameter is held constant, the regression model is given by

$$T \sim \text{WEIB}(\theta(\mathbf{s}), \kappa),$$

and

$$\log \theta = \beta_0 + \mathbf{s}^T \boldsymbol{\beta}.$$

Assuming each ALT experiment will last to time *t*, the probability of failure by the end of testing period is

$$F(t) = 1 - e^{-(t/\theta(s))^\kappa}.$$

As shown in Section 9.2.4, for the case of Type I censoring and a fixed shape parameter, the Weibull regression can be formulated by a GLM such as

- c_i, which is the censoring indicator variable for the ith test unit, can be treated as from a Poisson distribution with mean μ_i;
- the linear predictor is given by $\eta_i = \beta_0 + \beta_1 x_1 + \cdots + \beta_k x_k$, where $x_j, j = 1, ..., k$, are natural stress variables and $\beta_0 = \log \lambda_0$;
- the link function is given by $\log \mu_i = \eta_i +$ an offset term, and this offset term is $\kappa \log t_i$.

The expected information matrix of a GLM is given by

$$I = \mathbf{X}^T \mathbf{W} \mathbf{X}, \tag{9.6}$$

where $\mathbf{X}$ is the design matrix with dimension of $n \times p$, and each of its row corresponds to a test unit and each of its column corresponds to a stress variable. $\mathbf{W}$ is the weight matrix, which is a $n \times n$ diagonal matrix. For a Poisson regression, the diagonal elements are μ_i's, $i = 1, ..., n$. Because

$$\mu_i = e^{\eta_i} t_i^{\kappa}, \tag{9.7}$$

and the actual testing time of the ith test unit, t_i, has not been observed yet at the stage of test planning, we use the expected testing time in place of t_i in Eq. (9.7). Given that the censoring time is t_c, the expected testing time is

$$E(t_i) = E(T)F(t_c) + t_c S(t_c), \tag{9.8}$$

where $E(T)$ is the mean of Weibull failure time distribution, $F(t_c)$ is the probability of failure by the censoring time t_c, and S_c is the probability of survival at t_c; specifically,

$$E(T) = \theta \, \Gamma \left(1 + \frac{1}{\kappa} \right),$$

$$F(t_c) = 1 - e^{-(t_c/\theta)^{\kappa}}$$

and

$$S(t_c) = e^{-(t_c/\theta)^{\kappa}},$$

where

$$\theta = e^{\beta_0 + \beta_1 s_1 + \cdots + \beta_k s_k}.$$

Therefore, we can maximize the determinant of the information matrix defined by Eq. (9.6) to obtain the best design matrix, which is the D-optimal test plan. Note that, to find the optimal test plan, we need to know the values of the regression coefficients, $\beta_0, \beta_1, ..., \beta_k$. These values are called planning values. Thus, the optimal test plan depends on the parameter setting of the regression model; i.e. it is only locally optimal. This is a common feature of OED for a nonlinear model.

To derive the U_c-optimal test plan, we first compute the prediction variance of failure time prediction at the use condition. It is given by

$$V(\mathbf{x}_{\text{use}}) = \mathbf{x}_{\text{use}}^T (\mathbf{X}^T \mathbf{W} \mathbf{X})^{-1} \mathbf{x}_{\text{use}}, \tag{9.9}$$

where

$$\mathbf{x}_{\text{use}} = [1, x_{1,\text{use}}, ..., x_{k,\text{use}}]^T$$

is the vector of use stress variables. The design matrix that minimizes Eq. (9.9) is the U_c-optimal test plan.

For the I-optimal test plan, we minimize the average prediction variance over the ranges of use stress variables, which is given by

$$\text{Average Variance} = \frac{\displaystyle\int_{\mathbf{x}_{use}} \mathbf{x}^T (\mathbf{X}^T \mathbf{W} \mathbf{X})^{-1} \mathbf{x} \, d\mathbf{x}}{\displaystyle\int_{\mathbf{x}_{use}} d\mathbf{x}}. \tag{9.10}$$

One can see that a design matrix that minimizes the numerator of Eq. (9.10) will be sufficient, and it is the I-optimal test plan.

The R package `ALTopt` is developed for finding optimal ALT plans when a Weibull regression model is given. There are two functions, `altopt.rc` and `altopt.ic`, in the package for generating optimal designs for right censoring and interval censoring schemes, respectively. Other functions, such as `design.plot`, `pv.contour.rc` and `pv.contour.ic`, are used to visualize the optimal design and the prediction variance over the design and use region.

Example 9.6 Consider the adhesive bond ALT example from Meeker and Escobar (1998a). We will use this example to demonstrate the construction process of a *D*-optimal test plan with single stress factor by the GLM approach. The following information is given for the test:

- Stress factor: Temperature
- Failure time distribution: Weibull
- Use condition: 50°C
- Highest stress limit: 120°C
- Number of test units: 300 test units
- Testing time (Censoring time): 180 days
- Assumed shape parameter value: $\alpha = 1.667$
- Assumed regression coefficients: $\beta_0 = 27.89$, $\beta_1 = -1.21$ (These values were obtained by re-parameterization from those in the original example via $\beta_0 = (-1.667) \times (-16.733) = 27.89$ and $\beta_1 = (-1.667) \times 0.7265 = -1.21$.)

Besides the aforementioned information, the lowest stress limit is specified as 80°C in order to guarantee enough failure time data from the test. Applying the Arrhenius acceleration model, the natural stress variable is defined as

$$S = 1/kT = 11,605/(273.15 + T).$$

Accordingly, the use condition and the lowest and highest stress limits are transformed to be

$$S_U = \frac{11,605}{273.15 + 50} = 35.91$$

$$S_L = \frac{11,605}{273.15 + 80} = 32.86$$

$$S_H = \frac{11,605}{273.15 + 120} = 29.52.$$

In DOE, it is a common practice to use the coded variable instead of the original one. As such, the natural stress variable is further transformed to the coded variable via

$$x = \frac{S - S_H}{S_L - S_H}$$

so that the design region is constrained to $[0, 1]$. Note that $x = 0$ is the highest stress.

The following linear predictor model is obtained in terms of the coded variable,

$$\log \mu = \beta_0 + \beta_1 S = \beta_0 + \beta_1 \{(S_L - S_H)x + S_H\} = -7.855 - 4.049x.$$

Now, run the R package `ALTopt` to generate the *D*-optimal design. The software package basically runs the optimization routine. First, its algorithm assigns each of 300 design points randomly into the

Table 9.3 The D-optimal design for the adhesive bond ALT.

x	Number of test units
0.000	173
0.557	127

design region, and then the algorithm has these design points moved in the direction of an improved objective function value at each iteration. The iterative process stops when the objective function value has converged. Since the result depends on the initial random assignment, it is recommended to run the routine several times and pick the best one. Table 9.3 shows the *D*-optimal design generated by the following lines of code, which has the highest objective function value among 30 different runs of the optimization routine:

```
library("ALTopt")
set.seed(200)
altopt.rc("D", N = 300, tc = 180, nf = 1,
    alpha = 1.667, formula = ~x1,
    coef = c(-7.855, -4.049), nOpt = 30)
```

It shows the two distinct design points and the number of test units allocated to each point. We can transform the test locations back to the original variable, which results in 173 test units allocated to 120°C and 127 test units allocated to 96.7°C. We can show that 154 test units are expected to fail at 120°C, and so are 26 test units at 96.7°C before the end of the testing period if the assumed parameter values were correct. There numbers are large enough for model parameter estimation.

Note that this test plan places the lower test stress level at 96.7°C, instead of the allowable lowest testing temperature 80°C. When there is no censoring to be expected or all test units are expected to test to failure, the two extreme design points in the design region are indeed the best testing conditions and an optimal design is to place equal number of test units at these design points. However, due to Type I censoring, the expected number of failures is too few at the temperature of 80°C; thus the optimal test plan has to move the lower stress level up to 96.7°C. In addition, the allocations of test units at the two design points are not even. ∎

The JMP software has an Accelerated Life Test Plan function (in its DOE -> Special Purpose function group), where the user can plan Type I censored ALTs with one or two stress factors that will reach with *D*-optimal or *I*-optimal. However, the current function cannot automatically select the best lower stress level. It requires the user to specify the lowest stress level, and the software will find the best test unit allocation plan at both lower and higher stress level. For the previous example, if we specify the lowest test stress level to be 80°C, then the software uses the balanced design at temperatures 80°C and 120°C as the starting candidate design and, from there, it finds the *D*-optimal design to be allocating 116 test units to temperature 80°C and 184 units to 120°C. However, if we specify the lowest test stress level to be 95°C, the software gives the same design as in the previous example. See Figure 9.2.

The JMP software also provides a nonlinear design function (again, in its "DOE − > Special Purpose" function group), where the user may generate optimal designs for GLM models. However, this function

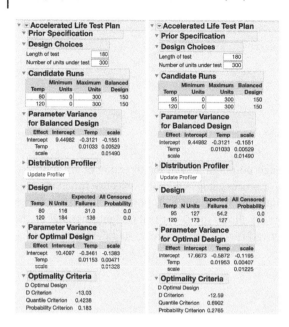

Figure 9.2 The JMP's optimal ALT plans depend on the specification of the initial candidate test plans. Source: SAS Institute Inc.

does not handle altered weight matrices that are induced by censored failure times, thus its usefulness to optimal ALT planning is limited.

9.3.2 Locality of Optimal ALT Plans

It is important to point out that optimal ALT plans are actually local optimal, because these plans are derived based on the assumed reliability model and model parameter value. In fact the weight matrix W depends on the regression coefficients and the shape parameter, as one can see from Eq. (9.7). Therefore, we need some *a priori* knowledge about these parameters, which is often obtained from engineering experience, domain experts' opinions, previous similar tests and literature. If the assumed model parameter values are in fact incorrect, we might end up generating a completely different and inefficient test plan.

Using the previous adhesive bond example, suppose the linear predictor model is still true, but we have assumed the following model, with β_1 being -2.428 instead of -4.049, incorrectly,

$$\log \hat{\mu} = -7.855 - 2.428x. \tag{9.11}$$

The *D*-optimal design based on the incorrect model of Eq. (9.11) is shown in Table 9.4. First, a substantial change of the lowest testing stress condition from 0.557 (96.7°C) to 0.928 (82.6°C) has been suggested. Second, there is a significant difference for the test unit allocations. Specifically, 46 more test units were allocated at the higher stress level in Table 9.3, while 22 more test units were allocated at the lower stress level in Table 9.4. As a result only 8 test units are expected to fail at the lower stress level (i.e. 82.6°C), which might not be enough number of failures for an accurate parameter estimation.

If at the ALT planning stage, the experimenter does not have a precise reliability model of the product under study, a range of model parameter could be assumed. In such case, we suggest trying out different model parameter values to generate multiple test plans. The experimenter may then combine them, possibly with weights, to produce a final plan. In Section 9.4, we will briefly introduce the concept

Table 9.4 The D-optimal design based on the incorrect linear predictor model.

x	Number of test units
0.000	139
0.928	161

of Bayesian OED, which is another way to accommodate model uncertainty at the experimental design stage.

9.3.3 Comparing Optimal ALT Plans

In the previous glass capacitor accelerated life testing example, the experimental design was a 2 by 4 factorial design and equal allocation of test units at all testing conditions. These tests had to be scheduled as failure censoring in order to observe enough failures at each testing condition. In fact, many tests were ended after a very long period, which is a drawback of this design.

This factorial design employs eight distinct design points, so eight testing conditions need to be set up. An optimal design need not require this many design points. For a linear regression function with two independent variables, only three design points are sufficient for estimating regression coefficients. If the interaction between these variables is taken into consideration, four design points are required. The highest testing stress condition is always selected, as it will generate the most failures. The selection of lower testing stress conditions is affected by the optimality criterion. Under D-optimality, we prefer the region enclosed by these selected design points to be large so that the variance of coefficient estimates can be reduced. Under U_c-optimality, we prefer the lower testing stress conditions to be close to the use stress condition, so as to reduce the reliability prediction variance at the use condition. However, it is also required that at these lower testing stress conditions a sufficient number of failures can be produced, so these stress levels cannot be too low and the number of test units to be allocated at these design points needs to be large enough.

Example 9.7 The Weibull regression model fitted for the glass capacitor data is given by

$$T \overset{\text{estimated}}{\sim} \text{WEIB}^*(\hat{\lambda}(\mathbf{s}), 2.814)$$

where

$$\log \hat{\lambda}(\mathbf{s}) = -5.409 - 1.507 s_1 + 4.568 s_2.$$

The testing temperature's range is from 170°C to 180°C and the testing voltage's range is from 200 volts to 350 volts. Find the D-optimal ALT plan for this model with $n = 64$ runs.

Solution:
First, we convert the temperature in Celsius and the voltage to the natural stress variables by the Arrhenius transformation and natural logarithm transformation, respectively. We have that s_1 varies from 26.19

to 25.61 and s_2 from 5.30 to 5.86. Again, according to the `ALTopt` design convention, we shift these ranges to 1 (lowest stress level) and 0 (highest stress level). The regression function becomes

$$\log \lambda(\mathbf{x}) = -17.244 - 0.871x_1 - 2.556x_2.$$

Let the maximum testing duration be 1000 time units and there are 64 test units. The *D*-optimal design is presented in Table 9.5.

Table 9.5 The *D*-optimal design for the glass capacitor ALT.

x_1	x_2	Number of test units
1	0.233	38
0	0.960	15
0	0	11

We can see that this design selects two relatively lower stress conditions and the numbers of test units being assigned to these conditions are higher than the number assigned to the highest stress condition $(0, 0)$. This is because the probability of no failures is higher at a lower stress condition, so it is better to allocate more test units there. Furthermore, this design has design points that are located on the border of the design region to enlarge the area that these design points can enclose, consequentially increasing the determinant of expected Fisher information matrix.

Now, if the testing duration is reduced to 500 time units, it can be anticipated that more censoring would happen at lower testing stress levels. Therefore, these lower stress conditions should move closer to the highest stress condition. The *D*-optimal design found for a shorter testing duration is given in Table 9.6. One can also see that the number of test units allocated to the highest stress level is increased from 11 to 14. This is due to the increasing probability of censoring at the highest stress level too. Figure 9.3 illustrates the two *D*-optimal designs with different lengths of testing time.

Table 9.6 The *D*-optimal design for the glass capacitor ALT with a shorter testing duration.

x_1	x_2	Number of test units
1	0.094	29
0	0.825	21
0	0	14

The U_c-optimal design for a use condition of 160°C and 100 volts is presented in Table 9.7. One can see that, in this design, the selected lower stress conditions are closer to the use stress condition $(2.046, 2.239)$. This is because the purpose of this experiment is to infer the reliability at the use condition as precisely as possible; thus the lower stress design points should be placed closer to the use condition point. Figure 9.4

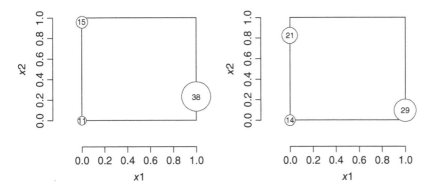

Figure 9.3 The two *D*-optimal ALT plans with different testing duration.

Table 9.7 The U_c-optimal design for the glass capacitor ALT.

x_1	x_2	Number of test units
0.937	0.936	27
0.764	0.996	15
0	0	22

Figure 9.4 The U_c-optimal ALT plan with expected prediction variance.

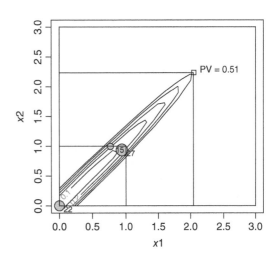

shows the expected reliability prediction variance of this design over the experimental design region, and further extending to the use stress condition point. The needed R code is:

```
library(ALTopt)

set.seed(200)
(m1<-altopt.rc(optType = "D", N = 64,
```

```
tc = 1000, nf = 2, alpha = 2.814,
formula = ~x1 + x2, coef = c(-17.244, -0.871, -2.556)))

(m11<-altopt.rc(optType = "D", N = 64,
tc = 500, nf = 2, alpha = 2.814,
formula = ~x1 + x2, coef = c(-17.244, -0.871, -2.556)))

design.plot(m1$opt.design.kmeans, x1, x2)

design.plot(m11$opt.design.kmeans, x1, x2)


(m2 <- altopt.rc(optType = "U", N = 64,
        tc = 1000, nf = 2, alpha = 2.814,
        formula = ~x1 + x2, coef = c(-17.244, -0.871, -2.556),
        useCond=c(2.046, 2.238)))

design.plot(m2$opt.design.kmeans, x1, x2)

pv.contour.rc(m2$opt.design.kmeans, x1, x2,
tc = 1000, nf = 2, alpha = 2.814,
formula = ~ x1 + x2, coef = c(-17.244, -0.871, -2.556),
useCond=c(2.046, 2.238))
```

■

9.4 Bayesian Design for GLM

In this section we discuss a more general approach to experimental designs for GLMs with the model parameter-dependent problem. For the GLMs the general form of the information matrix is the matrix

$$X'\Delta V\Delta X = X'WX$$

where W is the *Hessian weight matrix*. The matrix V is a diagonal matrix whose main diagonal elements are the variances of the individual response observations and is a matrix that depends on the link function specified. For certain choices of link functions, called *canonical links*, Δ is an identity matrix. The canonical link for the binomial distribution is the logistic link, the canonical link for the Poisson distribution is the log link, and the canonical link for the gamma distribution is the inverse link, although in practice we often use the log link with the gamma distribution because it ensures that all of the predicted responses will be non-negative. The elements of the V matrix will always contain the unknown model parameters. For example, in a logistic regression model for the variance of the ith observation is

$$\sigma_i^2 = n_i \frac{\exp(-x'\beta)}{(1 + \exp(-x'\beta))^2}$$

where there are n_i observations at the ith design point. As a result, for many GLMs, the optimal design cannot be found without knowledge of the parameters.

One approach to the optimal design problem for a GLM is to guess or estimate values of the unknown parameters. In this case the information matrix is a function of only the design points only and a D-optimal design can be found. More properly, this design should be called a *conditional D-optimal design*, because it depends on the estimated values of the unknown parameters. A two-stage or sequential procedure could also be employed, where the initial design is run and preliminary estimates of the model parameters obtained; then the initial design is augmented with additional runs to complete the experiment. A third alternative is to employ a Bayesian approach. This would involve assessing a prior distribution on the unknown parameters and integrating them out of the information matrix. If the prior distribution of β is $f(\beta)$ then an appropriate Bayesian D-optimality criterion is

$$\phi = \int \log(|I(\beta)|) f(\beta) \, d\beta$$

where $I(\beta)$ is the information matrix. Finding a design that maximizes this criterion is a computationally intensive process as typically it requires that a multidimensional integral be evaluated many times. Gotwalt et al. (2009) developed a clever quadrature scheme that provides extremely accurate values of this integral and is computationally efficient. JMP uses this procedure with a coordinate exchange procedure that can be used to find designs for a variety of nonlinear models, including GLMs. We now present several examples of this procedure.

Example 9.8 *Example: Logistic Regression with Two Design Variables*
An experimenter wants to fit a logistic regression model in two variables x_1 and x_2, where both variables are in coded units in the range $-1 \leq x_i \leq 1$. Therefore the model we plan to fit is

$$y_i \stackrel{\text{indep.}}{\sim} \text{BIN}\left(1, \frac{1}{1 + \exp(-(\beta_0 + \beta_1 x_{i1} + \beta_2 x_{i2}))}\right)$$

for which

$$E(y_i) = \frac{1}{1 + \exp(-(\beta_0 + \beta_1 x_{i1} + \beta_2 x_{i2}))}.$$

Our prior information about the model suggests that reasonable ranges for the model parameters are

$$1 \leq \beta_0 \leq 3$$
$$1.5 \leq \beta_1 \leq 4.5$$
$$-1 \leq \beta_2 \leq -3.$$

Solution:
The experimenter believes that a normal distribution defined over these ranges constitutes a reasonable summary of the prior information. Find a Bayesian D-optimal design with $n = 12$ runs.

Table 9.8 and Figure 9.5 present the Bayesian D-optimal design found by JMP. Notice that there are six distinct design points and the model has three parameters. Runs 2, 3, and 12 from Table 9.8 are not replicated, while runs $(5, 6, 9)$, $(1, 4, 7)$, and $(8, 10, 11)$ comprise the replicates for the other three distinct design points. ∎

Table 9.8 Bayesian D optimal design for the logistic regression in Example [crosref].

```
Nonlinear Design

Parameters

Name        Role         Values
b0          Continuous      1        3
b1          Continuous      1.5      4.5
b2          Continuous     -1       -3

Design
Run           x1              x2
1          -0.57255       1
2          -0.44157      -1
3          -1            -0.02109
4          -0.57255       1
5          -1            -1
6          -1            -1
7          -0.57255       1
8           0.402077      1
9          -1            -1
10          0.402077      1
11          0.402077      1
12          1             1
```

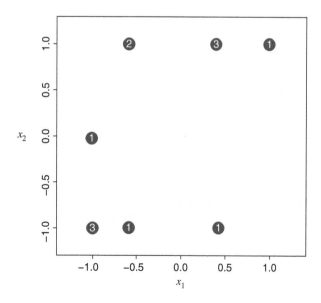

Figure 9.5 Bayesian D-optimal design for the first-order logistic regression model in Example 9.8.

Example 9.9 Reconsider the logistic regression model from Example 9.8 The model is

$$y \overset{\text{indep.}}{\sim} \text{BIN}\left(1, \frac{1}{1 + \exp(-(\beta_0 + \beta_1 x_1 + \beta_2 x_2 + \beta_{12} x_1 x_2))}\right).$$

As in Example 9.8, we assume that a normal prior distribution is appropriate. The likely ranges for the model parameters are

$$1 \leq \beta_0 \leq 3$$

$$1.5 \leq \beta_1 \leq 4.5$$

$$-1 \leq \beta_2 \leq -3$$

We want to use a design with 12 runs.

Table 9.9 and Figure 9.6 present the Bayesian D-optimal design. This design is rather different from the designs for the no-interaction model. It has seven distinct design points for a model with four parameters with runs 3, 6, and 10 from Table 9.9 unreplicated and runs $(1, 12)$, $(2, 11)$, $(4, 5, 8)$, and $(7, 9)$ forming the other four points.

Table 9.9 Bayesian D optimal design for the first-order logistic regression in example 9.9.

```
          Nonlinear Design

          Parameters

          Name       Role        Values
          b0      Continuous       1        3
          b1      Continuous      1.5      4.5
          b2      Continuous      -1       -3
          b12     Continuous     -0.5      1.5

          Design
          Run         x1            x2
          1         1            1
          2        -1            1
          3        -0.39451     -1
          4        -1           -1
          5        -1           -1
          6         0.261768     1
          7        -0.03877     -0.15756
          8        -1           -1
          9         0.03877     -0.15756
          10       -0.24769      1
          11       -1            1
          12        1            1
```

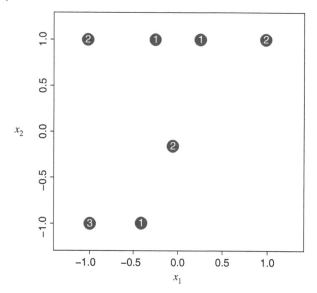

Figure 9.6 Bayesian D-optimal design for the logistic regression model in Example 9.9

9.5 Reliability Experiments with Design and Manufacturing Process Variables

Product reliability improvement can be achieved through an experimental study, where a lifetime variable is the main response of interest. Although some environmental stress variables are often involved in these experiments in order to accelerate the product failure process to obtain more failure observations, they are not the only factors in these experiments and, oftentimes, they are not the most important factors either. There are product design variables and manufacturing process variables to be studied by these experiments and these factors are often the ones of interest to engineers for product or process optimization. Therefore, these product design and process variables are treated as control factors, while the stress variables are treated as noise factors in these experiments, and failure time censoring is often ignored at the experimental design stage. In the literature, factorial designs and Taguchi's orthogonal array designs were employed for this type of experimental study, which are in general effective at finding the optimal product or process design; however, they may not be the most efficient method for model building. In this section, we will use several examples to demonstrate the use of experiments for product reliability improvement and the necessity of considering failure time censoring in such experiments.

The worsted yarn example discussed in Chapter 5 has three experimental factors: length, amplitude, and load. The product is tested to failure and the latter two factors are clearly the stress variables to cause failure to happen sooner. The original experiment uses a 3×3 factorial design. Our previous analysis has shown that an adequate model for the response is given by

$$Y \overset{\text{estimated}}{\sim} \text{LOGN}(\hat{\mu}, \hat{\sigma})$$

where

$$\log \hat{\mu} = 6.335 + 0.832x_1 - 0.631x_2 - 0.392x_3$$

and

$$\hat{\sigma} = 0.1713.$$

Note that with a loglinear relationship between scale parameter and experimental factors, choosing three levels for each factor is unnecessary, as a two-level per factor experiment is sufficient for model building. The actual response values from this experiment have a wide range. The largest value of cycles-to-failure is 3636, which is 40 times the smallest response value, 90. Therefore, if the experimenter has to terminate the experiment at 1000 cycles, there would be 9 runs being censored, which is one third of all runs, and most of them happen at the $x_2 = -1$ and $x_3 = -1$ level. With this test duration constraint, we may want to re-design this experiment so that a severe failure time censoring would not occur at a particular experimental setting.

Example 9.10 For the worsted yarn example, assume the estimated cycle-to-failure model is the true model. If the design variable, x_1 (length) is set at the level of -1, then the loglinear function of median life is

$$\log \mu = 5.503 - 0.631x_2 - 0.392x_3.$$

If x_1 is set at the level of 1, this function becomes

$$\log \mu = 7.167 - 0.631x_2 - 0.392x_3.$$

The experimenter wants to conduct a reliability experiment at both the low and high levels of the other two stress variables, but is afraid that heavy censoring may happen at some stress levels. Assume the test duration is 1000 time units and there are 12 units available for tests at the low and high levels of x_1. Re-design this experiment and find the proper sample size assignment to each testing condition.

Solution:
As there are 12 test units available for tests, we need to plan the assignment of these test units to each stress condition. Using JMP's Accelerated Life Plan function, we plan the optimal ALTs with factors x_2 and x_3. Figure 9.7 shows the JMP outputs for the cases of $x_1 = -1$ (left) and $x_1 = 1$ (right).

One can see that no matter whether we are at the low or high level of factor x_1, three stress conditions (combinations of x_2 andn x_3 levels) are chosen, because they are sufficient for estimating the regression coefficients of these stress variables. However, at $x_1 = 1$, $x_2 = -1$, and $x_3 = 1$, it is likely that the test will not produce any failures; thus, we need to adjust this stress condition. By setting the lower level of x_2 and x_3 to 0, instead of -1, we have the adjusted test plan for the case of $x_1 = 1$ shown in Figure 9.8. ∎

From this example, one can see that a regular factorial design may not be the best choice for reliability experiments when the test-to-failure scheme is impractical. Instead, we can separate the product design or manufacturing process factors from the environmental stress factors, and apply factorial designs and optimal ALT designs on different types of experimental design factors so as to increase the design efficiency. Obviously, this strategy requires knowledge of an accurate response model even before conducting experiments, which is not always possible in practice. Nevertheless, it is a good practice to evaluate the

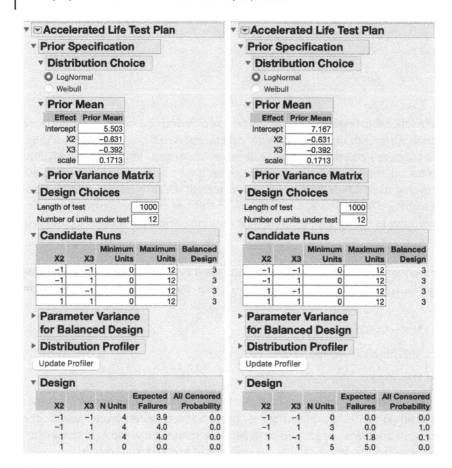

Figure 9.7 The optimal ALT plans for worst yarn experiment with $x_1 = -1$ and $x_1 = 1$. Source: SAS Institute Inc.

likelihood of censoring for any ALT experiments. If it is unlikely to observe a failure or there will be very few failures to be anticipated at a particular stress condition, the experimenter needs to either raise the stress level or assign more test units to that level.

Most industrial life tests are planned for studying the effects of product design variables on lifetime, while the stress variable is used to shorten the testing time and its effect is only a secondary consideration. The impact of censoring is often ignored, especially at the experimental design stage. With such experiments, the design variables are often categorical variables and the stress variables are continuous. We need to carefully analyze the data and detect any possible interaction between design factors and stress factors, so as to correctly select the best product design.

Taguchi argued that improvement of life characteristics required life testing to be conducted during the product design and prototype stages and it was better to cover as many factors as possible in a series of experiments. He recommended using a large inner array with many controllable factors and a smaller outer array for noise or stress factors. In the following two examples, we analyze two relatively large experiments with more design and process factors.

Figure 9.8 The optimal ALT plans for worst yarn experiment with adjusted stress conditions. Source: SAS Institute Inc.

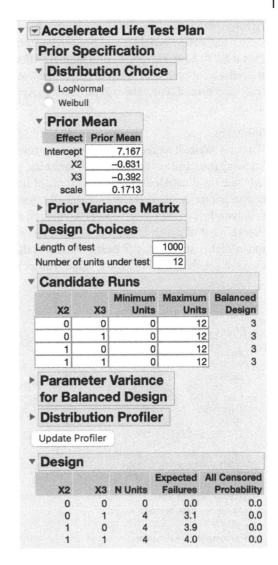

Accelerated Life Test Plan

▼ **Prior Specification**

▼ **Distribution Choice**

- ● LogNormal
- ○ Weibull

▼ **Prior Mean**

Effect	Prior Mean
Intercept	7.167
X2	−0.631
X3	−0.392
scale	0.1713

▶ **Prior Variance Matrix**

▼ **Design Choices**

Length of test	1000
Number of units under test	12

▼ **Candidate Runs**

X2	X3	Minimum Units	Maximum Units	Balanced Design
0	0	0	12	3
0	1	0	12	3
1	0	0	12	3
1	1	0	12	3

▶ **Parameter Variance for Balanced Design**

▶ **Distribution Profiler**

Update Profiler

▼ **Design**

X2	X3	N Units	Expected Failures	All Censored Probability
0	0	0	0.0	0.0
0	1	4	3.1	0.0
1	0	4	3.9	0.0
1	1	4	4.0	0.0

Example 9.11 In Taguchi (1986), a durability test of clutch springs is described. It has the following experimental factors:

- *A.* Shape, 3 levels
- *B.* Hole ratio, 2 levels
- *C.* Coining, 2 levels
- *D.* Stress σ_t, 90, 65, 40
- *E.* Stress σ_c, 200, 170, 140
- *F.* Shot preening, 3 levels
- *G.* Outer perimeter planning, 3 levels

Among these factors, factors *D* and *E* are stress variables and they are continuous variables, while other factors are design variables and are categorical. The original design also combines factors *B* and *C* to Factor *BC* to have 3 levels. To find all main effects and the DE and DF interactions, an L_{27} array design is used, and at each experimental run, three test units are assigned. The responses, cycles-to-failure, are interval censored. The dataset is provided in Appendix D. Analyze this dataset and find the best product design.

Solution:

We apply Weibull regression to the interval-censored clutch spring lifetime experimental data. Without knowing the nature of the two stress variables, we apply a log transformation on both of them.

Figure 9.9 illustrates the initial analysis of this data set. The analysis of main effects shows that the design factors, *A*, *BC*, and *F* are statistically significant while *G* is not. In particular, factor *F* has a strong positive effect when it is set at level 2. This can also be seen from the data, as when we examine survival events, most of which happen at *F* = 2. The effects of both stress variables are statistically significant too. A higher stress value of factor *D* reduces the product's lifetime, but factor *E* has an opposite effect. One can see that higher factor *E* value will increase the lifetime prediction. Because the nature of this

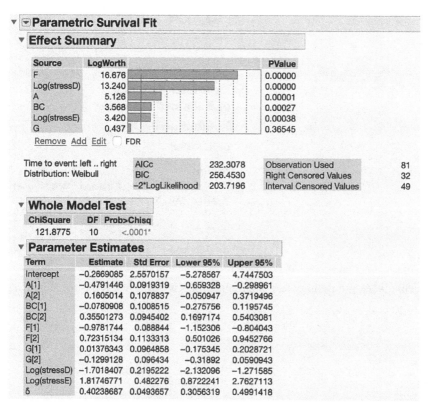

Figure 9.9 The Weibull regression parameter estimations for the clutch spring life testing experiment. Source: SAS Institute Inc.

stress variable is unknown to us, we cannot comment on this unconventional finding. Further analysis shows that the interaction effect between D and E is negligible and the estimate of the interaction effect between D and F is unstable, because when $F = 2$ and $D = 40$ or 65, almost all test units survived to the end of testing; thus no failure time information could be utilized.

Using the Quantile Profiler tool in the Parametric Survival Fit function of JMP (see Figure 9.10), we set the levels of A, BC, F and G to 3, 2, 2, and 1, respectively, and vary D and F over their test ranges. We can see that the estimated characteristic life of this product has a big variability, changing from 11.76 to 89.39. This again shows that these stress variables have a significant impact on product lifetime.

Based on these analyses, to improve reliability, we recommend setting factor F at its second level. This is the most important design factor to the product lifetime. As to other design factors, it is better to set $A = 3$ and $BC = 2$. While for G, as it will not affect the product lifetime much, engineers may choose its level based on other considerations, such as cost.

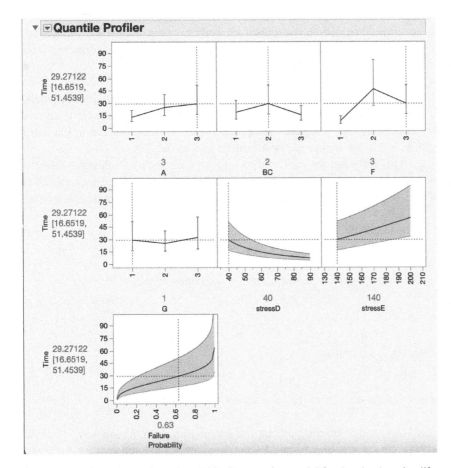

Figure 9.10 Quantile profiler of the Weibull regression model for the clutch spring life testing experiment. Source: SAS Institute Inc.

Reading the original testing data, we find that this product has very few failures over the entire testing duration (11 time units) unless the stress variable D is set at its most severe level, $D = 90$. As the Weibull regression model is available now, we can use it to plan a reliability demonstration test to show that the recommended product design is indeed as durable as we wanted. Assume that it is desired to demonstrate this product's 10th failure percentile is at least 10 time units with a 90% confidence level when it is used under the stress condition, $D = 40$ and $E = 140$. Then, again with the design factors set at $A = 3$, $BC = 2$, $F = 2$ and $G = 1$, we have the following acceleration model:

$$T \overset{\text{estimated}}{\sim} \text{WEIB}(\hat{\theta}(D,E), \hat{\kappa})$$

where

$$\log \hat{\theta} = 0.8250 - 1.7018 \log(D) + 1.8175 \log(E)$$

and

$$\hat{\kappa} = 2.4852.$$

A regular demonstration test would require 30 test units if a maximum of one failure is allowed. Now, conducting the test at an accelerated stress level with $D = 90$, it is easy to find that the acceleration factor is then 4 and only 8 test units are needed. See Figure 9.11. ∎

Example 9.12 The drill bit reliability experiment previously described in Example 6.9 (which originally appeared in Wu and Hamada (2000)) is re-analyzed here. Again, in this experiment there are 11 control factors (drill bit design factors), denoted by A through J and L, and 5 noise factors (stress factors), denoted by M through Q. All factors have 2 levels, except Factor A, which has 4 levels.

The inner array consists of 11 control factors, which follows a 16-run resolution III design, while the outer array has 5 noise factors, which follows a 2^{5-2} fractional factorial design. So, there are a total of 128 experimental runs. Each test unit is tested to failure until 3000 time units, which generates a few type I censored lifetime observations.

For this dataset (available in Appendix D), we find an exponential regression model provides a good fit. Factors A, C, D, J, O, and P are significant factors. The optimal levels of design factors are set at $A = 2$, $C = 1$, $D = 1$ and $J = 1$, where all other insignificant factors may be set to level -1 (see Figures 9.12 and 9.13). Then we have the following exponential regression model with noise factors:

$$T \overset{\text{estimated}}{\sim} \text{EXP}(\hat{\theta})$$

and

$$\log \hat{\theta} = 6.8005 + 0.0399 M_{(-)} + 0.1776 N_{(-)} + 0.9333 O_{(-)}$$
$$+ 0.6946 P_{(-)} - 0.13652 Q_{(-)}$$

where the coefficient values of M through Q are the values when these noise factors are set at their low levels.

Because the noise factors M, N, and Q are insignificant, we set their levels to the low level. Now we compare the MTTFs of this product at the highest and lowest stress levels of O and P. It can be found that the MTTF at $O = -1$ and $P = -1$ is 4960.86 and at $O = 1, P = 1$ it is 974.03. The acceleration factor is about 5.

Figure 9.11 Sample size determination for the demonstration test of clutch spring durability. Source: SAS Institute Inc.

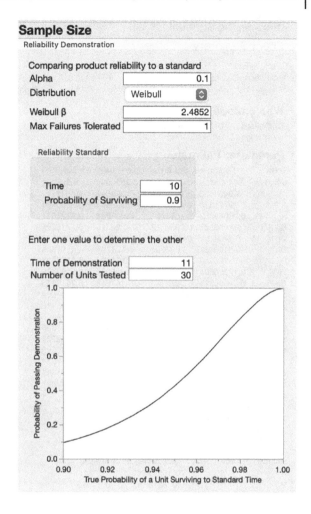

Suppose Factors O and P are indeed continuous stress variables and we are interested in further investigating this product's lifetime acceleration model caused by these two factors. We can plan a D-optimal ALT. Following the convention of R package `ALTopt`, we convert the highest testing stress condition to $(0, 0)$ and the lowest testing condition to $(1, 1)$, and search the optimal ALT plan according to the following link function:

$$\log \hat{\lambda} = -6.8814 - 1.8666O - 1.3891P.$$

The optimal ALT plan with 12 test units is given in Table 9.10.

Note that this is a 2×2 factorial design with varying sample size assigned to each testing condition. Since there are four design points planned, we may also use the experimental outcomes to investigate the interaction effect of these two stress variables. ∎

Parametric Survival Fit

Effect Summary

Time to event: T	AICc	1694.884	Observation Used	128
Distribution: Exponential	BIC	1742.035	Uncensored Values	114
Censored By: Failure	-2*LogLikelihood	1649.847	Right Censored Values	14

Whole Model Test

ChiSquare	DF	Prob>Chisq
161.7692	18	<.0001*

Parameter Estimates

Term	Estimate	Std Error	Lower 95%	Upper 95%
Intercept	6.44084539	0.0956016	6.2534697	6.628221
A[1]	-1.0404606	0.1629044	-1.359747	-0.721174
A[2]	0.29771348	0.1675168	-0.030613	0.6260404
A[3]	0.29045052	0.1666771	-0.036231	0.6171317
B[-1]	-0.016275	0.0968872	-0.20617	0.1736204
C[-1]	-0.2209485	0.0965327	-0.410149	-0.031748
D[-1]	-0.3100421	0.097804	-0.501734	-0.11835
E[-1]	0.07126648	0.099107	-0.12298	0.2655127
F[-1]	0.12368601	0.0965633	-0.065575	0.3129465
G[-1]	0.04998925	0.0979263	-0.141943	0.2419213
H[-1]	0.16083452	0.0968752	-0.029037	0.3507064
I[-1]	-0.1506025	0.0995907	-0.345797	0.0445916
J[-1]	-0.2470155	0.096272	-0.435705	-0.058326
L[-1]	-0.1769944	0.1005605	-0.374089	0.0201005
M[-1]	0.03994173	0.1050686	-0.165989	0.2458724
N[-1]	0.17755613	0.1037793	-0.025848	0.3809598
O[-1]	0.93332445	0.1073404	0.7229412	1.1437077
P[-1]	0.69456846	0.1053911	0.4880057	0.9011312
Q[-1]	-0.1365188	0.1055149	-0.343324	0.0702867
δ	1	0	1	1

Confidence Intervals are Wald

Variable Selection using Generalized Regression

Wald Tests

Source	Nparm	DF	Wald ChiSquare	Prob>ChiSq
A	3	3	41.1650374	<.0001*
B	1	1	0.02821698	0.8666
C	1	1	5.23881189	0.0221*
D	1	1	10.0491157	0.0015*
E	1	1	0.51708456	0.4721
F	1	1	1.6406547	0.2002
G	1	1	0.26058789	0.6097
H	1	1	2.75634353	0.0969
I	1	1	2.28679518	0.1305
J	1	1	6.58337672	0.0103*
L	1	1	3.0978788	0.0784
M	1	1	0.14451336	0.7038
N	1	1	2.92718374	0.0871
O	1	1	75.6030247	<.0001*
P	1	1	43.4332306	<.0001*
Q	1	1	1.6740055	0.1957

Figure 9.12 The exponential regression parameter estimations for the drill bit life testing experiment. Source: SAS Institute Inc.

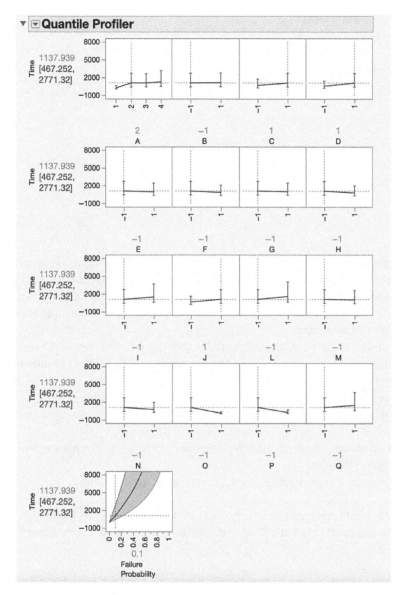

Figure 9.13 Quantile profiler of the exponential regression model for the drill bit life testing experiment. Source: SAS Institute Inc.

Table 9.10 The *D*-optimal design for the drill bit ALT.

O	P	Number of test units
0	0	2
1	0	4
0	1	3
1	1	3

Problems

9.1 Consider the scenario presented in Example 8.8, but the material's activation energy is in fact unknown. Here, 25°C is the normal use temperature and 100°C is the highest temperature that this material can be tested on without altering its failure mechanism. There are 200 test units. Assume the exponential lifetime distribution, *EXP*(1000), for this material at its normal-use temperature and let the test duration be 100 time units.

   (a) Set the lower test temperature to be 60°C. What is the higher test temperature? Why? How would you allocate these test units to the lower and higher test temperature so as to estimate the activation energy with the highest precision?

   (b) Design a *D*-optimal experiment for better quantifying the life-stress relationship.

   (c) Does the *D*-optimal design assign an equal number of test units to the lower and higher test temperature? Why?

   (d) Design an *I*-optimal experiment for predicting the material's MTTF at the normal-use temperature with the highest precision.

   (e) Compare the *D*-optimal and *I*-optimal test plans, what are the differences? Why?

9.2 Consider the glass capacitor data presented in Example 9.1. Fit the data with a Weibull regression model and then use this model to plan a *D*-optimal reliability test with a test duration of 1000 time units.

9.3 Compare the *D*-optimal test plan derived in the previous problem with the factorial design originally used by the researcher. What are the differences? Why?

9.4 Using the Weibull regression model derived from Problem 9.2 and the *D*-optimal test plan and the original factorial design, simulate 1000 experiments for each test plan. Each experiment should have 64 test units and a test duration of 1000 time units. Use these simulation data to estimate the characteristic life of Weibull distribution. What are the differences between the results from the *D*-optimal experiments and the factorial experiments? Which design gives you a better result? Why?

9.5 Suppose that you want to design an experiment to fit a logistic regression model in two predictors. A first-order model is assumed. Based on prior information about the experimental situation, reasonable ranges for the model parameters are

$$0.5 \leq \beta_0 \leq 5$$
$$2 \leq \beta_1 \leq 6$$
$$1 \leq \beta_2 \leq 4$$

and the normal distribution is a reasonable prior. Find a Bayesian D-optimal design with 10 runs for this experiment.

9.6 Reconsider the situation in Problem 9.5. Suppose that a uniform prior distribution is selected. Find a Bayesian D-optimal design with 10 runs for this experiment. Compare this with the design found in Problem 6.1.

9.7 Suppose that you want to design an experiment to fit a logistic regression model in two predictors. A first-order model with interaction is assumed. Based on prior information about the experimental situation, reasonable ranges for the model parameters are

$$0.5 \leq \beta_0 \leq 8$$
$$1 \leq \beta_1 \leq 5$$
$$0.5 \leq \beta_2 \leq 1.5$$

and the normal distribution is a reasonable prior. Find a Bayesian D-optimal design with 16 runs for this experiment.

9.8 Reconsider the situation in Problem 6.3. Suppose that a uniform prior distribution is selected. Find a Bayesian D-optimal design with 16 runs for this experiment. Compare this with the design found in Problem 6.3.

9.9 You are studying the failure rate of a component as a function of three stress factors; time, temperature, and relative humidity. The response variable is the proportion of units that fail at each set of test conditions. You want to construct a Bayesian D-optimal design for this experiment. You have some prior information about the anticipated results, which can be summarized with the following ranges of the parameters:

$$-1 \leq \beta_0 \leq 0.5$$
$$0 \leq \beta_1 \leq 0.5$$
$$-1 \leq \beta_2 \leq 0.5$$
$$0 \leq \beta_3 \leq 0.5$$
$$0.5 \leq \beta_{12} \leq 1.5$$

$$0 \le \beta_{13} \le 0.75$$

$$0 \le \beta_{13} \le 0.5.$$

Assume that the normal distribution is a reasonable prior.

(a) Find a Bayesian D-optimal design with 8 runs for this experiment. How does this design compare with a standard 23 factorial design?

(b) Find a Bayesian D-optimal design with 16 runs for this experiment.

9.10 Rework Problem 9.9 using a uniform prior. What difference does this make in the designs obtained?

9.11 Suppose that you want to design an experiment to fit a Poisson regression model in two predictors. A first-order model is assumed. Based on prior information about the experimental situation, reasonable ranges for the model parameters are

$$0.5 \le \beta_0 \le 3$$

$$1 \le \beta_1 \le 3$$

$$0.25 \le \beta_2 \le 0.75$$

and the normal distribution is a reasonable prior. Find a Bayesian D-optimal design with 16 runs for this experiment.

9.12 Reconsider the situation in Problem 6.7. Suppose that a uniform prior distribution is selected. Find a Bayesian D-optimal design with 16 runs for this experiment. Compare this with the design found in Problem 6.7.

A

The Survival Package in R

R (R Core Team 2020) is a free and open source software system for doing statistical analysis. RStudio (RStudio Team 2020) is designed to allow users to write code in the R language in a supportive environment. While R commands can be entered one-at-a-time at the command prompt, the real power of R lies in the ability to put lines of code in an R script and execute them all at once. We will assume that the reader is familiar with the basics of R, including the ability to write an R script and execute it. There are a number of excellent materials that describe R programming. A solid introduction is given in Matloff (2011) and a deeper and more modern treatment is given in Grolemund and Wickham (2018).

An R package is collection of functions, variables, and sometimes data sets that are designed around a set of related tasks. R contains thousands of packages. In 2017, the number of R packages available at the Comprehensive R Archive Network (CRAN)

```
https://cran.r-project.org/
```

surpassed 10,000. There are also a number of R packages available at other sites, such as `https://github.com/`. The package that we focus on here is the survival package (Therneau 2015) that was designed to analyze reliability or survival data and contains functions to accomplish this task.

When you start R, directly with the R graphical user interface (GUI) or with R Studio, you are loading base R, the set of R commands that perform basic functions. Any packages that are used must first be installed on your computer (which you must do just once) and then loaded into your system each time you load R. To install the survival package, type

```
install.packages( "survival" , dependencies = TRUE )
```

The `dependencies = TRUE` argument tells R to load any other packages that are required by the package you are installing. This step needs to be done just once on each computer. Once the package is installed, it must be loaded every time you begin R. This can be done with the command

```
library( survival )
```

We will illustrate R's survival package using the toaster snubber data set, which gives the failure or censoring times for the snubber component on a toaster (which muffles the ejection of bread so that the bread doesn't fly into the air) for an old design and a new design. This data set was discussed

Design of Experiments for Reliability Achievement, First Edition.
Steven E. Rigdon, Rong Pan, Douglas C. Montgomery, and Laura J. Freeman.
© 2022 John Wiley & Sons, Inc. Published 2022 by John Wiley & Sons, Inc.
Companion website: www.wiley.com/go/rigdon/designexperiments

in Example 3.3 and the raw data are presented in Table 3.2. Data are available in a `csv` file named `snubber.csv`, which we can load into R using the `read.csv` command. Note that the first row of the file contains the variables names. This is called the header line and the command within `read.csv` that says `header=TRUE` indicates this:

```
snubber = read.csv( "snubber.csv" , header=TRUE )
```

We can look at the first few lines of the data using the `head` command in R. The statement

```
head( snubber )
```

produces the output

```
  Design Event Cens
1      1    90    1
2      1    90    1
3      1    90    0
4      1   190    0
5      1   218    0
6      1   218    0
```

The first column of numbers simply provides the observation number and is not really a variable in the data frame. The next three columns give the `Design` (1 = Old, 2 = New), the `Event` time (which can be either a failure time or a censoring time, and `Cens` an indicator variable (1 = failure, 0 = censored observation).

For illustration, we'll focus on the old design. We can select only those observations for which `Design` = 1 using the filter command in the `dplyr` package, a popular package for data management and manipulation:

```
library( dplyr )
designOld = filter( snubber , Design == 1 )
```

Note that the logical operator in R is the double equal sign (==). The single equal sign is used for named arguments in function calls, as in `header=TRUE` when we executed the `read.csv` command. The equal sign is also used as the assignment operator as in `designOld = filter(snubber, Design==1)`.[1]

Many of the functions in the `survival` package require the input to be a survival object. This is a particular type of object in R that involves the event time along with a failure indicator. This indicator is often called a censoring indicator, but it is coded so that a failure yields a 1 and a censored observation yields a 0, so it is really an indicator on whether an observation is a failure or not, so it should properly be called a failure indicator. We can create a survival object using the `Surv` function in the `survival` package. First, let's pull the variables `Event` and `Cens` from the `designOld` data frame; this can be done with the commands

```
Event = designOld[,"Event"]
Cens = designOld[,"Cens"]
```

---

1 The assignment operator can also be written as <-, and some R programmers prefer this. We will use the single equal sign to make assignments.

The survival object `designOldSurv` can then be created and printed by

```
designOldSurv = Surv( Event, Cens )
print( designOldSurv )
```

There are other optional arguments to Surv in addition to those used earlier. For example, one argument describes the type of censoring, e.g. left-censoring, right-censoring, interval-censoring, etc. The default is right-censoring. The output from this `print` statement is

```
 [1]   90    90    90+ 190+ 218+ 218+ 241+ 268   349+ 378+
[11] 378+ 410   410   410+ 485   508   600+ 600+ 600+ 600+
[21] 631   631   631   635   658   658+ 731   739   739+ 739+
[31] 739+ 739+ 790   790+ 790+ 790+ 790+ 790+ 790+ 790+
[41] 790+ 790+ 790+ 790+ 855   980   980   980+ 980+ 980+
[51] 980+ 980+
```

We can obtain the Kaplan–Meier estimate of the survival function using the survfit command available in the survival package.

```
KM.designOld = survfit(designOldSurv~1,conf.type="log-log")
plot( KM.designOld, lwd=c(2,1.4,1,4),
      col=c("blue","red","red") )
```

The `conf.type="log-log"` indicates the method of finding confidence intervals for the survival probabilities. The first argument to `survfit` is a model, as indicated by the tilde ~. In this case there are no predictor variables so the only regressor is the constant, which is denoted by "1." The line widths and colors refer, respectively, to the middle, lower, and upper curves, that is, the point estimate, the lower limit, and the upper limit. This code produces the plot shown in Figure A.1. If we were to ask for a summary of the object `KM.designOld`, we would enter

```
summary( KM.designOld )
```

which yields

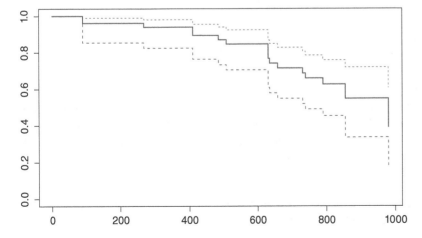

Figure A.1 Kaplan–Meier estimate of the old snubber design.

```
Call: survfit(formula = designOldSurv ~ 1, conf.type = "log-log")
 time n.risk n.event survival std.err lower 95% CI upper 95% CI
   90     52       2    0.962  0.0267        0.855        0.990
  268     45       1    0.940  0.0336        0.826        0.980
  410     41       2    0.894  0.0449        0.764        0.955
  485     38       1    0.871  0.0495        0.734        0.940
  508     37       1    0.847  0.0535        0.705        0.924
  631     32       3    0.768  0.0652        0.609        0.869
  635     29       1    0.741  0.0682        0.579        0.849
  658     28       1    0.715  0.0707        0.550        0.828
  731     26       1    0.687  0.0731        0.520        0.807
  739     25       1    0.660  0.0752        0.491        0.784
  790     20       1    0.627  0.0783        0.454        0.758
  855      8       1    0.549  0.1003        0.336        0.719
  980      7       2    0.392  0.1179        0.172        0.607
```

The first column gives the failure times (not the censoring times) while the second and third columns give the number of items at risk (i.e. still operating) and the number of failures at that time, respectively. The next column gives the estimated survival probability while the last three columns give the standard error and the limits for a 95% confidence interval.

Parametric distributions, such as the exponential or Weibull, can be fit to reliability data with censoring. This is done through the survreg function. Let's fit the exponential and the Weibull to the snubber data described earlier. The following code

```
designOldExponential = survreg( designOldSurv ~ 1 ,
                                dist="exponential" )
summary( designOldExponential )
```

fits the exponential distribution to the snubber data and yields the following output

```
Call:
survreg(formula = designOldSurv ~ 1, dist = "exponential")
            Value Std. Error    z       p
(Intercept) 7.502      0.236 31.8 <2e-16

Scale fixed at 1

Exponential distribution
Loglik(model)= -153    Loglik(intercept only)= -153
Number of Newton-Raphson Iterations: 4
n= 52
```

The output from R must be analyzed with care. The internal calculations use the logarithms of the lifetimes, and as you recall the logarithm of an $EXP(\theta)$ variable is an $SEV(\log \theta, 1)$ distribution, and the log of a $WEIB(\kappa, \theta)$ distribution is an $SEV(\log \theta, 1/\kappa)$ distribution. The terminology in R's output reflects the use of logarithms. For example, the intercept parameter is the location parameter μ of the SEV distribution, so it is $\log \theta$. Thus, in the aforementioned example 7.502 is the estimate of $\log \theta$, so

$$\hat{\theta} = \exp(7.502) = 1811.662.$$

For the case of the exponential distribution, we have a formula for the MLE; it is given in Eq. (3.20). The MLE is the total time on test divided by the number of failures:

$$\hat{\theta} = \frac{32598}{18} = 1811.$$

The discrepancy comes from the fact that R rounded the estimate to 7.502 when it printed the output from `designOldExponential`. We can get at a more exact number by typing `designOld Exponential$icoef`. This is the object that contains all of the estimated coefficients, but for the exponential distribution there is only one parameter. The more accurate estimate is 7.501634, which gives

$$\hat{\theta} = \exp(7.501634) = 1811$$

making it agree with Eq. (3.20).

We can then estimate the reliability or survival function by

$$S(t) = \exp\left(-t/\hat{\theta}\right).$$

We could then compare the estimated exponential distribution's reliability or survival curve with the Kaplan–Meier curve defined previously by plotting them on the same axes. This can be done with the following code

```
EventF = Event[ Cens == 1 ]
EventC = Event[ Cens == 0 ]
TTT = sum( EventF ) + sum( Event C )
thetahat = TTT / length( EventF )
t = seq(0,1000,1)
S = exp( - t / thetahat )
windows( 9 , 6 )
plot( KM.designOld, lwd=c(2,1.4,1.4) ,
      col=c("blue","gray","gray") )
  lines( t , S , col="black" , lwd=2 )
```

which yields the graph in Figure A.2. The exponential curve (the smooth curve) seems to underestimate survival early on, and then over estimate survival later. Thus, the exponential distribution may not be a good fit.

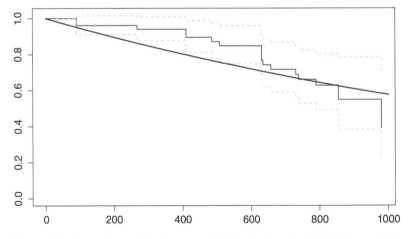

Figure A.2 Comparison of fitted exponential distribution fit with the Kaplan–Meier fit when applied to the Snubber data.

Let's next fit the Weibull distribution to the snubber data. The only change to the `survreg` function is to use `dist="weibull"`:

```
designOldWeibull = survreg( designOldSurv ~ 1 ,
                            dist="weibull" )
summary( designOldWeibull )
```

This code yields the following output:

```
Call:
survreg(formula = designOldSurv ~ 1,
dist = "weibull")
              Value Std. Error    z      p
(Intercept)   7.002   0.128    54.90  <2e-16
Log(scale)   -0.811   0.207    -3.92  9e-05

Scale= 0.444

Weibull distribution
Loglik(model) = -147.4    Loglik(intercept only) = -147.4
Number of Newton-Raphson Iterations: 8
n= 52
```

Once again, we must keep in mind that R works with the logarithm of the failure times. Thus, the "scale" parameter is the scale parameter σ of the SEV distribution, which in this case is estimated to be 0.444. We can get more accurate estimates of the parameters than R prints out with the `summary` command by typing

```
designOldWeibull$icoef
```

which produces

```
(Intercept)   Log(scale)
  7.0024980  -0.8111157
```

The estimates of the SEV parameters are then the exponentiated values of these estimates:

$$\hat{\mu} = 7.0024980.$$

and

$$\log \hat{\sigma} = -0.8111157.$$

Thus,

$$\hat{\sigma} = \exp(\log \hat{\sigma}) = \exp(-0.8111157) = 0.444362.$$

The estimates of the parameters are therefore

$$\hat{\theta} = \exp(\hat{\mu}) = \exp(7.0024980) = 1099.376$$

and

$$\hat{\kappa} = \frac{1}{\hat{\sigma}} = \frac{1}{0.444362} = 2.250417.$$

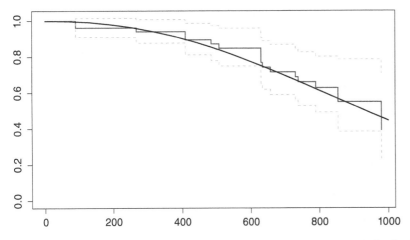

Figure A.3 Comparison of fitted Weibull distribution fit with the Kaplan–Meier fit when applied to the Snubber data.

The estimate for the shape parameter κ can also be obtained from directly from the output of `survreg` using `1/designOldWeibull$scale`.[2]

As we did for the exponential distribution, we could calculate the reliability function for the Weibull and plot it with the Kaplan–Meier function to assess the fit. The following code accomplishes this task, and the plot is shown in Figure A.3.

```
thetahat = exp(designOldWeibull$coefficients[1])
kappahat = 1/designOldWeibull$scale

t = seq(0,1000,1)
S = exp( - ( t / thetahat )^kappahat    )

windows( 9 , 6 )
plot( KM.designOld, lwd=c(2,1.4,1.4) ,
      col=c("blue","gray","gray") )
  lines( t , S , col="black" , lwd=2 )
```

The Weibull fit is much closer than with the exponential distribution.

A similar procedure exists for other distributions, including

- `"gaussian"` (what we've called the normal distribution)
- `"logistic"`
- `"lognormal"` also called `"loggaussian"`
- `"loglogistic"`
- `"rayleigh"`

---

2 Note that the Weibull distribution has a scale parameter θ and a shape parameter κ. The logarithm of the failure time has an SEV distribution where the location parameter is μ and the scale parameter is σ. These parameters are functionally related, but the scale parameter of the Weibull is related to the location parameter of the SEV ($\mu = \log\theta$) and the shape parameter of the Weibull is related to the scale parameter of the SEV ($\sigma = 1/\kappa$). Care must be taken when reading the output from any software system, especially when the term "scale" parameter is used, since this could refer either to θ or σ.

The `flexsurv` package can handle additional distributions, including the gamma distribution (`"gamma"`) and the generalized gamma distribution (`"gengamma"`). These distributions can be fit using the `flexsurvreg` function, which is analogous to the `survreg` function in the survival function.

To illustrate lifetime regression using R's `survival` package, let's look at the nickel super alloy data discussed in 6.10. In the following code, we load in the data file, extract the variables, and create the survival object.

```
Nickel = read.csv("Nickel.csv")
kCycles = Nickel$kCycles
Pseudostress = Nickel$Pseudostress
Cens = Nickel$Cens
kCyclesSurv = Surv( kCycles , Cens )
```

We could plot the data in a scatter plot using the following commands:

```
windows( 9 , 7 )
plot( Pseudostress[Cens==1] , kCycles[Cens==1] , pch=19 ,
      xlab="Pseudostress" , ylab="Lifetime in Kilocycles" )
  points( Pseudostress[Cens==0] , kCycles[Cens==0] , pch=1 ,
          col="gray" )
  for ( i in 1:length( Pseudostress[Cens==0] ) )
  {
   lines( rep( Pseudostress[Cens==0][i] , 2 ) ,
          c( kCycles[Cens==0][i] , rep(9999999) ) , col="gray" )
  }
```

The plot is shown in Figure A.4.

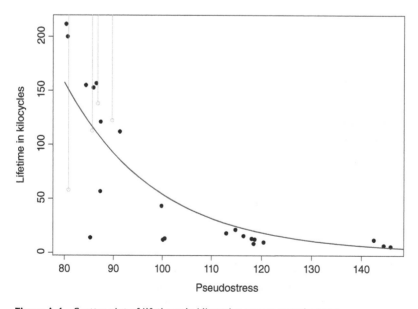

Figure A.4 Scatter plot of lifetimes in kilocycles versus pseudostress.

This data set has four censored observations, indicating that we do not know the exact value of the *y*-coordinate, kCycles for these four values, although a lower bound is given by the censoring time. This uncertainty is indicated by the gray line extending upward from the censoring time. Failure times are indicated by a solid circle, and censored times are indicated by an open circle. As the pseudostress increases, the lifetimes tend to get shorter but not in a linear relationship. Even with a first-order model where kCycles ~ Pseudostress would not produce a linear model on the original scales because the relationship between the Pseudostress and the characteristic life in kilocycles is

$$\theta = \exp(\beta_0 + \beta_1 \text{ Pseudostress})$$

which is nonlinear.

We can now apply the survreg function, but this time we put the response variable to the left of the formula operator (~) and we put the predictor variable Pseudostress to the right. (If there were multiple predictors we would enter each one separated by a plus sign.) We assign the results of the lifetime regression in the variable kCyclesWeibulReg, and then we give a summary:

```
kCyclesWeibullReg = survreg( kCyclesSurv ~ Pseudostress ,
                               dist="weibull" )
summary( kCyclesWeibullReg )
```

This produces the lifetime regression output

```
Call:
survreg(formula = kCyclesSurv ~ Pseudostress, dist = "weibull")
                Value Std. Error      z       p
(Intercept)   9.45774    0.45980  20.57 <2e-16
Pseudostress -0.05348    0.00426 -12.54 <2e-16
Log(scale)   -0.68306    0.17633  -3.87 0.00011

Scale= 0.505

Weibull distribution
Loglik(model)= -99.5   Loglik(intercept only)= -118.4
        Chisq= 37.76 on 1 degrees of freedom, p= 8e-10
Number of Newton-Raphson Iterations: 7
n= 26
```

From this output, notice the regression-like output in the table, where the first column names the parameter, (Intercept), Pseudostress, or Log(scale), the second column gives the point estimate, the third gives the standard error, the fourth gives the *z* statistic, and the last column gives the *P*-value for testing whether the true underlying parameter is zero. In this case we see that the very low *P*-value for Pseudostress indicates that there is strong evidence that the true coefficient is nonzero. In other words, there is strong evidence that Pseudostress affects the characteristic life, but this was to be expected.

We can extract estimates of the parameters β_0, β_1, and σ (the scale parameter of the SEV distribution), which are contained in the object kCyclesWeibullReg:

```
beta0hat = kCyclesWeibullReg$coef[1]
beta1hat = kCyclesWeibullReg$coef[2]
kappahat = 1/kCyclesWeibullReg$scale
```

We can then compute the estimated characteristic life over a grid of Pseudostress:

```
PS = seq(80,150,1)
thetahat = exp( beta0hat + beta1hat*PS )
```

The estimated expected life curve $E(T|\hat{\kappa}, \hat{\beta}_0, \hat{\beta}_1) = \hat{\theta}\Gamma(1 + 1/\hat{\kappa})$ can then be added to the scatter plot defined previously:

```
windows ( 9 , 7 )
plot ( Pseudostress[Cens==1] , kCycles[Cens==1] , pch=19 ,
      xlab="Pseudostress" , ylab="Lifetime in Kilocycles" )
  points ( Pseudostress[Cens==0] , kCycles[Cens==0] , pch=1,
          col="gray" )
  for ( i in 1:length( Pseudostress[Cens==0] ) )
  {
    lines ( rep( Pseudostress[Cens==0][i] , 2 ) ,
            c( kCycles[Cens==0][i], rep(9999999) ) , col="gray" )
  }
  lines ( PS , thetahat*gamma(1+1/kappahat) , lwd=2 , col="blue" )
```

The expected life curve does not seem to fit well. It overpredicts all outcomes between pseudostress levels 110 and 130, and then it underpredicts all values over 140. This suggests that a second-order model might be an improvement. The following code defines a new variable Pseudostress2 to be the square of Pseudostress, runs the survival regression, extracts the parameter estimates from the kCyclesWeibullReg object, and computes the estimate of the characteristic life θ over a grid of Pseudostress values:

```
Pseudostress2 = Pseudostress^2
kCyclesWeibullReg = survreg( kCyclesSurv ~ Pseudostress +
                                           Pseudostress2 ,
                           dist="weibull" )
summary( kCyclesWeibullReg )

beta0hat = kCyclesWeibullReg$coef[1]
beta1hat = kCyclesWeibullReg$coef[2]
beta2hat = kCyclesWeibullReg$coef[3]
kappahat = 1/kCyclesWeibullReg$scale
PS = seq(80,150,1)
thetahat = exp( beta0hat + beta1hat*PS + beta2hat*PS^2 )
```

Note that you must define a new variable for the square of Pseudostress; the command

```
kCyclesWeibullReg = survreg( kCyclesSurv ~ Pseudostress +
                                           Pseudostress^2,
                           dist="weibull" )
```

does not work. The output from summary(kCyclesWeibullReg) is shown below:

```
Call:
survreg(formula = kCyclesSurv ~ Pseudostress + Pseudostress2,
   dist = "weibull" )
                 Value Std. Error    z       p
(Intercept)   20.805347   2.607281  7.98 1.5e-15
```

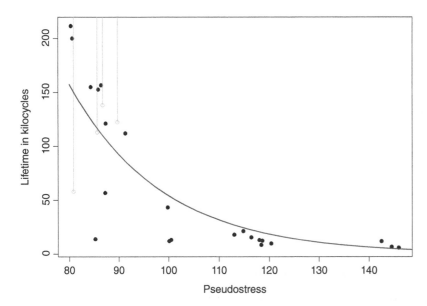

Figure A.5 Scatter plot of lifetimes in kilocycles versus pseudostress with the estimated expected lifetime curve from a first-order model.

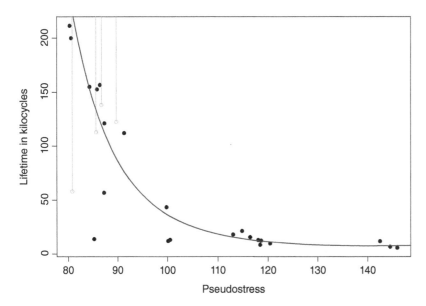

Figure A.6 Scatter plot of lifetimes in kilocycles versus pseudostress with the estimated expected lifetime curve from a second-order model.

```
Pseudostress   -0.264811    0.047931 -5.52 3.3e-08
Pseudostress2  0.000939     0.000213  4.40 1.1e-05
Log(scale)     -1.015148    0.181386 -5.60 2.2e-08

Scale= 0.362
Weibull distribution
Loglik(model)= -92.9   Loglik(intercept only)= -118.4
        Chisq= 50.98 on 2 degrees of freedom, p= 8.5e-12
Number of Newton-Raphson Iterations: 8
n= 26
```

We see that both the variables `Pseudostress` and `Pseudostress2` are significant, with very low *P*-values. We can extract the point estimates and estimate θ over a grid of `Pseudostress` with the following code:

```
beta0hat = kCyclesWeibullReg$coef[1]
beta1hat = kCyclesWeibullReg$coef[2]
beta2hat = kCyclesWeibullReg$coef[3]
kappahat = 1/kCyclesWeibullReg$scale
PS = seq(80,150,1)
thetahat = exp( beta0hat + beta1hat*PS + beta2hat*PS^2 )
```

With the variables defined this way (rather than with the result from the first-order model), similar code that was used to produce Figure A.5 can be used to produce the curve shown in Figure A.6. The second-order model seems to fit the data much better than the first-order model.

B

Design of Experiments using JMP

This appendix provides a brief overview of the capabilities of the JMP software to facilitate designing and analyzing an experiment. The JMP website (www.jmp.com) has many additional, more detailed resources to help a user become familiar with the software. The JMP menu has a tab labeled "DOE" that is the entry point for JMP's experimental design tools.

When clicking on the DOE tab, several options for creating experimental design are opened. The first option is the Custom Designer. The Custom Designer creates optimal designs using several criteria, including the D and I criteria discussed in this book. Generally, if the experimenter is considering a first-order model the Custom Designer defaults to the D-criterion and if the model of interest is second order the Custom Designer defaults to the I-criterion. This can be changed by the user. To update the optimality criteria, launch the Custom Designer by DOE → Custom Design, then select the triangle next to Custom Design, and scroll down to Optimality Criterion to select the appropriate option.

Another option is Definitive Screening. These are the definitive screening designs (DSD) discussed in Chapter 5. This option will construct the basic DSD for continuous three-level factors. It will also allow two-level categorical factors to be added to the basic DSD, and it will allow the user to incorporate the blocking feature for DSDs. There is also an option to analyze a DSD that has already been run. This feature utilizes a specialized analysis algorithm developed for the DSD.

The Classical option opens up a variety of standard designs, many of which have been discussed in Chapters 4 and 5. Screening designs include two-level fractional factorial designs for continuous factors and categorical factors with an arbitrary number of levels. Response Surface Designs include several variations of the central composite design and the Box–Behnken design. There are also options for full factorial designs, designs for experiments with mixtures, and Taguchi orthogonal arrays. These latter two types of designs ae not discussed in this book.

The Design Diagnostics option provides capability to evaluate designs and to compare two or more candidate designs for a specific application. There are also options for Augment Design that provide methodology for adding runs to an existing design using an optimal augmentation criterion and for Consumer Studies that provides discrete choice experiments.

The special purpose option includes designs for accelerated life test experiments and a capability to design experiments for fitting nonlinear models. Since the generalized linear model involves a nonlinear

Design of Experiments for Reliability Achievement, First Edition.
Steven E. Rigdon, Rong Pan, Douglas C. Montgomery, and Laura J. Freeman.
© 2022 John Wiley & Sons, Inc. Published 2022 by John Wiley & Sons, Inc.
Companion website: www.wiley.com/go/rigdon/designexperiments

model, this design tool is useful for designing experiments where several types of non-normal responses are anticipated.

Designed experiments can be analyzed in JMP using the tools available in the Analyze tab. The Fit Model option is useful for factorial and fractional factorial designs and for most response surface designs. The Reliability and Survival option provides several specialized techniques for the analysis of reliability data.

Figure B.1 shows an example of using JMP to create a design, using the six-factor optimal design problem in Chapter 4. Recall that this example involves an experiment with six continuous two-level factors, and we are unsure about the two-factor interactions as well as the main effects. A design that allows us to estimate all of 21 of these effects is a one-half fraction of the 2^6, or a 2^{6-1} fractional factorial with 32 runs. However, the experimenter cannot afford to do 32 runs. The maximum possible number of runs is 28. The smallest possible design that could estimate all of these effects would have 22 runs. In Chapter 4, the experimenters decided to use a *D*-optimal design with 28 runs.

The middle panel in Figure B.1 shows how to use the JMP Custom Designer tool to construct a *D*-optimal design with 24 runs. Other design tools in the DOE platform are used similarly. The following are step-by-step instructions to construct the *D*-optimal design. At the bottom of the instructions there is an illustration of the input display from JMP.

1. Select DOE → Custom Design.
2. Under Responses, specify the response(s):
   - Double-click to rename the response (default is Y).
   - Change the response goal or desired value (default is Maximize).
   - Click Add Response to add additional responses.
3. Under Factors, specify the experimental factors:
   - Click Add Factor, and select the factor type and number of levels. To add several factors of the same type and number of levels use Add N Factors (enter a number) and click Add Factor.
   - Double-click to rename the factors.
   - Enter the factor values (the experimental settings). You can use the default levels if you wish. For example, −1 and +1 for the low and high levels of a 2-level continuous factor.
   - If it is challenging to fully randomize an experiment you can use the platform to generate a split-plot or a split-split plot design. To do this, Change Easy under Changes to Hard or Very Hard. This will generate a split-plot or split-split plot design.
4. Click Continue.
5. Under Model, specify the statistical model to be estimated:
   - To add all interaction or power terms up to a given degree, click the corresponding button.
   - To add terms needed to perform a response surface analysis, click RSM (or Scheffe Cubic for mixture designs).
   - To add specific power terms, highlight one or more factors in the Factors window and click Powers. To add an interaction term, highlight one or more factors in the Factors window and select the factors in the Model window that you wish to cross with those chosen factors.
   - To remove a term, highlight it and click Remove Term.
   - To reduce the number of runs needed (at the expense of effect aliasing), click Necessary for a term (under Estimability) and change to If Possible.

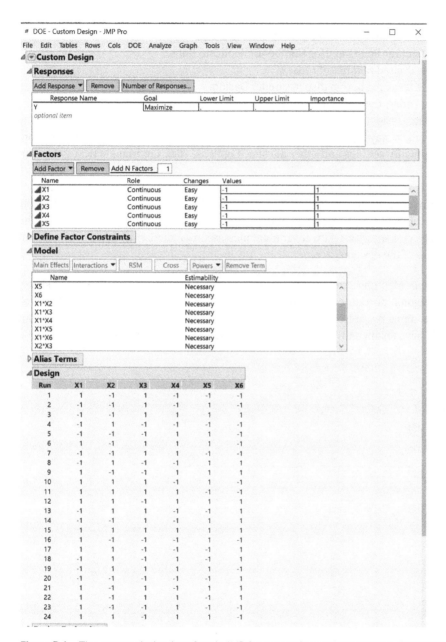

Figure B.1 The custom design interface in JMP for generating optimal designs. Source: SAS Institute Inc.

6. Under Design Generation, fine-tune the design (as needed):
   - Specify the block size or number of whole plots (with Hard to change factors).
   - Enter the Number of Center Points and/or Number of Replicate Runs.
   - Select (or specify) the desired number of runs.
7. Click Make Design. The resulting design displays under Design.
8. Select the desired Run Order, then click Make Table to generate the design table (or Back) to make changes. Model, Evaluate Design, and DOE Dialog scripts are saved to the data table (top left), and the design specification window stays open to change or regenerate the design if needed. Notes: For more options, including optimality settings and other advanced options, click the triangle next to Custom Design.
9. The Design Evaluation panel houses a variety of diagnostics. Figure B.2 shows the prediction variance profile plots and the fraction of design space for this design. The prediction variance profile plots show that the variance of the predicted response at the center of the design region is 0.050642 and the fraction of design space plot shows a relatively gradual increase in prediction variance over a region that includes about 80% of the design space.

JMP also produces a Color Map on Correlations between every pair of model terms. Figure B.3 shows the shading in the off-diagonal portion of the matrix indicates the size of the correlation coefficients. For this design, the correlations range from 0 to approximately 0.175. The design is not orthogonal; an orthogonal design would have resulted in all off-diagonal elements of the matrix being 0. However, these

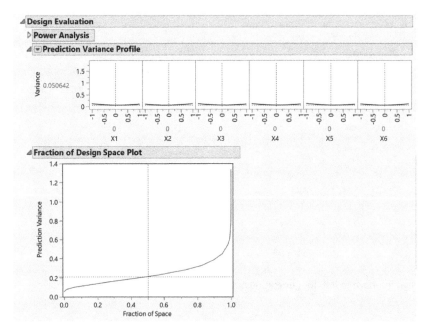

Figure B.2 Evaluation of the 24 run *D*-optimal design using prediction as a function of the design factors and across the full design space. Source: SAS Institute Inc.

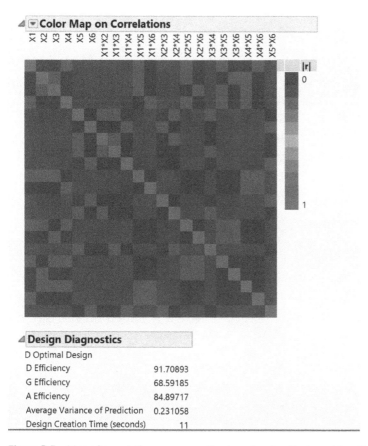

Figure B.3 Map of correlations among estimates of main effects and two-factor interactions. Source: SAS Institute Inc.

are very small correlations and unlikely to cause significant problems in obtaining reliable estimates of all 6 main effects and the 15 two-factor interactions.

The Design Diagnostics display in the following text shows the D, A, and I efficiency of the 24-run design. Notice that the D-efficiency is approximately 91.7. If the design had been orthogonal the D-efficiency would have been 100%. There is no 24-run design for this model that is orthogonal.

C

The Expected Fisher Information Matrix

If the lifetime variable T follows a log-location-scale distribution, then the natural logarithm of this variable follows a location-scale distribution; i.e.

$$Y = \log T \sim F(y; \mu, \sigma).$$

Here, μ is the location parameter and σ the scale parameter of the location-scale distribution. We can standardize this variable by defining $Z = (Y - \mu)/\sigma$. The CDF of Y is then given by

$$F(y) = P(Y < y) = P\left(Z < \frac{y - \mu}{\sigma}\right) = P(Z < z) = \Phi(z) = \Phi\left(\frac{y - \mu}{\sigma}\right),$$

and the density function is

$$f(y) = \frac{1}{\sigma}\phi(z) = \frac{1}{\sigma}\phi\left(\frac{y - \mu}{\sigma}\right)$$

where $\phi(z) = \frac{d}{dz}\Phi(Z)$. Note that, for the log-normal lifetime distribution, $\Phi(z)$ is the standard normal CDF, and for the Weibull lifetime distribution, $\Phi(z)$ is the standard smallest extreme value CDF.

Assume the experiment will test all specimens to failure, so there are no censored observations. The likelihood contribution of specimen i is given by its density function, i.e. $\frac{1}{\sigma}\phi(z_i)$. Its log-likelihood contribution is given by

$$l_i(\mu, \sigma) = -\log \sigma + \log \phi(z_i).$$

The total log-likelihood function of n independent specimens is

$$\ell(\mu, \sigma) = \sum_{i=1}^{n} l_i(\mu, \sigma).$$

As these specimens are i.i.d. and their random variable expectations are identical, we will drop the subscript i in the following derivation; i.e.

$$E[\ell(\mu, \sigma)] = nE[l(\mu, \sigma)] = nE\left[-\log \sigma + \log \phi(Z)\right] = -n \log \sigma - E\left[\log \phi(Z)\right].$$

Here $l(\mu, \sigma) = -\log \sigma + \phi(Z)$, which has the same expectation across all i. The first derivatives of $l(\mu, \sigma)$ with respect to its location and scale parameters are

$$\frac{\partial l}{\partial \mu} = \frac{\phi'(z)}{\phi(z)}\frac{\partial z}{\partial \mu} = -\frac{1}{\sigma}\frac{\phi'(z)}{\phi(z)},$$

Design of Experiments for Reliability Achievement, First Edition.
Steven E. Rigdon, Rong Pan, Douglas C. Montgomery, and Laura J. Freeman.
© 2022 John Wiley & Sons, Inc. Published 2022 by John Wiley & Sons, Inc.
Companion website: www.wiley.com/go/rigdon/designexperiments

and

$$\frac{\partial l}{\partial \sigma} = -\frac{1}{\sigma} + \frac{\phi'(z)}{\phi(z)} \frac{\partial z}{\partial \sigma} = -\frac{1}{\sigma} - \frac{1}{\sigma} \frac{z\phi'(z)}{\phi(z)},$$

where $\phi'(z) = \frac{d}{dz}\phi(z)$. Keep in mind that z is a function of y; specifically, $z = (y - \mu)/\sigma$ so that

$$\frac{\partial z}{\partial \mu} = -\frac{1}{\sigma}$$

$$\frac{\partial z}{\partial \sigma} = -\frac{y - \mu}{\sigma^2} = -\frac{z}{\sigma}.$$

Taking the second derivatives gives

$$\frac{\partial^2 l}{\partial \mu^2} = -\frac{1}{\sigma} \frac{\phi''(z)\phi(z) - [\phi'(z)]^2}{\phi^2(z)} \frac{\partial z}{\partial \mu} = \frac{1}{\sigma^2}\left[\frac{\phi''(z)}{\phi(z)} - \frac{[\phi'(z)]^2}{\phi^2(z)}\right],$$

$$\frac{\partial^2 l}{\partial \mu \, \partial \sigma} = \frac{1}{\sigma^2}\frac{\phi'(z)}{\phi(z)} - \frac{1}{\sigma}\frac{\phi''(z)\phi(z) - [\phi'(z)]^2}{\phi^2(z)}\frac{\partial z}{\partial \sigma}$$

$$= \frac{1}{\sigma^2}\left[\frac{\phi'(z)}{\phi(z)} + \frac{z\phi''(z)}{\phi(z)} - \frac{z[\phi'(z)]^2}{[\phi(z)]^2}\right],$$

and

$$\frac{\partial^2 l}{\partial \sigma^2} = \frac{1}{\sigma^2} + \frac{1}{\sigma^2}\frac{z\phi'(z)}{\phi(z)} - \frac{1}{\sigma}\frac{\phi'(z)\phi(z) + z\phi''(z)\phi(z) - z[\phi'(z)]^2}{[\phi(z)]^2}\frac{\partial z}{\partial \sigma}$$

$$= \frac{1}{\sigma^2}\left[1 + \frac{2z\phi'(z)}{\phi(z)} + \frac{z^2\phi''(z)}{\phi(z)} - \frac{z^2[\phi'(z)]^2}{[\phi(z)]^2}\right].$$

From the probabilistic distribution theory (Casella and Berger 2002), it is known that the expectation of the first derivative of the log-likelihood function is equal to zero; i.e. $E[\frac{d}{d\theta}\ell(X;\theta)] = 0$. In addition, for an i.i.d. sample, the expectation of the negative second derivative of the log-likelihood function is proportional to the expectation of the square of first derivative of log density function; i.e.

$$E\left[-\frac{d^2}{d\theta^2}\ell(X|\theta)\right] = nE\left[\left(\frac{d}{d\theta}\ell(X|\theta)\right)^2\right].$$

The entries of expected Fisher information matrix are the negative expected second derivatives of the log likelihood function. In terms of a sample of log-lifetime variable Y, these entries are

$$I_{11} = E\left[-\frac{\partial^2}{\partial \mu^2}\ell(Y|\mu,\sigma)\right] = \frac{n}{\sigma^2}E\left[\left(\frac{\phi'(Z)}{\phi(Z)}\right)^2\right],$$

and

$$I_{22} = E\left[-\frac{\partial^2}{\partial \sigma^2}\ell(Y|\mu,\sigma)\right] = \frac{n}{\sigma^2}E\left[1 + 2Z\frac{\phi'(Z)}{\phi(Z)} + Z^2\left(\frac{\phi'(Z)}{\phi(Z)}\right)^2\right]$$

$$= \frac{n}{\sigma^2}E\left[Z^2\left(\frac{\phi'(Z)}{\phi(Z)}\right)^2\right] - \frac{n}{\sigma^2}.$$

The second equation of the formula above utilizes the identity of $-E\left[\frac{Z\phi'(Z)}{\phi(Z)}\right] = 1$. Furthermore, it can be shown that $E\left[\frac{\phi'(Z)}{\phi(Z)}\right] = 0$ and $E\left[\frac{Z\phi''(Z)}{\phi(Z)}\right] = 0$. Thus, the expected negative mixed partial derivative is given by

$$I_{12} = E\left[-\frac{\partial^2}{\partial\mu\,\partial\sigma}\,\ell(Y|\mu,\sigma)\right] = \frac{n}{\sigma^2}\,E\left[Z\left(\frac{\phi'(Z)}{\phi(Z)}\right)^2\right].$$

Finally, the expected Fisher information matrix is

$$\mathbf{I} = \begin{bmatrix} I_{11} & I_{12} \\ I_{12} & I_{22} \end{bmatrix}.$$

The expected variance-covariance matrix of parameter estimators is the inverse of expected Fisher information matrix, i.e., $\mathbf{I}^{-1}$.

C.1 Lognormal Distribution

Let the lifetime variable T follows a lognormal distribution, then its natural logarithm transformation Y follows a normal distribution; i.e. $Y \sim N(\mu, \sigma^2)$. The standard normal density function is given by

$$\phi(z) = \frac{1}{2}e^{-z^2/2},$$

and its first derivative is given by

$$\phi'(z) = -\frac{z}{2}e^{-z^2/2}.$$

Now we have $\phi'(z)/\phi(z) = -z$, so

$$E\left[\left(\frac{\phi'(Z)}{\phi(Z)}\right)^2\right] = E[Z^2] = 1,$$

$$E\left[Z\left(\frac{\phi'(Z)}{\phi(Z)}\right)^2\right] = E[Z^3] = 0,$$

and

$$E\left[Z^2\left(\frac{\phi'(Z)}{\phi(Z)}\right)^2\right] = E[Z^4] = 3.$$

Finally, the expected Fisher information matrix is given by

$$\mathbf{I} = \frac{n}{\sigma^2}\begin{bmatrix} 1 & 0 \\ 0 & 2 \end{bmatrix}.$$

C.2 Weibull Distribution

If the lifetime variable T follows a Weibull distribution, then its natural logarithm transformation Y follows a smallest extreme value distribution; i.e., $Y \sim \text{SEV}(\mu, \sigma)$, where $\mu = \log\theta$ and θ is the characteristic life of the Weibull distribution, and $\sigma = 1/\kappa$, the inverse of the shape parameter. The standard SEV density function is given by

$$\phi(z) = e^{z - e^z},$$

and its first derivative is given by

$$\phi'(z) = e^{z - e^z}(1 - e^z).$$

Now we have $\phi'(z)/\phi(z) = 1 - e^z$, so

$$E\left[\left(\frac{\phi'(Z)}{\phi(Z)}\right)^2\right] = E\left[(1 - e^Z)^2\right] = 1,$$

$$E\left[Z\left(\frac{\phi'(Z)}{\phi(Z)}\right)^2\right] = E\left[Z(1 - e^Z)^2\right] = 0.422784,$$

and

$$E\left[Z^2\left(\frac{\phi'(Z)}{\phi(Z)}\right)^2\right] = E[Z^2(1 - e^Z)^2] = 2.82368.$$

The integrations needed for these expectations are computed by the *Mathematica* online calculator (https://www.wolframalpha.com/examples/mathematics/calculus-and-analysis/).

Finally, the expected Fisher information matrix is given by

$$\mathbf{I} = \frac{n}{\sigma^2}\begin{bmatrix} 1 & 0.422784 \\ 0.422784 & 1.82368 \end{bmatrix}.$$

In a lifetime regression where only the lifetime scale parameter is affected by covariates, for a single specimen i, the common regression model is given by

$$Y_i = \log T_i \sim F(y; \mu_i, \sigma)$$

where

$$\mu_i = \beta_0 + \mathbf{x}_i^T \boldsymbol{\beta}.$$

Here β_0 is the intercept parameter and $\boldsymbol{\beta} = [\beta_1, \ldots, \beta_k]^T$ is the vector of covariate effects of the covariate vector $\mathbf{x} = [x_1, \ldots, x_k]^T$.

If, for example, there is only one covariate, then

$$\mu_i = \beta_0 + \beta_1 x_i.$$

To obtain the expected Fisher information for the coefficients β_0 and β_1, we need to attach the derivative of μ with respect to these coefficients to the previous formulas. Again, using the standardized variable Z, the following can be derived:

$$\frac{\partial \ell_i}{\partial \beta_0} = \frac{\phi'(z_i)}{\phi(z_i)} \frac{\partial z_i}{\partial \mu_i} \frac{\partial \mu_i}{\partial \beta_0} = -\frac{1}{\sigma} \frac{\phi'(z_i)}{\phi(z_i)},$$

$$\frac{\partial \ell_i}{\partial \beta_1} = \frac{\phi'(z_i)}{\phi(z_i)} \frac{\partial z_i}{\partial \mu_i} \frac{\partial \mu_i}{\partial \beta_1} = -\frac{1}{\sigma} \frac{\phi'(z_i)}{\phi(z_i)} x_i,$$

and

$$\frac{\partial \ell_i}{\partial \sigma} = -\frac{1}{\sigma} - \frac{1}{\sigma} \frac{z_i \phi'(z_i)}{\phi(z_i)}.$$

The mixed partial derivatives are given by

$$\frac{\partial^2 \ell_i}{\partial \beta_0 \partial \beta_1} = \frac{1}{\sigma^2} \left[\frac{\phi''(z_i)}{\phi(z_i)} - \left(\frac{\phi'(z_i)}{\phi(z_i)} \right)^2 \right] x_i,$$

$$\frac{\partial^2 \ell_i}{\partial \beta_0 \partial \sigma} = \frac{1}{\sigma^2} \left[\frac{\phi'(z_i)}{\phi(z_i)} + z_i \frac{\phi''(z_i)}{\phi(z_i)} - z_i \left(\frac{\phi'(z_i)}{\phi(z_i)} \right)^2 \right],$$

and

$$\frac{\partial^2 \ell_i}{\partial \beta_1 \partial \sigma} = \frac{1}{\sigma^2} \left[\frac{\phi'(z_i)}{\phi(z_i)} + z_i \frac{\phi''(z_i)}{\phi(z_i)} - z_i \left(\frac{\phi'(z_i)}{\phi(z_i)} \right)^2 \right] x_i.$$

Similarly, we compute the entries of expected Fisher information matrix as

$$I_{11} = E\left[-\frac{\partial^2}{\partial \beta_0^2} \ell(Y|\beta_0, \beta_1, \sigma) \right] = \frac{n}{\sigma^2} E\left[\left(\frac{\phi'(Z)}{\phi(Z)} \right)^2 \right],$$

$$I_{22} = E\left[-\frac{\partial^2}{\partial \beta_1^2} \ell(Y|\beta_0, \beta_1, \sigma) \right] = \frac{n}{\sigma^2} E\left[\left(\frac{\phi'(Z)}{\phi(Z)} \right)^2 \right] \sum_{i=1}^{n} x_i^2,$$

$$I_{12} = E\left[-\frac{\partial^2}{\partial \beta_0 \partial \beta_1} \ell(Y|\beta_0, \beta_1, \sigma) \right] = \frac{n}{\sigma^2} E\left[\left(\frac{\phi'(Z)}{\phi(Z)} \right)^2 \right] \sum_{i=1}^{n} x_i,$$

$$I_{33} = E\left[-\frac{\partial^2}{\partial \sigma^2} \ell(Y|\beta_0, \beta_1, \sigma) \right] = \frac{n}{\sigma^2} E\left[Z^2 \left(\frac{\phi'(Z)}{\phi(Z)} \right)^2 \right] - \frac{n}{\sigma^2},$$

$$I_{13} = E\left[-\frac{\partial^2}{\partial \beta_0 \partial \sigma} \ell(Y|\beta_0, \beta_1, \sigma) \right] = \frac{n}{\sigma^2} E\left[Z \left(\frac{\phi'(Z)}{\phi(Z)} \right)^2 \right],$$

$$I_{23} = E\left[-\frac{\partial^2}{\partial \beta_1 \partial \sigma} \ell(Y|\beta_0, \beta_1, \sigma) \right] = \frac{n}{\sigma^2} E\left[Z \left(\frac{\phi'(Z)}{\phi(Z)} \right)^2 \right] \sum_{i=1}^{n} x_i.$$

C.3 Lognormal Distribution

Following the previous lognormal distribution example, let the parameter μ be a linear function of the covariate x. Then the expected Fisher information matrix is given by

$$\mathbf{I}(\mathbf{X}) = \frac{n}{\sigma^2} \begin{bmatrix} 1 & \sum_{i=1}^{n} x_i & 0 \\ \sum_{i=1}^{n} x_i & \sum_{i=1}^{n} x_i^2 & 0 \\ 0 & 0 & 2 \end{bmatrix}.$$

C.4 Weibull Distribution

Following the previous Weibull distribution example, let the parameter μ be a linear function of the covariate x. Then the expected Fisher information matrix is given by

$$\mathbf{I(X)} = \frac{n}{\sigma^2} \begin{bmatrix} 1 & \sum_{i=1}^{n} x_i & 0.422784 \\ \sum_{i=1}^{n} x_i & \sum_{i=1}^{n} x_i^2 & 0.422784 \sum_{i=1}^{n} x_i \\ 0.422784 & 0.422784 \sum_{i=1}^{n} x_i & 1.82368 \end{bmatrix}.$$

The development of the information matrices in this appendix has so far assumed that there is no censoring. In practice, censoring is common and methods are needed to obtain the information matrix for the parameters. These methods are beyond the scope of this book, but the interested reader is referred to Escobar and Meeker (1986, 1994, 1998).

D

Data Sets

Design of Experiments for Reliability Achievement, First Edition.
Steven E. Rigdon, Rong Pan, Douglas C. Montgomery, and Laura J. Freeman.
© 2022 John Wiley & Sons, Inc. Published 2022 by John Wiley & Sons, Inc.
Companion website: www.wiley.com/go/rigdon/designexperiments

Table D.1 Electrical appliance lifetime data.

Lifetime in cycles	Failure code
11	1
2223	9
4329	9
3112	9
13403	0
6367	0
2451	5
381	6
1062	5
1594	2
329	6
2327	6
958	10
7846	9
170	6
3059	6
3504	9
2568	9
2471	9
3214	9
3034	9
2694	9
49	15
6976	9
35	15
2400	9
1167	9
2831	2
2702	10
708	6
1925	9
1990	9
2551	9
2761	6
2565	0
3478	9

Source: Data from Nelson (1970).

Table D.2 Glass capacitor accelerated life testing data.

Temperature	Voltage	T	Failure	Temperature	Voltage	T	Failure
170	200	439	1	180	200	959	1
170	200	904	1	180	200	1065	1
170	200	1092	1	180	200	1065	1
170	200	1105	1	180	200	1087	1
170	200	1105	0	180	200	1087	0
170	200	1105	0	180	200	1087	0
170	200	1105	0	180	200	1087	0
170	200	1105	0	180	200	1087	0
170	250	572	1	180	250	216	1
170	250	690	1	180	250	315	1
170	250	904	1	180	250	455	1
170	250	1090	1	180	250	473	1
170	250	1090	0	180	250	473	0
170	250	1090	0	180	250	473	0
170	250	1090	0	180	250	473	0
170	250	1090	0	180	250	473	0
170	300	315	1	180	300	241	1
170	300	315	1	180	300	315	1
170	300	439	1	180	300	332	1
170	300	628	1	180	300	380	1
170	300	628	0	180	300	380	0
170	300	628	0	180	300	380	0
170	300	628	0	180	300	380	0
170	300	628	0	180	300	380	0
170	350	258	1	180	350	241	1
170	350	258	1	180	350	241	1
170	350	347	1	180	350	435	1
170	350	588	1	180	350	455	1
170	350	588	0	180	350	455	0
170	350	588	0	180	350	455	0
170	350	588	0	180	350	455	0
170	350	588	0	180	350	455	0

Source: Data from Zelen (1959).

Table D.3 Solder joint survival times in cycles for 3 PCB types and 3 levels of temperatures.

Temp. (°C)	Cycles to failure		
	PCB type 1	PCB type 2	PCB type 3
20	218	685	791
20	265	899	1140
20	279	1020	1169
20	282	1082	1217
20	336	1207	1267
20	469	1396	1409
20	496	1411	1447
20	507	1417	1476
20	685	1470	1488
20	685	1999	1545
60	185	593	704
60	242	722	827
60	254	859	925
60	280	863	930
60	305	956	984
60	353	1017	984
60	381	1038	1006
60	504	1107	1166
60	556	1264	1258
60	697	1362	1362
100	7	188	98
100	46	248	154
100	52	266	193
100	82	269	230
100	90	291	239
100	100	345	270
100	101	352	295
100	105	381	332
100	112	385	491
100	151	445	532

Source: Based on Lau et al. (1988).

Table D.4 Clutch spring durability testing data.

A	BC	F	G	D	E	Left	Right	Censoring
1	1	1	1	90	200	1	2	Interval
1	1	1	1	90	200	1	2	Interval
1	1	1	1	90	200	1	2	Interval
2	2	2	2	90	200	4	5	Interval
2	2	2	2	90	200	5	6	Interval
2	2	2	2	90	200	11	.	Right
3	3	3	3	90	200	2	3	Interval
3	3	3	3	90	200	2	3	Interval
3	3	3	3	90	200	11	.	Right
1	1	2	3	90	170	2	3	Interval
1	1	2	3	90	170	3	4	Interval
1	1	2	3	90	170	3	4	Interval
2	2	3	1	90	170	5	6	Interval
2	2	3	1	90	170	11	.	Right
2	2	3	1	90	170	11	.	Right
3	3	1	2	90	170	1	2	Interval
3	3	1	2	90	170	1	2	Interval
3	3	1	2	90	170	1	2	Interval
1	1	3	2	90	140	1	2	Interval
1	1	3	2	90	140	1	2	Interval
1	1	3	2	90	140	3	4	Interval
2	2	1	3	90	140	1	2	Interval
2	2	1	3	90	140	1	2	Interval
2	2	1	3	90	140	2	3	Interval
3	3	2	1	90	140	3	4	Interval
3	3	2	1	90	140	3	4	Interval
3	3	2	1	90	140	4	5	Interval
1	2	1	1	65	200	1	2	Interval
1	2	1	1	65	200	1	2	Interval
1	2	1	1	65	200	2	3	Interval
2	3	2	2	65	200	11	.	Right
2	3	2	2	65	200	11	.	Right
2	3	2	2	65	200	11	.	Right
3	1	3	3	65	200	6	7	Interval

(Continued)

Table D.4 (Continued)

A	BC	F	G	D	E	Left	Right	Censoring
3	1	3	3	65	200	11	.	Right
3	1	3	3	65	200	11	.	Right
1	2	2	3	65	170	11	.	Right
1	2	2	3	65	170	11	.	Right
1	2	2	3	65	170	11	.	Right
2	3	3	1	65	170	2	3	Interval
2	3	3	1	65	170	2	3	Interval
2	3	3	1	65	170	2	3	Interval
3	1	1	2	65	170	1	2	Interval
3	1	1	2	65	170	2	3	Interval
3	1	1	2	65	170	2	3	Interval
1	2	3	2	65	140	2	3	Interval
1	2	3	2	65	140	3	4	Interval
1	2	3	2	65	140	4	5	Interval
2	3	1	3	65	140	2	3	Interval
2	3	1	3	65	140	2	3	Interval
2	3	1	3	65	140	2	3	Interval
3	1	2	1	65	140	11	.	Right
3	1	2	1	65	140	11	.	Right
3	1	2	1	65	140	11	.	Right
1	3	1	1	40	200	3	4	Interval
1	3	1	1	40	200	4	5	Interval
1	3	1	1	40	200	4	5	Interval
2	1	2	2	40	200	11	.	Right
2	1	2	2	40	200	11	.	Right
2	1	2	2	40	200	11	.	Right
3	2	3	3	40	200	11	.	Right
3	2	3	3	40	200	11	.	Right
3	2	3	3	40	200	11	.	Right
1	3	2	3	40	170	11	.	Right
1	3	2	3	40	170	11	.	Right
1	3	2	3	40	170	11	.	Right
2	1	3	1	40	170	11	.	Right
2	1	3	1	40	170	11	.	Right
2	1	3	1	40	170	11	.	Right

(Continued)

Table D.4 (Continued)

A	BC	F	G	D	E	Left	Right	Censoring
3	2	1	2	40	170	5	6	Interval
3	2	1	2	40	170	11	.	Right
3	2	1	2	40	170	11	.	Right
1	3	3	2	40	140	4	5	Interval
1	3	3	2	40	140	4	5	Interval
1	3	3	2	40	140	6	7	Interval
2	1	1	3	40	140	2	3	Interval
2	1	1	3	40	140	2	3	Interval
2	1	1	3	40	140	3	4	Interval
3	2	2	1	40	140	11	.	Right
3	2	2	1	40	140	11	.	Right
3	2	2	1	40	140	11	.	Right

Source: Data from Taguchi (1986).

Table D.5 Drill bit life testing data.

A	B	C	D	E	F	G	H	I	J	L	M	N	O	P	Q	Cycle	Failure
1	−1	−1	−1	−1	−1	−1	−1	−1	−1	−1	−1	−1	−1	−1	−1	1280	1
1	−1	−1	−1	−1	−1	−1	−1	−1	−1	−1	−1	−1	−1	1	1	44	1
1	−1	−1	−1	−1	−1	−1	−1	−1	−1	−1	1	1	−1	1	150	1
1	−1	−1	−1	−1	−1	−1	−1	−1	−1	−1	1	1	1	−1	20	1
1	−1	−1	−1	−1	−1	−1	−1	−1	−1	1	−1	1	−1	1	60	1
1	−1	−1	−1	−1	−1	−1	−1	−1	−1	1	−1	1	1	−1	2	1
1	−1	−1	−1	−1	−1	−1	−1	−1	−1	1	1	−1	−1	−1	65	1
1	−1	−1	−1	−1	−1	−1	−1	−1	−1	1	1	−1	1	1	25	1
1	−1	−1	−1	1	−1	1	1	1	1	1	−1	−1	−1	−1	−1	2680	1
1	−1	−1	−1	1	−1	1	1	1	1	1	−1	−1	−1	1	1	125	1
1	−1	−1	−1	1	−1	1	1	1	1	1	−1	1	1	−1	1	120	1
1	−1	−1	−1	1	−1	1	1	1	1	1	−1	1	1	1	−1	2	1
1	−1	−1	−1	1	−1	1	1	1	1	1	1	−1	1	−1	1	165	1
1	−1	−1	−1	1	−1	1	1	1	1	1	1	−1	1	1	−1	100	1
1	−1	−1	−1	1	−1	1	1	1	1	1	1	1	−1	−1	−1	795	1
1	−1	−1	−1	1	−1	1	1	1	1	1	1	1	−1	1	1	307	1
1	1	1	1	−1	1	−1	−1	−1	1	1	−1	−1	−1	−1	−1	2670	1
1	1	1	1	−1	1	−1	−1	−1	1	1	−1	−1	−1	1	1	480	1
1	1	1	1	−1	1	−1	−1	−1	1	1	−1	1	1	−1	1	762	1

(Continued)

Table D.5 (Continued)

A	B	C	D	E	F	G	H	I	J	L	M	N	O	P	Q	Cycle	Failure
1	1	1	1	-1	1	-1	-1	-1	1	1	-1	1	1	1	-1	130	1
1	1	1	1	-1	1	-1	-1	-1	1	1	1	-1	1	-1	1	1422	1
1	1	1	1	-1	1	-1	-1	-1	1	1	1	-1	1	1	-1	280	1
1	1	1	1	-1	1	-1	-1	-1	1	1	1	1	-1	-1	-1	670	1
1	1	1	1	-1	1	-1	-1	-1	1	1	1	1	-1	1	1	130	1
1	1	1	1	1	1	1	1	1	-1	-1	-1	-1	-1	-1	-1	2655	1
1	1	1	1	1	1	1	1	1	-1	-1	-1	-1	-1	1	1	90	1
1	1	1	1	1	1	1	1	1	-1	-1	-1	1	1	-1	1	7	1
1	1	1	1	1	1	1	1	1	-1	-1	-1	1	1	1	-1	27	1
1	1	1	1	1	1	1	1	1	-1	-1	1	-1	1	-1	1	3	1
1	1	1	1	1	1	1	1	1	-1	-1	1	-1	1	1	-1	15	1
1	1	1	1	1	1	1	1	1	-1	-1	1	1	-1	-1	-1	90	1
1	1	1	1	1	1	1	1	1	-1	-1	1	1	-1	1	1	480	1
2	-1	1	-1	1	1	-1	-1	1	-1	1	-1	-1	-1	-1	-1	3000	0
2	-1	1	-1	1	1	-1	-1	1	-1	1	-1	-1	-1	1	1	440	1
2	-1	1	-1	1	1	-1	-1	1	-1	1	-1	1	1	-1	1	480	1
2	-1	1	-1	1	1	-1	-1	1	-1	1	1	1	1	1	-1	10	1
2	-1	1	-1	1	1	-1	-1	1	-1	1	1	-1	1	-1	1	1260	1
2	-1	1	-1	1	1	-1	-1	1	-1	1	1	-1	1	1	-1	5	1
2	-1	1	-1	1	1	-1	-1	1	-1	1	1	1	-1	-1	-1	1720	1
2	-1	1	-1	1	1	-1	-1	1	-1	1	1	1	-1	1	1	3000	0
2	-1	1	-1	-1	1	1	1	-1	1	-1	-1	-1	-1	-1	-1	2586	1
2	-1	1	-1	-1	1	1	1	-1	1	-1	-1	-1	-1	1	1	6	1
2	-1	1	-1	-1	1	1	1	-1	1	-1	-1	1	1	-1	1	370	1
2	-1	1	-1	-1	1	1	1	-1	1	-1	-1	1	1	1	-1	45	1
2	-1	1	-1	-1	1	1	1	-1	1	-1	1	-1	1	-1	1	2190	1
2	-1	1	-1	-1	1	1	1	-1	1	-1	1	-1	1	1	-1	36	1
2	-1	1	-1	-1	1	1	1	-1	1	1	1	-1	-1	-1	-1	1030	1
2	-1	1	-1	-1	1	1	1	-1	1	1	1	-1	1	1	1	16	1
2	1	-1	1	1	-1	-1	-1	1	1	-1	-1	-1	-1	-1	-1	3000	0
2	1	-1	1	1	-1	-1	-1	1	1	-1	-1	-1	-1	1	1	2580	1
2	1	-1	1	1	-1	-1	-1	1	1	-1	-1	1	1	-1	1	20	1
2	1	-1	1	1	-1	-1	-1	1	1	-1	1	1	1	1	-1	320	1
2	1	-1	1	1	-1	-1	-1	1	1	-1	1	-1	1	-1	1	425	1
2	1	-1	1	1	-1	-1	-1	1	1	-1	1	-1	1	1	-1	85	1
2	1	-1	1	1	-1	-1	-1	1	1	-1	1	1	-1	-1	-1	950	1
2	1	-1	1	1	-1	-1	-1	1	1	-1	1	1	-1	1	1	3000	0

(Continued)

Table D.5 (Continued)

A	B	C	D	E	F	G	H	I	J	L	M	N	O	P	Q	Cycle	Failure
2	1	−1	1	−1	−1	1	1	−1	−1	1	−1	−1	−1	−1	−1	800	1
2	1	−1	1	−1	−1	1	1	−1	−1	1	−1	−1	−1	1	1	45	1
2	1	−1	1	−1	−1	1	1	−1	−1	1	−1	1	1	−1	1	260	1
2	1	−1	1	−1	−1	1	1	−1	−1	1	−1	1	1	1	−1	250	1
2	1	−1	1	−1	−1	1	1	−1	−1	1	1	−1	1	−1	1	1650	1
2	1	−1	1	−1	−1	1	1	−1	−1	1	1	−1	1	1	−1	470	1
2	1	−1	1	−1	−1	1	1	−1	−1	1	1	1	−1	−1	−1	1250	1
2	1	−1	1	−1	−1	1	1	−1	−1	1	1	1	−1	1	1	70	1
3	1	−1	−1	1	1	−1	1	−1	−1	−1	−1	−1	−1	−1	−1	3000	0
3	1	−1	−1	1	1	−1	1	−1	−1	−1	−1	−1	−1	1	1	190	1
3	1	−1	−1	1	1	−1	1	−1	−1	−1	−1	1	1	−1	1	140	1
3	1	−1	−1	1	1	−1	1	−1	−1	−1	−1	1	1	1	−1	2	1
3	1	−1	−1	1	1	−1	1	−1	−1	−1	1	−1	1	−1	1	100	1
3	1	−1	−1	1	1	−1	1	−1	−1	−1	1	−1	1	1	−1	3	1
3	1	−1	−1	1	1	−1	1	−1	−1	−1	1	1	−1	−1	−1	450	1
3	1	−1	−1	1	1	−1	1	−1	−1	−1	1	1	−1	1	1	840	1
3	1	−1	−1	−1	1	1	−1	1	1	1	−1	−1	−1	−1	−1	3000	0
3	1	−1	−1	−1	1	1	−1	1	1	1	−1	−1	−1	1	1	638	1
3	1	−1	−1	−1	1	1	−1	1	1	1	−1	1	1	−1	1	440	1
3	1	−1	−1	−1	1	1	−1	1	1	1	−1	1	1	1	−1	145	1
3	1	−1	−1	−1	1	1	−1	1	1	1	1	−1	1	−1	1	690	1
3	1	−1	−1	−1	1	1	−1	1	1	1	1	−1	1	1	−1	140	1
3	1	−1	−1	−1	1	1	−1	1	1	1	1	1	−1	−1	−1	1180	1
3	1	−1	−1	−1	1	1	−1	1	1	1	1	1	−1	1	1	1080	1
3	−1	1	1	1	−1	−1	1	−1	1	1	−1	−1	−1	−1	−1	3000	0
3	−1	1	1	1	−1	−1	1	−1	1	1	−1	−1	−1	1	1	180	1
3	−1	1	1	1	−1	−1	1	−1	1	1	−1	1	1	−1	1	870	1
3	−1	1	1	1	−1	−1	1	−1	1	1	−1	1	1	1	−1	310	1
3	−1	1	1	1	−1	−1	1	−1	1	1	1	−1	1	−1	1	2820	1
3	−1	1	1	1	−1	−1	1	−1	1	1	1	−1	1	1	−1	240	1
3	−1	1	1	1	−1	−1	1	−1	1	1	1	1	−1	−1	−1	2190	1
3	−1	1	1	1	−1	−1	1	−1	1	1	1	1	−1	1	1	1100	1
3	−1	1	1	−1	−1	1	−1	1	−1	−1	−1	−1	−1	−1	−1	3000	0
3	−1	1	1	−1	−1	1	−1	1	−1	−1	−1	−1	−1	1	1	612	1
3	−1	1	1	−1	−1	1	−1	1	−1	−1	−1	1	1	−1	1	1611	1
3	−1	1	1	−1	−1	1	−1	1	−1	−1	−1	1	1	1	−1	625	1

(Continued)

Table D.5 (Continued)

A	B	C	D	E	F	G	H	I	J	L	M	N	O	P	Q	Cycle	Failure
3	−1	1	1	−1	−1	1	−1	1	−1	−1	1	−1	1	−1	1	1720	1
3	−1	1	1	−1	−1	1	−1	1	−1	−1	1	−1	1	1	−1	165	1
3	−1	1	1	−1	−1	1	−1	1	−1	−1	1	1	−1	−1	−1	1881	1
3	−1	1	1	−1	−1	1	−1	1	−1	−1	1	1	−1	1	1	2780	1
4	1	1	−1	−1	−1	−1	1	1	−1	1	−1	−1	−1	−1	−1	3000	0
4	1	1	−1	−1	−1	−1	1	1	−1	1	−1	−1	−1	1	1	1145	1
4	1	1	−1	−1	−1	−1	1	1	−1	1	−1	1	1	−1	1	1060	1
4	1	1	−1	−1	−1	−1	1	1	−1	1	1	−1	1	1	−1	198	1
4	1	1	−1	−1	−1	−1	1	1	−1	1	1	−1	1	−1	1	1340	1
4	1	1	−1	−1	−1	−1	1	1	−1	1	1	−1	1	1	−1	95	1
4	1	1	−1	−1	−1	−1	1	1	−1	1	1	1	−1	−1	−1	2509	1
4	1	1	−1	−1	−1	−1	1	1	−1	1	1	1	−1	1	1	345	1
4	1	1	−1	1	−1	1	−1	−1	1	−1	−1	−1	−1	−1	−1	3000	0
4	1	1	−1	1	−1	1	−1	−1	1	−1	−1	−1	−1	1	1	970	1
4	1	1	−1	1	−1	1	−1	−1	1	−1	−1	1	1	−1	1	180	1
4	1	1	−1	1	−1	1	−1	−1	1	−1	−1	1	1	1	−1	220	1
4	1	1	−1	1	−1	1	−1	−1	1	−1	1	−1	1	−1	1	415	1
4	1	1	−1	1	−1	1	−1	−1	1	−1	1	−1	1	1	−1	70	1
4	1	1	−1	1	−1	1	−1	−1	1	−1	1	1	−1	−1	−1	2630	1
4	1	1	−1	1	−1	1	−1	−1	1	−1	1	1	−1	1	1	3000	0
4	−1	−1	1	−1	1	−1	1	1	1	−1	−1	−1	−1	−1	−1	3000	0
4	−1	−1	1	−1	1	−1	1	1	1	−1	−1	−1	−1	1	1	3000	0
4	−1	−1	1	−1	1	−1	1	1	1	−1	−1	1	1	−1	1	794	1
4	−1	−1	1	−1	1	−1	1	1	1	−1	−1	1	1	1	−1	40	1
4	−1	−1	1	−1	1	−1	1	1	1	−1	1	−1	1	−1	1	160	1
4	−1	−1	1	−1	1	−1	1	1	1	−1	1	−1	1	1	−1	50	1
4	−1	−1	1	−1	1	−1	1	1	1	−1	1	1	−1	−1	−1	495	1
4	−1	−1	1	−1	1	−1	1	1	1	−1	1	1	−1	1	1	3000	0
4	−1	−1	1	1	1	1	−1	−1	−1	1	−1	−1	−1	−1	−1	680	1
4	−1	−1	1	1	1	1	−1	−1	−1	1	−1	−1	−1	1	1	140	1
4	−1	−1	1	1	1	1	−1	−1	−1	1	−1	1	1	−1	1	809	1
4	−1	−1	1	1	1	1	−1	−1	−1	1	−1	1	1	1	−1	275	1
4	−1	−1	1	1	1	1	−1	−1	−1	1	1	−1	1	−1	1	1130	1
4	−1	−1	1	1	1	1	−1	−1	−1	1	1	−1	1	1	−1	145	1
4	−1	−1	1	1	1	1	−1	−1	−1	1	1	1	−1	−1	−1	2025	1
4	−1	−1	1	1	1	1	−1	−1	−1	1	1	1	−1	1	1	125	1

Source: Data from Wu and Hamada (2000).

Table D.6 Low-cycle fatigue test of nickel super alloy.

Pseudostress (ksi)	Cycles (1000s)	Censored
80.3	211.626	0
80.6	200.027	0
80.8	57.923	1
84.3	155	0
85.2	13.949	0
85.6	112.968	1
85.8	152.68	0
86.4	156.725	0
86.7	138.114	1
87.2	56.723	0
87.3	121.075	0
89.7	122.372	1
91.3	112.002	0
99.8	43.331	0
100.1	12.076	0
100.5	13.181	0
113	18.067	0
114.8	21.3	0
116.4	15.616	0
118	13.03	0
118.4	8.489	0
118.6	12.434	0
120.4	9.75	0
142.5	11.865	0
144.5	6.705	0
145.9	5.733	0

Source: Data from Nelson (1984).

Table D.7 Failure and censoring times for diesel generator fans.

Hours	Status	Hours	Status
450	Failed	4300	Censored
460	Censored	4600	Failed
1150	Failed	4850	Censored
1150	Failed	4850	Censored
1560	Censored	4850	Censored
1600	Failed	4850	Censored
1660	Censored	5000	Censored
1850	Censored	5000	Censored
1850	Censored	5000	Censored
1850	Censored	6100	Censored
1850	Censored	6100	Failed
1850	Censored	6100	Censored
2030	Censored	6100	Censored
2030	Censored	6300	Censored
2030	Censored	6450	Censored
2070	Failed	6450	Censored
2070	Failed	6700	Censored
2080	Failed	7450	Censored
2200	Censored	7800	Censored
3000	Censored	7800	Censored
3000	Censored	8100	Censored
3000	Censored	8100	Censored
3000	Censored	8200	Censored
3100	Failed	8500	Censored
3200	Censored	8500	Censored
3450	Failed	8500	Censored
3750	Censored	8750	Censored
3750	Censored	8750	Failed
4150	Censored	8750	Censored
4150	Censored	9400	Censored
4150	Censored	9900	Censored
4150	Censored	10100	Censored
4300	Censored	10100	Censored
4300	Censored	10100	Censored
4300	Censored	11500	Censored

Source: Modified from Nelson (1969).

E

Distributions Used in Life Testing

In this appendix, we summarize many of the distributions that are useful in studying lifetimes. We present several of the properties of each.

Exponential Distribution:

Probability Density Function:

$$f(t|\theta) = \frac{1}{\theta} \exp(-t/\theta), \qquad t > 0$$

Survival Function:

$$S(t|\theta) = \exp(-t/\theta), \qquad t > 0$$

Hazard Function:

$$h(t|\theta) = \frac{1}{\theta}, \qquad t > 0$$

Abbreviation:

$$T \sim \text{EXP}(\theta)$$

Mean:

$$E(T) = \theta$$

Variance:

$$V(T) = \theta^2$$

Exponential Distribution (Alternate Parameterization):

Probability Density Function:

$$f(t|\theta) = \lambda \exp(-\lambda t), \qquad t > 0$$

Design of Experiments for Reliability Achievement, First Edition.
Steven E. Rigdon, Rong Pan, Douglas C. Montgomery, and Laura J. Freeman.
© 2022 John Wiley & Sons, Inc. Published 2022 by John Wiley & Sons, Inc.
Companion website: www.wiley.com/go/rigdon/designexperiments

Survival Function:

$$S(t|\theta) = \exp(-\lambda t), \qquad t > 0$$

Hazard Function:

$$h(t|\theta) = \lambda, \qquad t > 0$$

Abbreviation:

$$T \sim \text{EXP}^*(\lambda)$$

Mean:

$$E(T) = \frac{1}{\lambda}$$

Variance:

$$V(T) = \frac{1}{\lambda^2}$$

Weibull Distribution:

Probability Density Function:

$$f(t|\theta, \kappa) = \frac{\kappa}{\theta} \left(\frac{t}{\theta} \right)^{\kappa-1} \exp\left(-\left(\frac{t}{\theta} \right)^{\kappa} \right), \qquad t > 0$$

Survival Function:

$$S(t|\theta, \kappa) = \exp\left(-\left(\frac{t}{\theta} \right)^{\kappa} \right), \qquad t > 0$$

Hazard Function:

$$h(t|\theta, \kappa) = \frac{\kappa}{\theta} \left(\frac{t}{\theta} \right)^{\kappa}, \qquad t > 0$$

Abbreviation:

$$T \sim \text{WEIB}(\theta, \kappa)$$

Mean:

$$E(T) = \theta \, \Gamma\left(1 + \frac{1}{\kappa} \right)$$

Variance:

$$V(X) = \theta^2 \left[\Gamma\left(1 + \frac{2}{\kappa} \right) - \left(\Gamma\left(1 + \frac{1}{\kappa} \right) \right)^2 \right]$$

Weibull Distribution (Alternative Parameterization):

The Weibull distribution is sometimes parameterized in this way. WinBUGS parameterizes the Weibull distribution this way.

Probability Density Function:

$$f(t|\lambda, \phi) = \lambda \phi t^{\phi-1} \exp(-\lambda t^{\phi}), \qquad t > 0$$

Survival Function:

$$F(t|\lambda) \exp(-\lambda t^{\phi}), \qquad t > 0.$$

Hazard Function:

$$h(t|\lambda, \phi) = \phi \lambda t^{\phi-1}, \qquad t > 0$$

Abbreviation:

$$T \sim \text{WEIB}^*(\lambda, \phi)$$

Mean:

$$E(T) = \lambda^{-(1/\phi)} \Gamma\left(1 + \frac{1}{\phi}\right)$$

Variance:

$$V(T) = \lambda^{-(2/\phi)} \left[\Gamma\left(1 + \frac{2}{\phi}\right) - \left(\Gamma\left(1 + \frac{1}{\phi}\right)\right)^2\right]$$

The parameters are of the two Weibull parameterizations are related by

$$\phi = \kappa$$
$$\lambda = \theta^{-\kappa}$$

or, equivalently

$$\kappa = \phi$$
$$\theta = \lambda^{-(1/\phi)}$$

Gamma Distribution:

Probability Density Function:

$$f(t|\theta, \alpha) = \frac{t^{\alpha-1}}{\theta^\alpha \Gamma(\alpha)} \exp(-t/\theta), \qquad t > 0$$

Survival Function:

$$S(t|\theta, \alpha) = \int_t^\infty f(u) \, du = \frac{\Gamma(\kappa, t/\theta)}{\Gamma(\kappa)}, \qquad t > 0.$$

where $\Gamma(\cdot)$ is the usual gamma function and $\Gamma(\cdot, \cdot)$ is the incomplete gamma function.

Abbreviation:

$$T \sim \text{GAM}(\theta, \alpha)$$

Mean:

$$E(T) = \theta \alpha$$

Variance:

$$V(T) = \theta^2 \alpha$$

Gamma Distribution (Alternative parameterization):

The gamma distribution is sometimes parameterized in this way. WinBUGS and many other MCMC engines parameterize the gamma distribution this way.

Probability Density Function:

$$f(t|\lambda, \alpha) = \frac{\lambda^\alpha}{\Gamma(\alpha)} t^{\alpha-1} \exp(-\lambda t), \qquad t > 0$$

Survival Function:

$$S(t|\lambda, \alpha) = \int_t^\infty f(u) \, du = \frac{\Gamma(\kappa, \lambda t)}{\Gamma(\kappa)}$$

where $\Gamma(\cdot)$ is the usual gamma function and $\Gamma(\cdot, \cdot)$ is the incomplete gamma function.

Abbreviation:

$$T \sim \text{dgamma}(\lambda, \alpha)$$

Mean:

$$E(T) = \frac{\alpha}{\lambda}$$

Variance:

$$V(T) = \frac{\alpha}{\lambda^2}$$

Generalized Gamma Distribution

The generalized gamma distribution is a three-parameter model (with parameters θ, κ, and α) and is a generalization of both the gamma distribution and the Weibull distribution. If $\alpha = 1$, then the distribution reduces to the $\text{WEIB}(\theta, \kappa)$ and if $\kappa = 1$ then the distribution reduces to the $\text{GAM}(\theta, \alpha)$ distribution.

Probability Density Function:

$$f(t|\theta, \alpha, \kappa) = \frac{1}{\Gamma(\alpha)} \frac{\kappa}{\theta} \left(\frac{t}{\theta}\right)^{\alpha\kappa-1} \exp\left[-\left(\frac{t}{\theta}\right)^\kappa\right] \qquad t > 0$$

Abbreviation:

$$T \sim \text{GGAM}(\theta, \kappa, \alpha)$$

Mean:

$$E(T) = \frac{\theta\, \Gamma\left(\alpha + \frac{1}{\kappa}\right)}{\Gamma(\alpha)}$$

Variance:

$$V(T) = \theta^2 \left[\frac{\Gamma\left(\alpha + \frac{2}{\kappa}\right)}{\Gamma(\alpha)} - \left(\frac{\Gamma\left(\alpha + \frac{1}{\kappa}\right)}{\Gamma(\alpha)} \right)^2 \right]$$

Standardized Smallest Extreme Value Distribution

Probability Density Function:

$$\phi_{\text{SSEV}}(z) = \exp(z - \exp(z)), \quad -\infty < z < \infty$$

Cumulative Distribution Function:

$$\Phi_{\text{SSEV}}(z) = 1 - \exp(-\exp(z)), \quad -\infty < z < \infty$$

Abbreviation:

$$Z \sim \text{SEV}(0, 1),$$

Mean:

$$E(Z) = -\gamma$$

where γ is Euler's constant ($\gamma \approx 0.577216$).

Variance:

$$V(X) = \frac{\pi^2}{6}$$

Smallest Extreme Value Distribution

Probability Density Function:

$$f(x) = \frac{1}{\sigma} \phi_{\text{SSEV}}\left(\frac{x - \mu}{\sigma}\right), \quad -\infty < z < \infty$$

Cumulative Distribution Function:

$$F(x) = \Phi_{\text{SSEV}}\left(\frac{x - \mu}{\sigma}\right), \quad -\infty < z < \infty$$

Abbreviation:

$$X \sim \text{SEV}(\mu, \sigma)$$

Note that μ and σ are not the mean and standard deviation, although they are the location and scale parameters, respectively.

Mean:

$$E(X) = \mu - \gamma$$

where γ is Euler's constant ($\gamma \approx 0.577216$).

Variance:

$$V(X) = \frac{\pi^2}{6}\sigma^2.$$

Bibliography

Aarset, M.V. (1987). How to identify a bathtub hazard rate. *IEEE Transactions on Reliability* 36 (1): 106–108.

Aitkin, M. and Clayton, D. (1980). The fitting of exponential, Weibull and extreme value distributions to complex censored survival data using glim. *Applied Statistics* 29: 156–163.

ASQ (2020). What is reliability? https://asq.org/quality-resources/reliability (accessed 17 June 2020).

Bain, L.J. and Engelhardt, M. (1992). *Introduction to Probability and Mathematical Statistics*. Duxbury Press.

Balakrishnan, N. and Kundu, D. (2013). Hybrid censoring: models, inferential results and applications. *Computational Statistics and Data Analysis* 57: 166–209.

Barbosa, E.P., Colosimo, E.A., and Louzada-Neto, F. (1996). Accelerated life tests analyzed by a piecewise exponential distribution via generalized linear models. *IEEE Transactions on Reliability* 45: 619–623.

Bartholomew, D.J. (1957). A problem in life testing. *Journal of the American Statistical Association* 52: 350–355.

Bergquist, B., Vanhatalo, E., and Nordenvaad, M.L. (2011). A Bayesian analysis of unreplicated two-level factorials using effects sparsity, hierarchy, and heredity. *Quality Engineering* 23 (2): 152–166.

Box, J.F. (1978). *RA Fisher, the Life of a Scientist*. Wiley.

Box, G. (1988). Signal-to-noise ratios, performance criteria, and transformations. *Technometrics* 30 (1): 1–17.

Box, G.E.P. (1999). Statistics as a catalyst to learning by scientific method part II–A discussion. *Journal of Quality Technology* 31 (1): 16–29.

Box, G.E.P. and Behnken, D.W. (1960). Some new three level designs for the study of quantitative variables. *Technometrics* 2 (4): 455–475.

Box, G.E.P. and Draper, N.R. (1987). *Empirical Model-Building and Response Surfaces*. Wiley.

Box, G.E.P. and Hunter, J.S. (1957). Multi-factor experimental designs for exploring response surfaces. *The Annals of Mathematical Statistics* 28 (1): 195–241.

Box, G.E.P. and Wilson, K.G. (1951). On the experimental attainment of optimum conditions. *Journal of the Royal Statistical Society. Series B (Methodological)* 13 (1): 1–45.

Box, G., Bisgaard, S., and Fung, C. (1988). An explanation and critique of Taguchi's contributions to quality engineering. *Quality and Reliability Engineering International* 4 (2): 123–131.

Brown, P.F. and Potts, J.R. (1977). Evaluation of powder processed turbine engine ball bearings [Interim Report, May 1975- December 1976]. Aero - Propulsion Laboratory Wright Patterson Air Force Base, *AFAPL-TR-77-26*.

Casella, G. and Berger, R.L. (2002). *Statistical Inference*, 2e. Duxbury, MA: Thomson Learning.

Design of Experiments for Reliability Achievement, First Edition.
Steven E. Rigdon, Rong Pan, Douglas C. Montgomery, and Laura J. Freeman.
© 2022 John Wiley & Sons, Inc. Published 2022 by John Wiley & Sons, Inc.
Companion website: www.wiley.com/go/rigdon/designexperiments

Coffin, L.F. (1954). A study of the effects of cyclic thermal stresses on the ductile metal. *Transactions of the ASME* 76: 931–950.

Cohen, A.C. (1963). Progressively censored samples in life testing. *Technometrics* 5: 327–399.

Coleman, D.E. and Montgomery, D.C. (1993). A systematic approach to planning for a designed industrial experiment. *Technometrics* 35 (1): 1–12.

Condra, L. (2001a). *Reliability Improvement with Design of Experiments*, 2e. CRC Press.

Condra, L.W. (2001b). *Reliability Improvement with Design of Experiments: Revised and Expanded*, 2e. Marcel Dekker.

Coolen, F.P.A. and Coolen-Schrijner, P. (2006). On zero-failure testing for Bayesian high-reliability demonstration. *Proceedings of the Institute of Mechanical Engineers Part O: Journal of Risk and Reliability* 220: 35–44.

Coolen, F.P.A., Coolen-Schrijner, P., and Rahrouh, M. (2005). Bayesian reliability demonstration for failure-free periods. *Reliability Engineering & System Safety* 88: 81–91.

Dahmen, K., Burkschat, M., and Cramer, E. (2012). A- and D-optimal progressive Type-II censoring designs based on fisher information. *Journal of Statistical Computation and Simulation* 82: 879–905.

Daniel, C. (1959). Use of half-normal plots in interpreting factorial two-level experiments. *Technometrics* 1 (4): 311–341.

Davis, T.P. (1995). Analysis of an experiment aimed at improving the reliability of transmission centre shafts. *Lifetime Data Analysis* 1 (3): 275–306.

Efron, B. and Hastie, T. (2016). *Computer Age Statistical Inference*. Cambridge: Cambridge University Press.

Epaarachchi, J. and Clausen, P. (2003). An empirical model for fatigue behavior prediction of glass fibre-reinforced plastic composites for various stress ratios and test frequencies. *Composites Part A Applied Science and Manufacturing* 34: 313–326.

Escobar, L.A. and Meeker, W.Q (1986). Algorithm as 218: elements of the Fisher information matrix for the smallest extreme value distribution and censored data. *Journal of the Royal Statistical Society. Series C (Applied Statistics)*, 35 (1): 80–86.

Escobar, L.A. and Meeker, W.Q. Jr. (1994). Algorithm AS 292: Fisher information matrix for the extreme value, normal and logistic distributions and censored data. *Applied Statistics* 43 (3): 533–540.

Escobar, L.A. and Meeker, W.Q. (1995). Planning accelerated life tests with two or more experimental factors. *Technometrics* 37: 411–427.

Escobar, L.A. and Meeker, W.Q. (1998). Fisher information matrices with censoring, truncation, and explanatory variables. *Statistica Sinica* 8 (1): 221–237.

Eyring, H. (1935). The activated complex in chemical reactions. *Journal of Chemical Physics* 3: 107–115.

Findelstein, D.M. (1986). A proportional hazards model for interval-censored failure time data. *Biometrics* 42: 845–854.

Fisher, R.A. (1958). *Statistical Methods for Research Workers*. Oliver and Boyd.

Fisher, R.A. (1966). *The Design of Experiments*, 8e. Hafner.

Gelman, A., Carlin, J.B., Stern, H.S. et al. (2013). *Bayesian Data Analysis*. CRC Press.

Geman, S. and Geman, D. (1984). Stochastic relaxation, gibbs distributions, and the Bayesian restoration of images. *IEEE Transactions on Pattern Analysis and Machine Intelligence* PAMI-6 (6): 721–741.

Gotwalt, C.M., Jones, B.A., and Steinberg, D.M. (2009). Fast computation of designs robust to parameter uncertainty for nonlinear settings. *Technometrics* 51 (1): 88–95.

Grolemund, G. and Wickham, H. (2018). *R for Data Science: Import, Tidy, Transform, Visualize, and Model Data*. O'Reilly.

Hastings, W.K. (1970). Monte Carlo sampling methods using Markov chains and their applications. *Biometrika* 57 (1): 97–109.

Hellstrand, C. (1989). The necessity of modern quality improvement and some experience with its implementation in the manufacture of rolling bearings. *Philosophical Transactions of the Royal Society of London, Series A: Mathematical, Physical and Engineering Sciences* 327 (1596): 529–537.

Hiergeist, P., Spitzer, A., and Rohl, S. (1989). Lifetime of thin oxide and oxide-nitride-oxide dielectrics within trench capacitors for DRAMs. *IEEE Transactions on Electron Devices* 36 (5): 913–919.

Hill, W.J. and Demler, W.R. (1970). More on planning experiments to increase research efficiency. *Industrial & Engineering Chemistry* 62 (10): 60–65.

Hobbs, G.K. (2005). *Accelerated Reliability Engineering*. Westminster, CO: Hobbs Engineering Corporation.

Hunter, J.S. (1985). Statistical design applied to product design. *Journal of Quality Technology* 17 (4): 210–221.

Hunter, J.S. (1989). Let's all beware the Latin square. *Quality Engineering* 1 (4): 453–465.

Islam, A. and Ahmad, N. (1994). Optimal design of accelerated life tests for Weibull distribution under periodic inspection and type I censoring. *Microelectronics Reliability* 34: 1459–1468.

JEDEC (2001). Solid State Reliability Assessment and Qualification Methodologies. *Technical report JEP143*. JEDEC Solid State Technology.

JEDEC (2003a). Failure Mechanisms and Models for Semiconductor Devices. *Technical report JEP122B*. JEDEC Solid State Technology.

JEDEC (2003b). Method for Developing Acceleration Models for Electronic Component Failure Mechanism. *Technical report JESD91A*. JEDEC Solid State Technology.

JEDEC (2004). Reliability Qualification and Semiconductor Devices Based on Physics of Failure Risk and Opportunity Assessment. *Technical report JEP148*. JEDEC Solid State Technology.

Johnson, L. and Burrows, S. (2011). For Starbucks, it's in the bag. *Quality Progress* 44 (3): 18.

Johnson, N.L., Kotz, S., and Balakrishnan, N. (1994). *Continuous Univeriate Distributions*, vol. 1. New York: Wiley.

Jones, B. and Nachtsheim, C.J. (2011a). A class of three-level designs for definitive screening in the presence of second-order effects. *Journal of Quality Technology* 43 (1): 1–15.

Jones, B. and Nachtsheim, C.J. (2011b). Efficient designs with minimal aliasing. *Technometrics* 53 (1): 62–71.

Kackar, R.N. (1985). Off-line quality control, parameter design, and the Taguchi method. *Journal of Quality Technology* 17 (4): 176–188.

Kaplan, E.L. and Meier, P. (1958). Nonparametric estimation from incomplete observations. *Journal of the American Statistical Association* 53 (282): 457–481.

Khuri, A.I. and Cornell, J.A. (1987). *Response Surfaces: Designs and Analyses*. Marcel Dekker.

Kiefer, J. (1959). Optimum experimental designs. *Journal of the Royal Statistical Society: Series B (Methodological)* 21 (2): 272–304.

Kiefer, J. (1961). Optimum designs in regression problems, II. *The Annals of Mathematical Statistics* 298–325. 32.

Kiefer, J. and Wolfowitz, J. (1959). Optimum designs in regression problems. *The Annals of Mathematical Statistics* 271–294. 30.

Lau, J.H., Harkins, G., Rice, D. et al. (1988). Experimental and statistical analyses of surface-mount technolgoy plcc solder-joint reliability. *IEEE Transactions on Reliability* 37: 524–530.

Lawless, J.F. (2003). *Statistical Models and Methods for Lifetime Data*, 2e. Wiley.

Leemis, L. (1995). *Reliability. Probabilistic Models and Statistical Methods*. Englewood Cliffs, NJ: Prentice-Hall.

Lenth, R.V. (1989). Quick and easy analysis of unreplicated factorials. *Technometrics* 31 (4): 469–473.

Li, X., Sudarsanam, N., and Frey, D.D. (2006). Regularities in data from factorial experiments. *Complexity* 11 (5): 32–45.

Li, P.C., Ting, W., and Kwong, D.L. (1989). Time-dependent dielectric breakdown of chemical-vapour-deposited SiO/sub2/gate dielectrics. *Electronics Letters* 25 (10): 665–666.

Lieblein, J. and Zelen, M. (1956). Statistical investigation of the fatigue life of deep-groove ball bearings. *Journal of Research of the National Bureau of Standards* 57 (5): 273–316.

Limon, S., Yadav, O.P., and Liao, H. (2017). A literature review on planning and analysis of accelerated testing for reliability assessment. *Quality and Reliaiblity Engineering International* 33: 2361–2383.

Lunn, D.J., Thomas, A., Best, N., and Spiegelhalter, D. (2000). Winbugs-a Bayesian modelling framework: concepts, structure, and extensibility. *Statistics and Computing* 10 (4): 325–337.

Manson, S.S. (1953). Behavior of materials under conditions of thermal stress. *Proceedings of the Heat Transfer Symposium*. Ann Arabor, MI: University of Michigan Engineering Research Institute, pp. 9–75.

Matloff, N. (2011). *The Art of R Programming: A Tour of Statistical Software Design*. No Starch Press.

McCool, J.I. (2012). *Using the Weibull Distribution: Reliability, Modeling, and Inference*, vol. 950. Wiley.

McCullagh, P. and Nelder, J. (1989). *Generalized Linear Models*, 2e. Oxfordshire: Routledge.

Meeker, W.Q. and Escobar, L.A. (1998a). *Statistical Methods for Reliability Data*. Wiley.

Meeker, W.Q. and Escobar, L.A. (1998b). Pitfalls of accelerated testing. *IEEE Transactions on Reliability* 47 (2): 114–118.

Meeker, W.Q., Sarakakis, G., and Gerokostopoulos, A. (2013). More pitfalls of accelerated tests. *Journal of Quality Technology* 45 (3): 213–222.

Metropolis, N., Rosenbluth, A.W., Rosenbluth, M.N. et al. (1953). Equation of state calculations by fast computing machines. *The Journal of Chemical Physics* 21 (6): 1087–1092.

Monroe, E.M., Pan, R., Anderson-Cook, C. et al. (2010). Sensitivity analysis of optimal designs for accelerated life testing. *Journal of Quality Technology* 42: 121–135.

Monroe, E.M., Pan, R., Anderson-Cook, C. et al. (2011). A generalized linear model approach to designing accelerated life test experiments. *Quality and Reliability Engineering International* 27: 595–607.

Montgomery, D.C. (2020). *Design and Analysis of Experiments*, 10e. Wiley.

Moore, D.F. (2016). *Applied Survival Analysis Using R*. Springer.

Morris, V.M., Hargreaves, C., Overall, K. et al. (1997). Optimization of the capillary electrophoresis separation of ranitidine and related compounds. *Journal of Chromatography A* 766 (1–2): 245–254.

Mueller, G. and Rigdon, S.E. (2015). The constant shape parameter assumption in Weibull regression. *Quality Engineering* 27 (3): 374–392.

Myers, R.H. and Montgomery, D.C. (1997). A tutorial on generalized linear models. *Journal of Quality Technology* 29 (3): 274–291.

Myers, R.H., Montgomery, D.C., Vining, G.G., and Robinson, T.J. (2012). *Generalized Linear Models: With Applications in Engineering and the Sciences*. Hoboken, NJ: Wiley.

Myers, R.H., Montgomery, D.C., and Anderson-Cook, C.M. (2016). *Response Surface Methodology: Process and Product Optimization Using Designed Experiments*. Hoboken, NJ: Wiley.

Nair, V.N. (1992). Taguchi's parameter design: a panel discussion. *Technometrics* 34 (2): 127–161.

Nasir, E.A. and Pan, R. (2017). Simulation-based Bayesian optimal experimental design for multi-factor accelerated life tests. *Communications in Statistics - Simulation and Computation* 46: 980–993.

Nelder, J.A. and Wedderburn, R.W.M. (1972). Generalized linear models. *Journal of the Royal Statistical Society: Series A (General)* 135 (3): 370–384.

Nelson, W.B. (1969). Hazard plotting for incomplete failure data. *Journal of Quality Technology* 1 (1): 27–52.

Nelson, W. (1970). Hazard plotting methods for analysis of life data with different failure modes. *Journal of Quality Technology* 2: 126–149.

Nelson, W.B. (1984). *Accelerated Testing: Statistical Models, Test Plans, and Data Analysis*. Wiley.

Nelson, W.B. (2005). A bibliography of accelerated test plans. *IEEE Transactions on Reliability* 54 (2): 194–197.

Nelson, W.B. (2015). An updated bibliography of accelerated test plans. *2015 Annual Reliability and Maintainability Symposium (RAMS)*. IEEE, pp. 1–6.

Nelson, W. and Kielpinski, T.J. (1976). Theory for optimum censored accelerated life tests for normal and lognormal life distributions. *Technometrics* 18: 105–114.

Nelson, W. and Meeker, W.Q. (1978). Theory for optimum accelerated censored life tests for Weibull and extreme value distributions. *Technometrics* 20: 171–177.

Ng, H.K.T., Chan, P.S., and Balakrishnan, N. (2004). Optimal progressive censoring plans for the Weibull distribution. *Technometrics* 46: 470–481.

Ng, H.K.T., Kundu, D., and Chan, P.S. (2009). Statistical analysis of exponential lifetimes under an adaptive Type-II progressive censoring scheme. *Naval Research Logistics* 56: 687–698.

Pan, R., Yang, T., and Seo, K. (2015). Planning constant-stress accelerated life tests for accelerated model selection. *IEEE Transactions on Reliability* 64: 1356–1366.

Park, J.-W. and Yum, B.-J. (1996). Optimal design of accelerated life tests with two stresses. *Naval Research Logistics* 43: 863–884.

Peck, D.S. (1986). Comprehensive model for humidity testing correlation. *Proceedings of the IEEE International Reliability Physics Symposium*. IEEE, pp. 44–50.

Pignatiello, J.J. Jr. and Ramberg, J.S. (1991). Top ten triumphs and tragedies of Genichi Taguchi. *Quality Engineering* 4 (2): 211–225.

Poon, G.K.K. and Williams, D.J. (1999). Characterization of a solder paste printing process and its optimization. *Soldering & Surface Mount Technology*. 11 (3). 23–26.

Radio Electronics Television Manufacturers Association (1955). *Eletronic Applications Reliability Review*. 3 18. Radio Electronics Television Manufacturers Association.

R Core Team (2020). *R: A Language and Environment for Statistical Computing*. Vienna, Austria: R Foundation for Statistical Computing. https://www.R-project.org/ (accessed 22 October 2021).

Rigdon, S.E. and Basu, A.P. (2000). *Statistical Methods for the Reliability of Repairable Systems*. New York: Wiley.

RStudio Team (2020). *RStudio: Integrated Development Environment for R*. Boston, MA: RStudio, Inc. http://www.rstudio.com/ (accessed 22 October 2021).

Seo, K. and Pan, R. (2015). ALTopt: An R package for optimal experimental design of accelerated life testing. *R Journal* 7: 2.

Seo, K. and Pan, R. (2018). Planning accelerated life tests with random effects of test chambers. *Applied Stochastic Models in Business and Industry* 34: 224–243.

Seo, K. and Pan, R. (2020). Planning accelerated life tests with multiple sources of random effects. *Journal of Quality Technology*. https://doi.org/10.1080/00224065.2020.1829214.

Seo, S.-K. and Yum, B.-J. (1991). Accelerated life test plans under intermittent inspection and Type-I censoring: the case of Weibull failure distribution. *Naval Research Logistics* 38: 1–22.

Sitter, R.R. and Torsney, B. (1995). Optimal designs for binary response experiments with two design variables. *Statistica Sinica* 5: 495–519.

Snee, R.D., Hare, L.B., and Trout, J.R. (1985). *Experiments in Industry: Design, Analysis, and Interpretation of Results*. American Society for Quality Control.

Sturtz, S., Ligges, U., and Gelman, A.E. (2005). R2WinBUGS: A package for running winbugs from R. https://cran.r-project.org/web/packages/R2WinBUGS/R2WinBUGS.pdf (accessed 22 October 2021).

Taguchi, G. (1986). *Introduction to Quality Engineering: Designing Quality into Products and Processes*. Asian Productivity Organization.

Taguchi, G. (1987). *System of Experimental Design: Engineering Methods to Optimize Quality and Minimize Cost: UNIPUB/Kraus Int*, vol. 2. UNIPUB.

Taguchi, G. (1991). *Introduction to Quality Engineering*. Asian productivity organization.

Taguchi, G. and Wu, Y. (1980). *Introduction to Off-line Quality*. Control Central Japan Quality Control Association.

Therneau, Terry M., Thomas, Lumley, Atkinson, Elizabeth, Crowson, Cynthia, (2015). A Package for Survival Analysis in S. version 2.38. https://cran.r-project.org/web/packages/survival/survival.pdf.

Tobias, P.A. and Trindade, D. (2011). *Applied Reliability*. CRC Press.

Wang, F.K. (2000). A new model with bathtub-shaped failure rate using an additive Burr XII distribution. *Reliability Engineering & System Safety* 70 (3): 305–312.

Wolfram Research Inc. (2010). Mathematica Edition: Version 8.0.

Wu, C.F.J. and Hamada, M. (2000). *Experiments: Planning, Analysis, and Parameter Design Optimization*. Wiley.

Xiao, L., Lin, D.K.J., and Bai, F. (2012). Constructing definitive screening designs using conference matrices. *Journal of Quality Technology* 44 (1): 2–8.

Yang, G. (2007). *Life Cycle Reliability Engineering*. Hoboken, NJ: Wiley.

Yang, T. and Pan, R. (2013). A novel approach to optimal accelerated life test planning with interval censoring. *IEEE Transactions on Reliability* 62: 527–536.

Yum, B.-J. and Choi, S.-C. (1989). Optimal design of accelerated life tests under periodic inspection. *Naval Research Logistics* 36: 779–795.

Zelen, M. (1959). Factorial experiments in life testing. *Technometrics* 1: 269–288.

Index

Design of Experiments for Reliability Achievement, First Edition.
Steven E. Rigdon, Rong Pan, Douglas C. Montgomery, and Laura J. Freeman.
© 2022 John Wiley & Sons, Inc. Published 2022 by John Wiley & Sons, Inc.
Companion website: www.wiley.com/go/rigdon/designexperiments